Radar Electronic Warfare

August Golden Jr.
Lieutenant Colonel, U.S. Air Force
Headquarters Air Force Systems Command
Andrews Air Force Base, Maryland

AIAA EDUCATION SERIES
J. S. Przemieniecki
Series Editor-in-Chief
Air Force Institute of Technology
Wright-Patterson Air Force Base, Ohio

Published by
American Institute of Aeronautics and Astronautics, Inc.
370 L'Enfant Promenade, SW, Washington, DC 20024

Texts Published in the AIAA Education Series

Re-Entry Vehicle Dynamics
Frank J. Regan, 1984
Aerothermodynamics of Gas Turbine and Rocket Propulsion
Gordon C. Oates, 1984
Aerothermodynamics of Aircraft Engine Components
Gordon C. Oates, Editor, 1985
Aircraft Combat Survivability Analysis and Design
Robert E. Ball, 1985
Intake Aerodynamics
J. Seddon and E.L. Goldsmith, 1985
Composite Materials for Aircraft Structures
Brian C. Hoskin and Alan A. Baker, Editors, 1986
Gasdynamics: Theory and Applications
George Emanuel, 1986
Aircraft Engine Design
Jack D. Mattingly, William H. Heiser, and
Daniel H. Daley, 1987
An Introduction to the Mathematics
and Methods of Aerodynamics
Richard H. Battin

American Institute of Aeronautics and Astronautics, Inc.
New York, New York

Library of Congress Cataloging in Publication Data

Golden, August.
Radar electronic warfare.

(AIAA education series)
Includes index.
1. Radar. 2. Electronics in military engineering.
3. Electronics in aeronautics. I. Title. II. Series.
UG1420.G65 1987 623′.043 87-1259
ISBN 0-930403-22-3

Printed in the United States of America.

FOREWORD

In designing aeronautical systems for military applications such as aircraft or missiles, the consideration of their performance in the electronic warfare (EW) environment is becoming of paramount importance. Strategic and tactical aircraft in combat operations are threatened by interceptor aircraft, surface-to-air missiles, and antiaircraft artillery. All combat aircraft need appropriate countermeasures to reduce or eliminate this threat. They need warning systems to ascertain when they are under attack, self-protection jammers, various devices to misdirect incoming weapons, and antiradiation missiles to destroy the emitters that guide those incoming missiles. It is therefore not surprising that electronic warfare has today become an integral part of any tactical or strategic air operation.

Lieutenant Colonel August Golden's textbook is based on lecture notes developed for various courses in electronic warfare he taught at the Air Force Institute of Technology (AFIT), USAF's graduate engineering school at Wright-Patterson Air Force Base, Ohio. Colonel Golden's experience with electronic warfare also includes work as a project engineer in the Electronic Warfare Division of the Air Force Avionics Laboratory, where he was responsible for the development of monopulse radar countermeasures and the evaluation of electronic countermeasures (ECM) against both ground and airborne tracking radars. He is presently assigned to Headquarters Air Force Systems Command, Andrews Air Force Base, Maryland.

Radar Electronic Warfare includes both the fundamental principles of radar electronic warfare and the specific analysis methods for EW systems. It can be used as a textbook for courses in electronic warfare as well as a reference text for EW engineers and EW officers (EWO's) operating electronic countermeasures (ECM) or counter-countermeasures (ECCM) systems.

J. S. PRZEMIENIECKI
Editor-in-Chief
Education Series

TABLE OF CONTENTS

PART II

PREFACE

In modern warfare, electronic communications and automatic control systems have become indispensable aids in the achievement of required weapon system performance. This is the era of smart bombs, infrared vision aids, and terrain avoidance radar, to name but a few modern systems introduced, at least in part, by electronic technology. From another point of view, however, this additional capability comes at the expense of more dependence on the proper performance of electronic devices. Electronic warfare (EW) seeks to insure proper performance of friendly electronic systems and deny proper performance of unfriendly systems. Leaders responsible for the nation's defense must recognize the potential effect of enemy EW systems on U.S. weapon systems. For example, Admiral Thomas H. Moorer, Chairman of the Joint Chiefs of Staff in 1970-1974, stated: "If there is a WW III, the winner will be the side that can best control and manage the electromagnetic spectrum [1]."

This text, an outgrowth of various courses in electronic warfare taught at the Air Force Institute of Technology, Wright-Patterson Air Force Base, Ohio, should aid anyone requiring an understanding of electronic warfare. Part I of the text is introductory in nature; therefore, it requires only minimum sophistication in mathematics. Included in this part are definitions of common terms used in the fields of radar and electronic warfare, discussion of radar and EW principles, and some simple analyses of interactions between radar systems and electronic countermeasures (ECM).

The first chapter in Part II presents the necessary background material in mathematics, electromagnetics, and probability and the second chapter develops tracking radar models. The remainder of the text is devoted to analysis of the effect of various ECM emissions on classes of radar systems.

The text will not make the reader an expert in such areas as electromagnetics, solid-state and thermionic devices, control systems, or signal processing. These areas have a major impact on electronic warfare, but they require intensive study. Additionally, this short text does not address the more practical aspects of electronic warfare, including ECM operations, radar system design and testing, and air operations. In other words, it focuses on radar and ECM principles rather than their operational use. Finally, it is oriented toward radars and countermeasures to radars (otherwise known as radar countermeasures), no doubt at the expense of proper coverage of other aspects of electronic warfare.

I am indebted to many people for their assistance in the development of this text. Drs. Robert Fontana and Jack D'Azzo, of the Electrical Engineering Department at the Air Force Institute of Technology, kindly provided the resources to accomplish the typing and graphic arts on this and earlier versions of the text. Captain Thomas Johnson reviewed the material and offered many helpful suggestions and Lt. Col. James Rutledge enhanced the capability of AFIT's word processor so it would properly handle the many equations in the text. Thanks go to Capt. James W. Howatt, Donna M. Smith, Becky Brumlow, and Juanita Hardin for typing this and earlier versions with diligence and perfection. A special mention must be given to John Smith, former editor of the Air Command and Staff College, for editing the text and to Stephen L. Johnston for reviewing portions of the text and for his encouragement.

Finally, I thank most of all my wife, Sharon, for her encouragement and love and my daughters, Tamara and Michelle, who received little time and attention over many past months.

LT. COL. AUGUST GOLDEN JR.
Headquarters Air Force Systems Command
Andrews Air Force Base, Maryland

PART I

1 INTRODUCTION TO ELECTRONIC WARFARE

1.1 HISTORICAL REVIEW

It is generally perceived that every action stemming the flow of information between components of opposing armed forces is an act of electronic warfare and many of these perceptions are logical. For example, the transmission of deceptive information by false signal fires certainly uses the optical portion of the electromagnetic spectrum. But such associations merely point out that the principles of war never change, only the tools for implementing them. After all, communications in future wars may conceivably use gravitational waves. The jamming of such signals will no doubt occur, but such action will not appear to be electromagnetic in nature. Thus, to aid in defining what is and what is not classified as electronic warfare, this chapter begins with an account of past activities generally considered to be acts of electronic warfare.

The first recorded electromagnetic actions related to warfare are different, depending on the writer cited. According to Refs. 1, 2 (Chap.2), and 3, on May 31, 1916, Admiral of the Fleet, Sir Henry Jackson, used coastal radio direction finders to observe movements of the German fleet and, on the basis of this information, positioned the British fleet so that it was able to oppose the Germans successfully. On the other hand, Shustov and Vakin [4] suggest an even earlier use of electronic warfare, claiming that the Russian torpedo boat *Gromky* and cruiser *Igumrad* jammed Japanese radio communications during the Russian-Japanese War at the Battle of Tsushima in May 1905. If this earlier claim is true, it attests to the fact that electronic warfare by itself does not win battles because the Russians were soundly defeated [5]. Finally, Johnston [68, p. 1] references a nonmilitary event involving Guglielmo Marconi, Lee DeForest, and John Greenleaf Pickard, all pioneers of radio. In 1901, Marconi and DeForest were contracted to use wireless to report international yacht races for the Associated Press and the Publishers Press Association. Their race reports, which were transmitted by wireless, were effectively jammed by Pickard of the American Wireless Telephone and Telegraph Company. This was accomplished by a simple procedure of continued transmittal of a series of interfering dashes that also transmitted Pickard's coded reports, while denying reception by the other two competitors. Thus, 1901 saw what was probably the first use of the electronic countermeasures.

The use of such electronic systems as radio and radar, as well as electronic warfare, became widespread in World War II. During the early stages of the war, the British concentrated on defeating various German navigational aids used to guide bombers to their targets in England [2]. Some of these aids—Lorenz, Headache, and Ruffian—used directional radio antennas that pointed their beams from the continent toward the desired target in England. German bombers flew within the beams to their destination. As shown in Fig. 1.1, the British countermeasure to these aids was deceptive in nature [91, p.127]. A jammer in the target area radiated signals similar to the German transmissions and these signals caused an apparent bending of the beam in which the aircraft was flying. The bomber crews thus dropped their loads away from the jammer position and flew home with an unjustified sense of accomplishment.

Navigational aids were not the only electronic warfare targets during World War II. In November 1941, the British jammed German tank communications during the Libyan campaign, but this successful application of airborne jamming was short-lived because no air cover was provided for the jamming aircraft. As the war moved closer to Germany, the Allies employed both active jamming of the radio control channels used for guiding interceptor aircraft and passive jamming in the form of chaff (or window) to confuse German Wurzburg and Freya radars [91]. Prior to the Normandy invasion, the suppression of Germany's defensive forces was of paramount importance to the Allies. German coastal radar sites were reduced by 82% through massive attacks with bombs and rockets. In addition, balloon-borne radar reflectors and deceptive transmissions were used to create "ghost" invasion fleets, which led the Germans to dilute their defenses [3]. Excellent histories of electronic warfare during World War II can be found in Refs. 91 and 92.

Equipment and tactics during the Korean War were essentially the same as those of World War II. Nevertheless, electronic warfare was indispensable by the end of 1951. According to some estimates, aircraft and crew losses

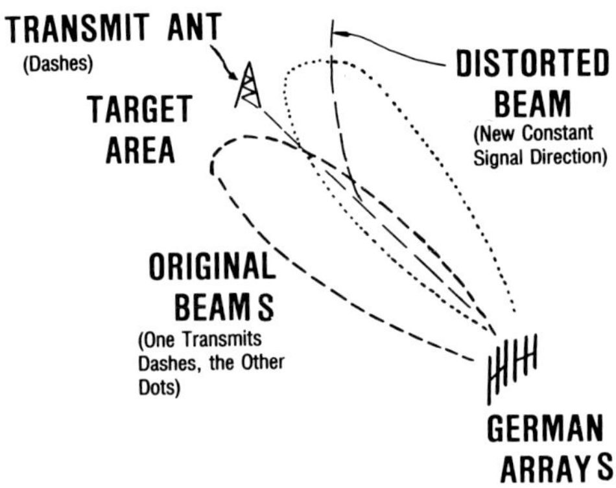

Fig. 1.1 Disruption of Lorenz navigational aid.

would have been triple the actual losses during the last two years of the war without the use of electronic warfare [1, p. 4].

Both before and after the Korean conflict, development of a capability for electronic warfare received low priority. As a result, few resources were available to counter the surface-to-air missile systems (SAM) and anti-aircraft artillery (AAA) facing the U.S. Air Force and Navy during the initial bombing of North Vietnam [1]. Only the EB-66C had an electronic countermeasure (ECM) capability during this first encounter with radar- and infrared (IR)-guided missiles. (The Germans experimented with both radio-guided and infrared-homing missiles during World War II [6], but they did not use them operationally. However, they did use radio-guided bombs.) Apparently, the EB-66C was insufficient, because many new EW systems were developed during the course of the war. Suppression of air defenses took the form of F-105G and F-4C "Wild Weasel" aircraft configured to seek out and destroy SAM radar sites. Individual fighters carried "PODS" equipped with flexible jamming systems to adapt to the ever-changing radar threat. These fighters also carried radar warning receivers to warn the crews of nearby SAM activity, newly designed dispensers of chaff, and infrared countermeasures in the form of flares. These efforts aided substantially in reducing aircraft losses [7].

But the radar defenses were surmounted for only a brief period. In the Yom Kippur War of October 1973, Israeli aircraft could not provide close air support until their ground forces had overrun the Egyptian defenses [8]. The proven EW defenses used against the SA-2 and SA-3 during the Vietnam War were inadequate against the SA-6 [9]. The SA-6, operating in a previously unused portion of the frequency spectrum, accounted for 30% of the Israeli aircraft losses. Other new weapons introduced during this war were the ZSU-23-4 (an AAA system mounted on a tracked vehicle with its own radar), a wheeled vehicle mounting numerous SA-7 infrared-guided missiles, and the Snapper and Saggar wire-guided antitank missiles. These new systems forced initial reliance on an old countermeasure (chaff) and a rapid scramble to develop new tactics and techniques [9]. This search for new responses to changing threats is a common pattern throughout the evolution of electronic warfare. Other patterns are the expanded use of the frequency spectrum and the increased complexity and power requirements of the jammer systems.

1.2 DEFINITION OF ELECTRONIC WARFARE

As shown in Fig. 1.2, electronic warfare is one element in a larger effort known as electronic combat (EC). In broad terms, EC is action taken in support of military operations against the enemy's electromagnetic capability. EC includes electronic warfare (EW) as well as elements of command, control, and communications countermeasures (C^3CM) and suppression of enemy air defense (SEAD) [10]. Electronic combat is a new term encompassing all activities historically considered to be electronic warfare. The activities of defense suppression and communication jamming, previously a part of

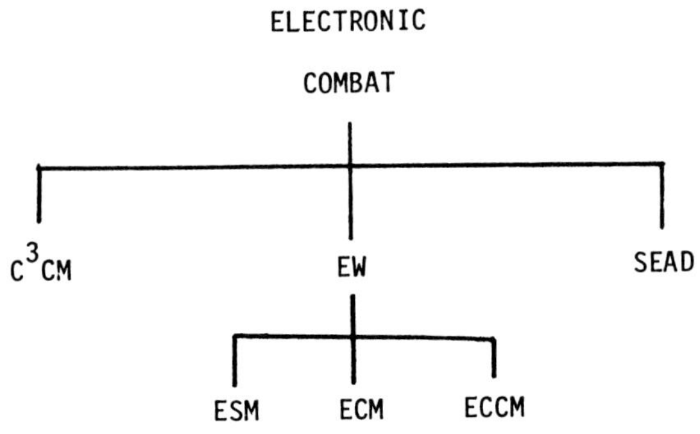

Fig. 1.2 Organization of electronic combat.

electronic warfare, are now considered as separate areas of SEAD and C^3CM, both of which are on an equal level with electronic warfare.

Electronic warfare is military action involving the use of electromagnetic energy to determine, exploit, reduce, or prevent hostile use of the electromagnetic spectrum and to maintain friendly use of the spectrum [1]. One distinction between electronic combat and electronic warfare is that electronic combat may involve action that does not use the electromagnetic spectrum. Electronic warfare consists of three principal elements:

Electronic warfare support measures (ESM)—gathering and immediate analysis of electronic emissions of weapon systems to determine a proper and immediate reaction.

Electronic countermeasures (ECM)—development and application of equipment and tactics to deny enemy use of electromagnetically controlled weapons.

Electronic counter-countermeasures (ECCM)—actions necessary to insure use of the electromagnetic spectrum by friendly forces.

Electronic warfare should be viewed as a tool for use with other tools available to the battlefield commander to achieve mission objectives. Proper application of electronic warfare should not be viewed as a panacea that can make an aircraft invulnerable to all electromagnetically controlled weapons. After all, the proper application of ordnance does not guarantee destruction of the target when the element of surprise is required. As shown in Fig. 1.3, for example, the objective of the forces to the right of the forward edge of the battle area (FEBA) is the destruction of a target on the left side of the figure. Certainly, the attacking commander will attempt to accomplish this objective with "acceptable" losses. His "tools" may include different types of aircraft, knowledge of environmental conditions, available armament, etc. In planning his mission, he will anticipate defenses and consider aircraft that can deliver the required ordnance to the target.

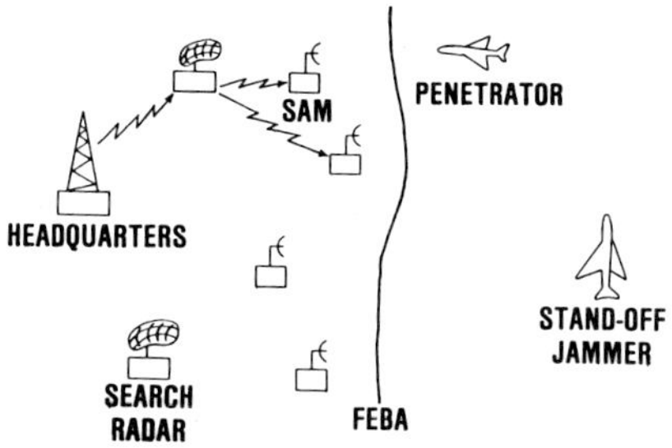

Fig. 1.3 Penetration problem.

A commander uses electronic warfare support measures to obtain information about enemy forces. Determination of the composition and posture of defensive forces actually requires use of all available intelligence information sources, but only certain specific activities in intelligence gathering can be considered as electronic support measures. ESM activities enable the commander to determine the enemy's EW capability, its effectiveness, current use of the capability, its location, etc.

The commander applies electronic countermeasures when he considers his route of penetration, altitude, and method of defense suppression. For example, will he employ standoff jammers (SOJ)? Will he carry ECM equipment on the penetrators? If so, how will he use the equipment? Obviously, electronic countermeasures are not merely "black boxes" that make penetrators impervious to surface-to-air missile systems. Electronic countermeasures are the intelligent consideration of all electromagnetic factors that can deny the enemy use of defensive systems and thus affect the outcome of a battle.

Opposing commanders must also consider electronic counter-countermeasures, a third element of electronic warfare. Both commanders may exercise emission control to deny knowledge of critical operating parameters in their equipment to opposing surveillance systems. The defensive forces must consider radar sighting for "optimum" protection of their resources against expected mixes of attack forces. The attackers need the ability to recognize malfunctioning of their terrain avoidance capability due to jamming by ground defenses. Again, another "black box" attached to the receiver, commonly called an ECCM "fix," may not be as important as knowing when to use it.

In other words, electronic warfare is not merely the search for the magical emission that will deny the enemy a functioning system. It is the constant gathering of information on the enemy's capabilities and his plans for using his capabilities. It is the development of technology to do more "things," but, in the final analysis, it is planning for "what things" to do "when" to achieve maximum enemy confusion and to gain time.

1.3 ORDER OF DISCUSSION

Since the objective of this text is to introduce the reader to radar electronic warfare, he must become familiar with both radar and electronic warfare. The radar system is treated first because it is considered to be the offensive element. Chapter 2 includes a discussion of radar principles, terminology, and performance capabilities. Chapter 3 contains a discussion of the ECM divisions, descriptions of generic ECM modulations, and an analysis of their effectiveness against various radars. Chapter 4 outlines various ECCM techniques and their use, and Chapter 5 concludes Part I with a description of the ESM receivers so essential for the control of modern ECM systems. Chapter 6 marks the beginning of Part II and contains background material used in the remainder of the text. Chapter 7 develops various radar models and Chapters 8–10 analyze the effects of certain ECM techniques on these radar models.

2
PRINCIPLES OF RADAR

Radar sensors are the eyes and ears of a large number of current weapon systems. An understanding of the capabilities and limitations of these sensors is essential to successful EW planning. This chapter reviews basic radar tasks, factors affecting achievement of these tasks, equations for determining radar range, classes of radars, and their principles of operation.

2.1 RADAR TASKS

A radar is tasked either to detect the presence of a target or to determine its location. Depending on the function emphasized, a radar system might be classified as a search or tracking radar, and it might be assigned an appropriate role within a larger weapon system. For example, determining the presence of a target is normally termed "target detection," while location of a target might be referred to as estimation of target parameters; i.e., the radar is tasked to determine the range of a target. Conceivably, a radar used as an office security system would detect an intruder (target) without giving any indication of the intruder's location within the office. The office could be called the radar's resolution cell. More typically, a radar is tasked either to detect intruders in many different resolution cells (or offices) or to determine an intruder's location within a particular resolution cell. Figure 2.1 shows one possible arrangement of resolution cells around a radar. The radar is tasked to determine the target's presence or absence within the large number of resolution cells designated by the location in range and angle of their centers and characterized by some finite range and angular extent. Thus, target "detection" naturally implies within a specific cell or a rough estimate of target location. Radar detection in other types of resolution cells is possible. If the radar in Fig. 2.1 is located on Earth, a three-dimensional resolution cell could be defined by considering the angle of elevation above the Earth as shown in Fig. 2.2. Other resolution cells are considered in later sections.

Once the rough estimate of target location is available, location of the target within the resolution cell or acquisition of other data is usually necessary. For example, most methods of target tracking require knowledge of the position of the target within the resolution cell for subsequent repositioning of the resolution cell and continued tracking. Other systems may observe the target reflections only for the purpose of identification. In this process, estimates of the size, orientation, and shape of the target may be required. The

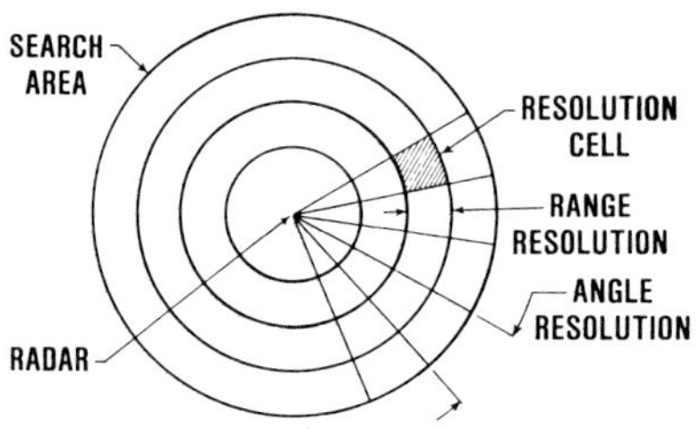

Fig. 2.1 Radar resolution.

same radar need not be tasked for both detection and estimation, although some capability in each area is the most common practice. It may be advantageous to use two radars, each with a specialized function in a given weapon system.

A weapon system may require many radars, each providing detection and estimation within a specific environment of target returns and noise. For example, a surveillance radar may detect a target and then use a secondary radar to identify the target. An airborne interceptor may be directed to the vicinity of the target from which its own radar can begin to acquire (detect) and then track (estimate position of) the target. A missile fired at the intruder may have its own radar, and so on.

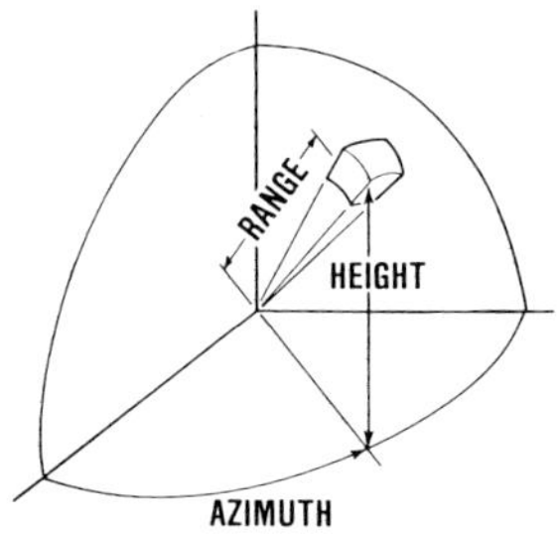

Fig. 2.2 Resolution in three dimensions.

2.2 FACTORS AFFECTING RADAR PERFORMANCE

Radar tasking has been described in terms of detecting the presence of, or estimating the position of, a target within a resolution cell. This section addresses factors that determine the size of the resolution cell, the level of target signal, the level of natural interfering signals, and, ultimately, radar performance.

Wave Propagation

Since the signals emitted by a radar are electromagnetic (EM) in form, discussion of EM wave propagation is necessary. As shown in Fig. 2.3, an EM wave in the atmosphere can be compared to a wave on the surface of a body of water. That is, a stone dropped in the center of a quiet pond is

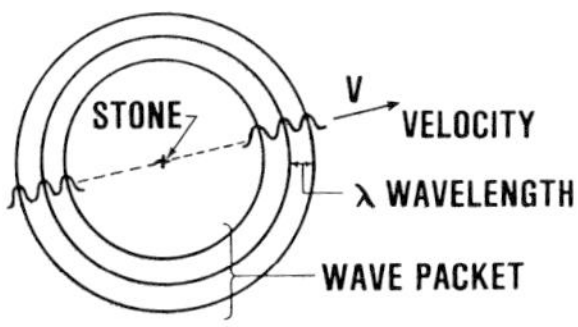

Fig. 2.3 Wave disturbance in water.

analogous to a radar transmitter in the sense that it disturbs (pulses) the surface of the water. This disturbance takes the form of a wave packet that travels away from the source at a fixed velocity V (for EM waves, this velocity is the speed of light or 3×10^8 m/s). The time interval over which the packet is observed as it passes a point can be designated by the arrival time of the packet T_i and the time the packet completely passes the point T_f. Passage of the wave packet at any particular point causes the surface of the pond to oscillate up and down. Passage of EM waves at a given point causes a similar oscillation of the electric or magnetic field. The time variation of the electric field at a point can be described by the equation (see Appendix).

$$E(t) = E_{\max} \cos (2\pi f_R t + \theta)$$

provided that the time t lies between times T_i and T_f previously defined. At other times, the electric field is zero. In this equation, the frequency of oscillation f_R is the inverse of the time between the occurrences of the peak value of the electric field. In the analogy of the water wave, the equivalent would be the inverse of the time between the peak displacements of the water. The frequency of oscillation is usually written in hertz (cycles/second) or in hertz to some power of 10, for example,

$$1{,}000{,}000 \text{ Hz} = 10^6 \text{ Hz} = 1 \text{ megahertz} = 1 \text{ MHz}$$

Also in the analogy of the water wave, a definite distance can be observed between subsequent wave crests. This distance is referred to as the "wavelength" λ of the disturbance (or signal) and is related to the frequency and velocity V of the wave by

$$f_R \lambda = V$$

If the signal lasts for an interval of time given by $\tau = T_f - T_i$, then the wave packet will occupy $V\tau$ meters of space. This distance is usually referred to as the pulse-width. Since the velocity of an EM wave in the atmosphere is usually 3×10^8 m/s, it is sufficient to designate the pulse width of a typical radar by its time duration τ.

The radio frequency (RF) used depends on the purpose of the radar. For typical radars, these frequencies vary from approximately 100 MHz to 20 GHz (1 GHz = 1000 MHz). If one assumes that the velocity is constant at

3×10^8 m/s, the wavelength of the radiated EM energy from typical radars varies over the values shown in Table 2.1.

Table 2.1 Wavelength vs Frequency

Frequency	Wavelength	Wavelength
f_R, MHz	λ, m	λ, in.
100	3.0	118
1,000	0.3	11.8
10,000	0.03	1.18
20,000	0.015	0.6

A number of conventions have arisen over the years for labeling various frequency ranges, for example, "X-band" and "super high frequencies." Figure 2.4 shows some of these conventions. In Fig. 2.4, "current U.S." refers to the present frequency band designation used by the Department of Defense in matters pertaining to electronic warfare [93]. "Old U.S." refers to the method of band designation used by most of the microwave industry in the United States [1,p.5-2]. Finally, ITU refers to the band designations set up by the International Telecommunications Union and in nearly universal use [63, pp. 18–60].

In the fluid analogy shown in Fig. 2.5, a block of wood or any other discontinuity at a distance R from the initial disturbance would create a reflected wave. Common experience with reflected waves in water suggests several observations that apply to EM waves. First, the wave reflected from the block starts at $T_1 = R/V$, the instant the original disturbance (or incident wave) reaches the block. Second, the reflected wave will reach the original position of the stone (the point where it created the disturbance) after a time

$$T_2 = \frac{2R}{V} = 2T_1$$

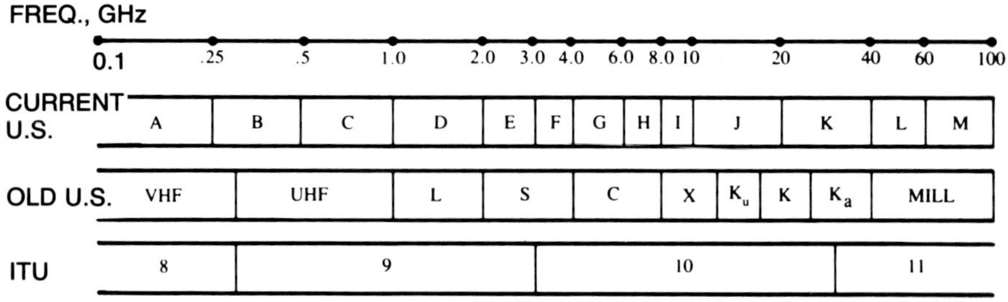

Fig. 2.4 Frequency bands.

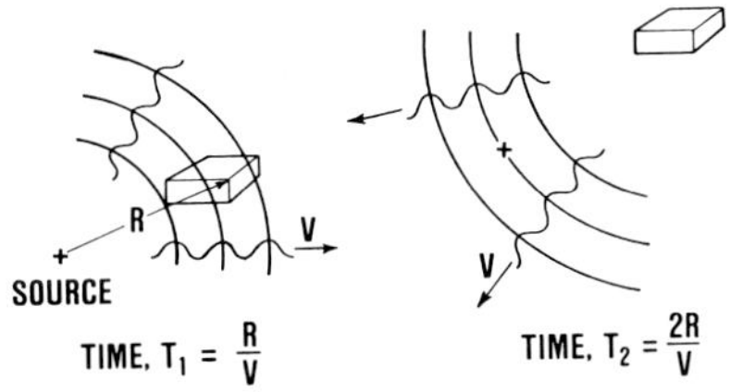

Fig. 2.5 Wave reflections.

Third, the strength of the reflected wave (i.e., its peak disturbance) decreases as the separation between the source (radar) and the block (target) increases. Finally, the reflected pulse width will be similar to the pulse width of the incident pulse. All times are measured from the time of the original disturbance.

The radiated and reflected EM waves are "polarized." In the water analogy, the water displacement always appears at right angles to the direction of wave propagation and this displacement is always vertical. The electric field also lies at right angles to the direction of wave propagation, but it can appear in any arbitrary direction or its direction can change with time. In the familiar analogy of polarized sunglasses in Fig. 2.6, for example, the view becomes alternately opaque and clear as one of two polarized filters is rotated because light is also an EM wave consisting of individual photons with different polarization. The filter furthest from the eye passes only vertical polarization and, depending on its orientation, the second filter then passes or stops this vertical component. The impact of polarization in electronic warfare is that, ideally, the radar antenna will be insensitive to a horizontally polarized jammer if the antenna is vertically polarized (the electric field oscillates vertically in some reference system).

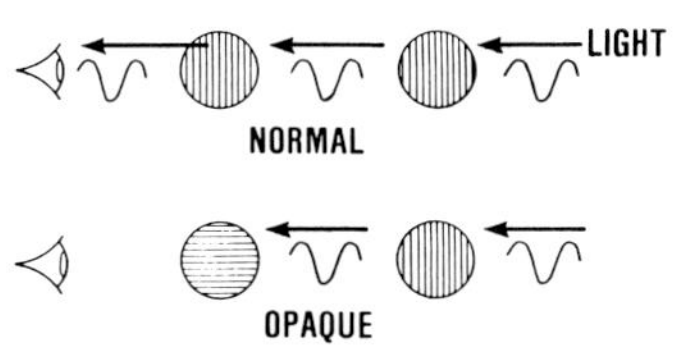

Fig. 2.6 Polarization.

Length of Range Resolution Cell

What would be the reflected signal from two targets if the radar and the two targets form a straight line? One can answer this question by assuming that multiple reflections can be ignored because the scattered signal from either target is small in relation to the incident signal. The total reflected signal is then simply the sum of the two individually reflected signals.

Table 2.2 shows the individual and combined signals for various separations of point targets. A point target is small relative to radar pulse width.

Table 2.2 Range Resolution

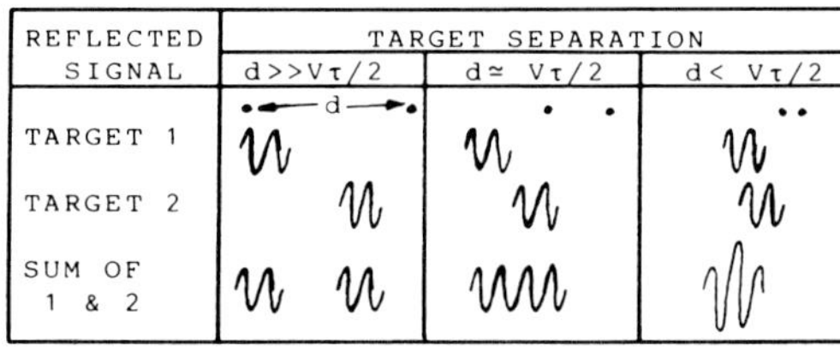

For target separations greater than $V\tau/2$, two distinct returns are observed. When target separation is slightly greater than $V\tau/2$, the individual returns almost merge into one and thus complicate the task of determining the number of targets. Resolution of individual targets becomes impossible if the target separation is less than $V\tau/2$. Thus, $V\tau/2$ is considered the smallest radar resolution cell that can be used to distinguish between two separate targets in a straight line.

Another perspective on the signal sensed by a radar at an instant in time T_1 is that it can be caused only by targets within $\pm V\tau/4$ of the distance associated with the instant of sampling, $R = VT_1/2$. Thus, the range interval within which a radar of pulse width τ searches for a target is $V\tau/2$ or its range resolution.

Angular Resolution and Antenna Gain

How can the reflected signal be increased without increasing the source strength? How can returns from targets at different angles but of the same range be isolated? This section provides an answer to the first question in a discussion of antenna gain. It then answers the second question by describing the close relationship between antenna gain and angular resolution. It concludes with a discussion of antenna performance measures.

The analogy of the water wave can be used to illustrate the concept of antenna gain. The stone can be considered a point source. In other words, lines drawn through any number of velocity vectors associated with the wave disturbance created by the stone would intersect at one point. Each of these lines is called a ray path. In antenna theory, such a source is considered isotropic since it radiates equal power in all directions. However, EM power must be concentrated in the desired direction for proper radar performance.

A widely used device for this purpose in optics, acoustics, and electromagnetics is the parabola. When a parabola is relatively large compared to the wavelength used, all rays directed toward it from a source at its focus emerge in parallel from its aperture. Thus, if a parabolic reflector is placed in the water at the proper distance from the dropped stone, the rays to the right of the reflector will increase in amplitude at the expense of the rays to the left of the reflector (Fig. 2.7).

The parabolic reflector antenna, then, is a device used to concentrate the available transmitter power in a given direction of space. If this transmitter

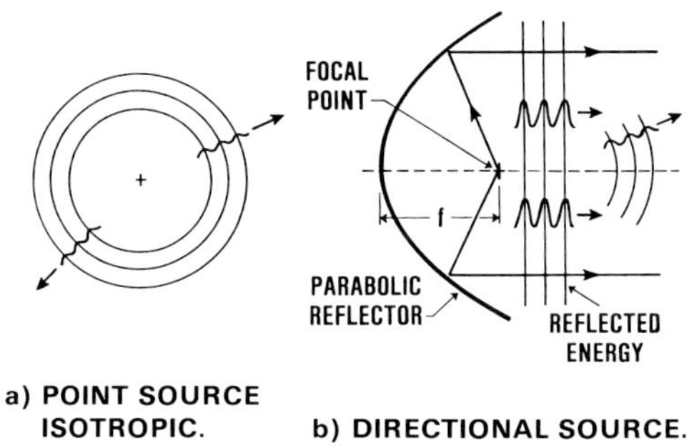

Fig. 2.7 Focusing principles.

power, denoted P_R, were radiated uniformly in all directions, the "power density" S in watts per square meter (W/m^2) at the target range R would be

$$S = \frac{P_R}{4\pi R^2}\ \text{W/m}^2$$

as the surface area of a sphere of radius R is $4\pi R^2$. If the antenna had a "gain" denoted G_R (purely numerical with no dimensions), the power density in *a particular direction* would increase in proportion to this amount. If the gain is 10, then the power density will be 10 times as large. If the antenna gain is in the direction of the target, the power density incident on the target becomes

$$S = \frac{P_R G_R}{4\pi R^2}\ \text{W/m}^2$$

As seen in the water analogy, higher gains are obtained from larger antennas. Actually, the gain of an antenna is related in a complicated manner to its physical size and the wavelength at which it is used. However, there is a simple relationship between the gain of an antenna and its "effective area" A_e. This relationship is

$$G_R = \frac{4\pi}{\lambda^2} A_e, \qquad A_e = \text{effective area}$$

The effective area is typically 50–75% of the actual physical area [13], excluding simple wire antennas.

Closely related to the achievement of antenna gain is the formation of very narrow antenna beams. If a fixed amount of power is focused in a given direction, this direction can be defined by angular limits. The power density in the defined direction is greater than that from an isotropic source and comes at the expense of the power density in other directions.

If all the power were radiated through the small shaded area shown in Fig. 2.8 rather than the total spherical surface, the power density (power per unit

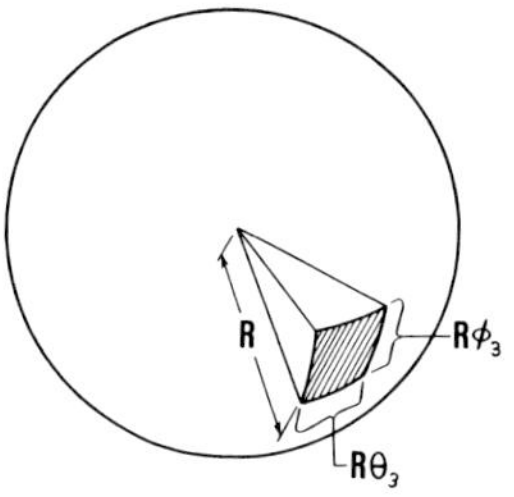

Fig. 2.8 Antenna gain.

area) must be $P_R/R^2\theta_3\varphi_3$ as opposed to $P_R/4\pi R^2$. Thus, the gain (angles measured in radians) would be

$$G_R = \frac{P_R/R^2\theta_3\varphi_3}{P_R/4\pi R^2} = \frac{4\pi}{\theta_3\varphi_3}$$

The region of the spherical surface through which power is radiated is uniquely defined by the angles θ_3 and φ_3. These angles can be considered the antenna beamwidth in the two directions. Thus, a relationship exists between antenna gain and beamwidth. Clearly, the size of the angle resolution cell is given by the antenna beamwidth (Fig. 2.1).

The ability to affect such focusing depends on the length L of the antenna in each dimension relative again to the wavelength of the radio frequency in use. The beamwidth achievable, typically expressed in degrees, is approximately $\theta_3 = 57.3\lambda/L$ (or $\theta_3 = \lambda/L$ radians) as shown in Fig. 2.9. If all power were concentrated in the shadowed regions of Figs. 2.8 and 2.9, the gain would be

$$G_R = \frac{\text{power density in beam}}{\text{isotropic power density}} = \frac{\dfrac{P_R}{R^2(\lambda^2/LaLe)}}{P_R/4\pi R^2} = 4\pi\frac{LaLe}{\lambda^2} = \frac{4\pi}{\lambda^2}A$$

Under these conditions, the gain is related directly to the physical aperture

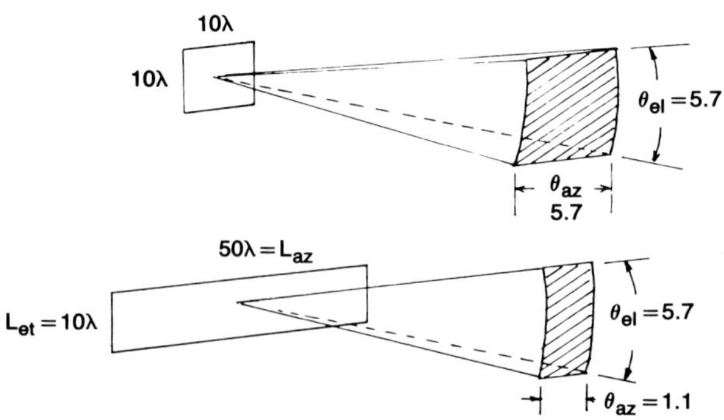

Fig. 2.9 Beamwidth dependence on antenna size.

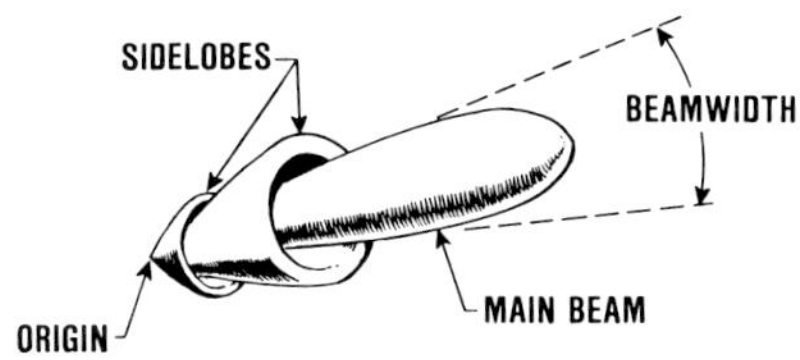

Fig. 2.10 Radiation pattern in three dimensions.

area A. In actual practice, the above gain is not achieved and an efficiency η is defined in such a manner that the effective area equals $\eta A = A_e$. Then $G_R = \eta(4\pi/\lambda^2)A = (4\pi/\lambda^2)A_e$ as before.

Another characteristic of directional antennas is that some of the RF power is distributed outside of the antenna main beam into so-called antenna sidelobes. The combination of the antenna main beam and its sidelobes is called the radiation pattern of the antenna. A representative radiation pattern is shown in Fig. 2.10. The distance from the origin to a point on the pattern is proportional to the power density of the radiation in the direction of the given point. An intersection of this pattern with a plane that includes the point of peak power density would have a plot as shown in Fig. 2.11.

Typically, the beamwidth is defined as the angular separation between the half-power points of the main lobe; separate values are quoted for azimuth and elevation plane cuts. Sidelobe values depend on design constraints and range from 1/20th to 1/100,000th of the peak value. If an antenna of a fixed dimension L is designed to yield the minimum possible beamwidth ($\theta_3 = 0.88\lambda/L$), it will also yield the highest sidelobes. Different designs will result in wider main beams ($\theta_3 = 1.2\lambda/L$) and much lower sidelobes from the same size antenna.

Some antennas common to radar or ECM systems are described in Table 2.3 [14,p.313] and Table 2.4 [15,Chap. 9,p.10]. In these tables, antenna gain and sidelobe levels are expressed in decibel (dB) form where

$$\text{Gain in dB} = 10 \log_{10} \text{gain}$$

The gain of an antenna in the direction of a sidelobe is often expressed relative to the maximum gain of the antenna. It is possible for an antenna to have a maximum gain of 40 dB and sidelobes at −30 dB relative to the

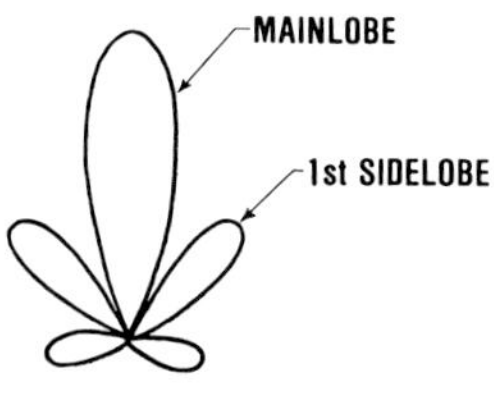

Fig. 2.11 Radiation pattern in a plane.

Table 2.3 Simple Antenna Characteristics

ANTENNA TYPE	TYPICAL RADIATION PATTERN	TYPICAL SPECIFICATIONS
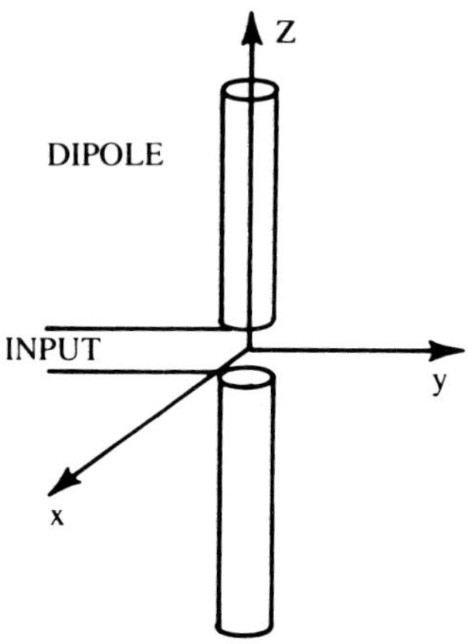	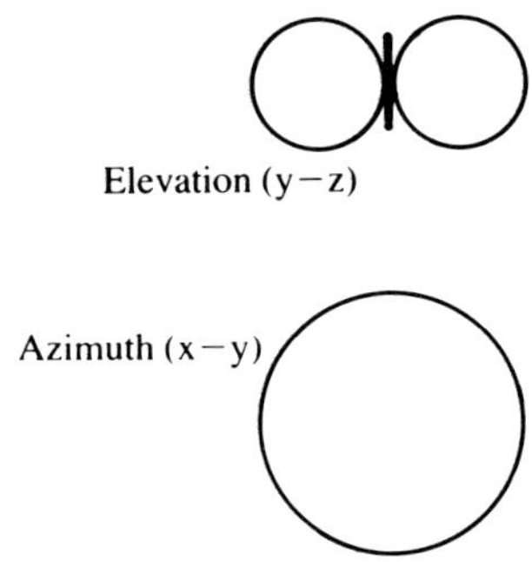	Polarization Linear Beamwidth 80° Gain 2 dB
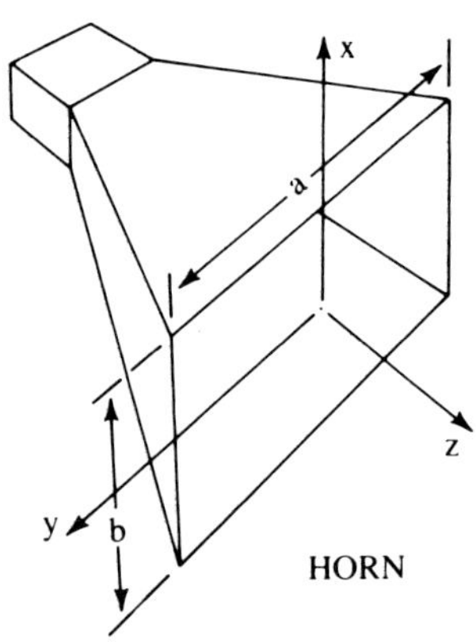	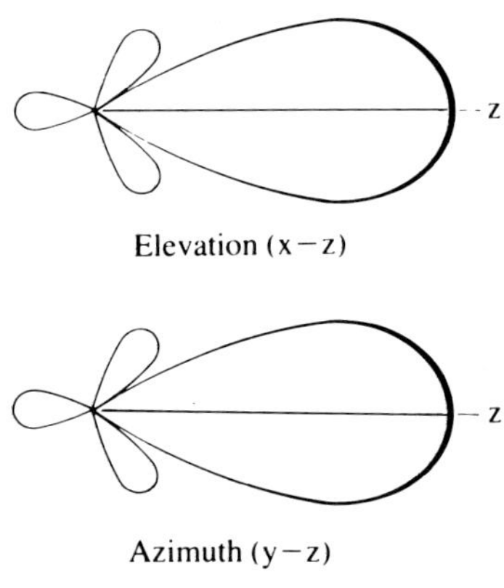	Polarization Linear Beamwidth: $a = b = \lambda$ $\theta_{el} \sim 50°$ $\theta_{az} \sim 72°$ Gain $\frac{2 \pi a b}{\lambda^2}$ [65, Chap 10]
CAVITY BACKED SPIRAL 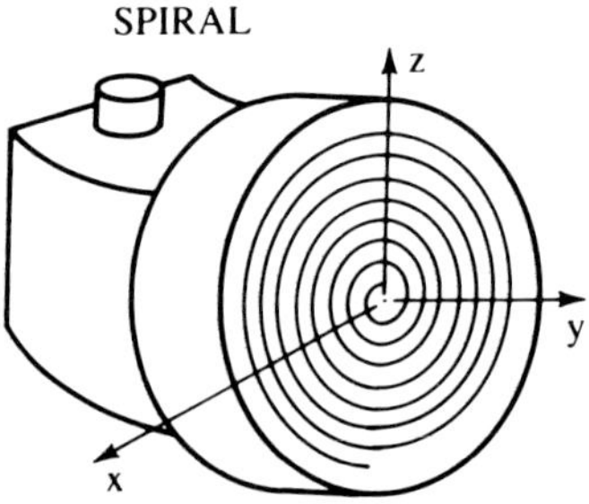	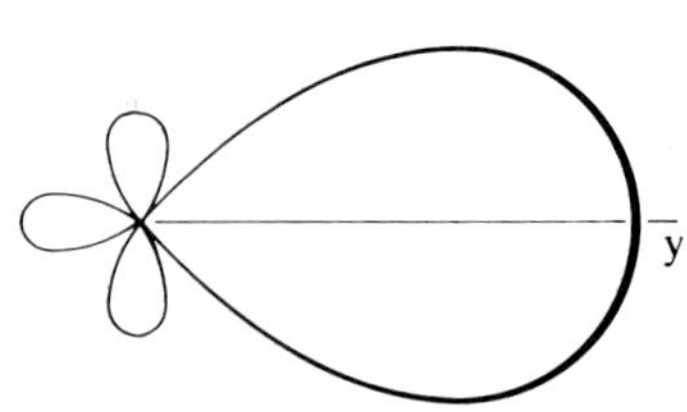 Elevation & Azimuth	Polarization Circular Beamwidth 90° Gain 2 to 4 dB Bandwidth 9 to 1 [40, p 426]

Table 2.4 Reflector Antenna Characteristics

ANTENNA TYPE	TYPICAL RADIATION PATTERN	TYPICAL SPECIFICATIONS
PARABOLA 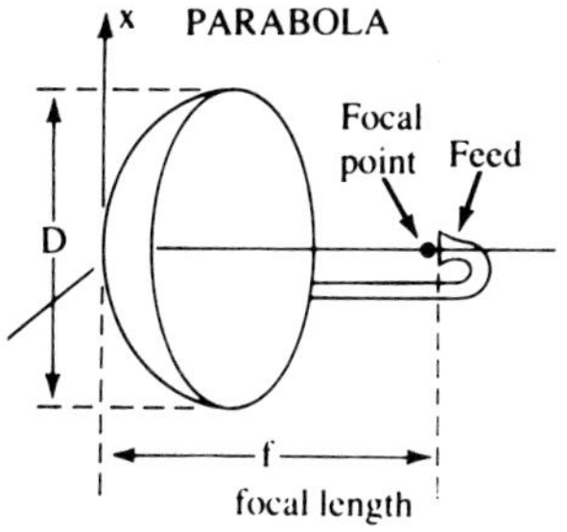	AZIMUTH & ELEVATION (1) 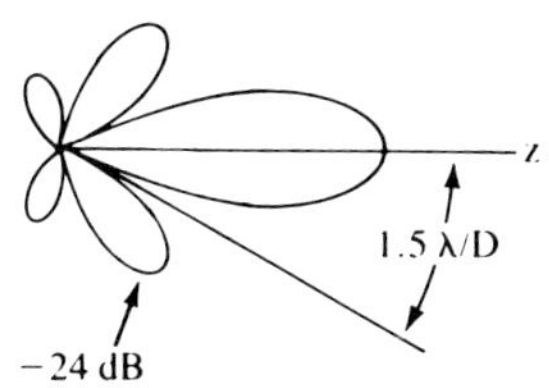	Polarization Same as feed on z − axis Beamwidth 1.17 λ/D Gain $0.65\left(\frac{\pi D}{\lambda}\right)^2$ Frequency Limit Lower D > 10 λ Upper Tolerances [65, Chap 12]
PARABOLIC CYLINDER 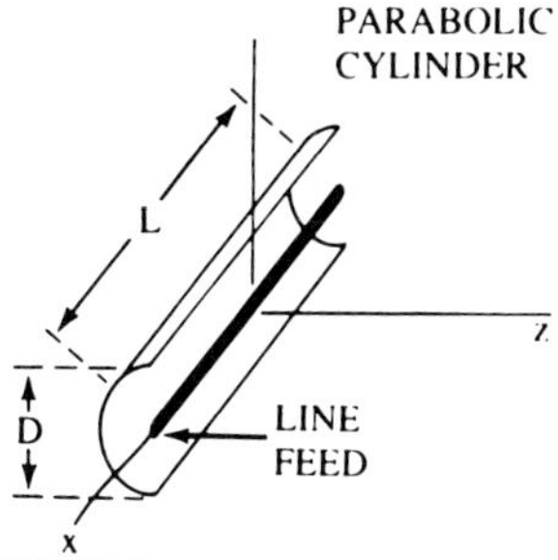	ELEVATION (y−z) 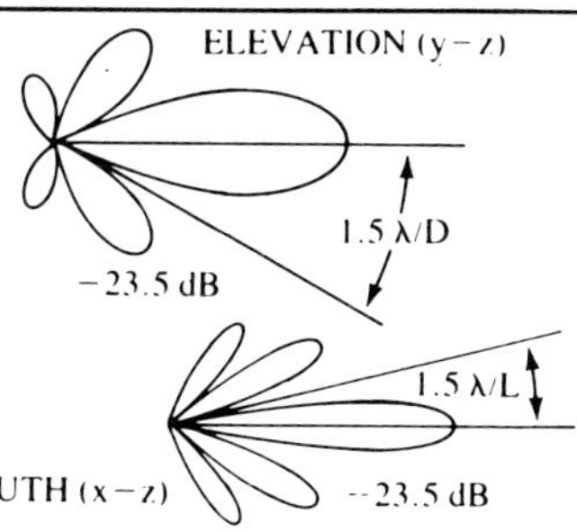 AZIMUTH (x−z) (2)	Polarization Same as feed on z-axis Beamwidth x-z plane 1.2 λ/L y-z plane 1.2 λ/D Gain $.664\frac{\pi}{\lambda^2}LD$ [13, p 338]
CASSEGRAIN 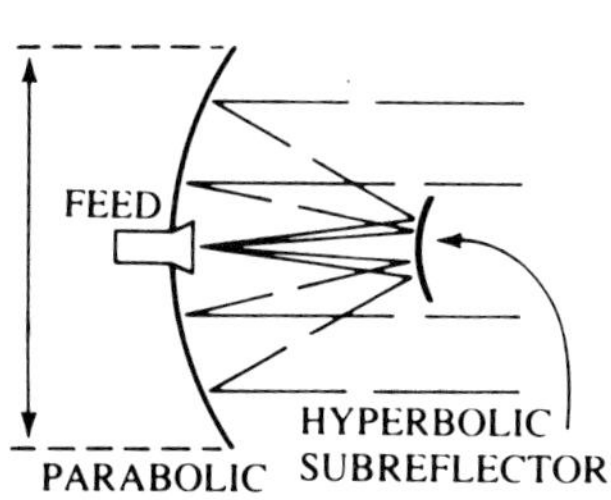	SIMILAR TO PARABOLA	

(1) Edge illumination of —10 dB. Measured.
(2) Theoretical values assuming cosine aperture distributions.

maximum gain. These sidelobes still have a 10 dB gain when compared to an isotropic source. Sometimes an average sidelobe level is given for an antenna. There is no common definition for the average sidelobe level, but often it is computed by averaging all but the largest sidelobes over a portion of the total spherical space surrounding the antenna.

Ratio of Target Signal to Receiver Noise

The signal-to-noise ratio is most important in quantifying radar detection and tracking performance. The received signal is an attenuated version of the emitted RF wave reflected by the target. The amplitude of this reflected pulse depends on many factors, such as target orientation or atmospheric conditions. There are various models for computing the probability associated with the occurrence of any range of values in amplitude of target returns. All such models contain an average or expected value of received power that can be written in terms of a "target cross section" σ_T as

$$P_r = S_i \frac{\sigma_T}{4\pi R^2} A_e$$

where

S_i = radiated power density at target = $P_R G_R / 4\pi R^2$
σ_T = target cross section, m^2
A_e = radar antenna effective area, m^2
P_r = power received by radar, watts

Reference to Fig. 2.12 should clarify various terms in this equation. The radar emits power of a certain density in the vicinity of the target. The target is modeled by an area σ_T, which intercepts $S_i\sigma_T$ watts and reradiates this power isotropically. The reradiated wave thus has a power density at the radar given by $S_i\sigma_T/4\pi R^2$. The radar antenna acts as a collector of RF energy over a given area, which is the same effective area discussed earlier. By using previous equivalent expressions, one can write this received power as

$$\begin{aligned} P_r &= S_i \frac{\sigma_T}{4\pi R^2} A_e \\ &= \frac{P_R G_R}{4\pi R^2} \frac{\sigma_T}{4\pi R^2} \frac{\lambda^2}{4\pi} G_R = \frac{P_R G_R^2 \sigma_T \lambda^2}{(4\pi)^3 R^4} \end{aligned}$$

This power is amplified to display the results to an operator or to be digitized for computer analysis. The basic processing steps in signal reception are shown in Fig. 2.13. The reflector focuses the returning EM wave on a small receiving antenna or feed. This signal may be amplified at RF and is almost always converted to a lower frequency, called the intermediate

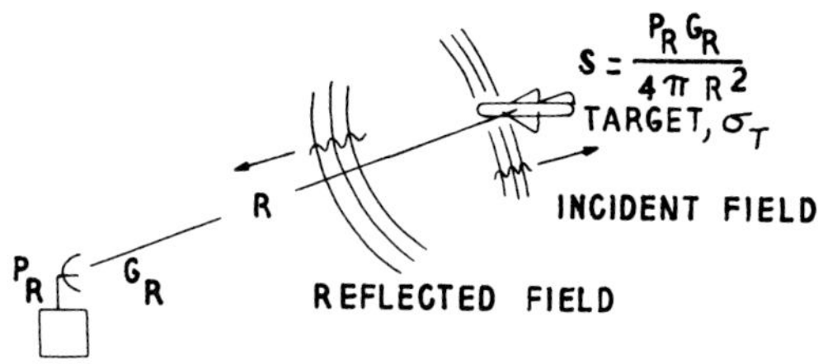

Fig. 2.12 Range equation geometry.

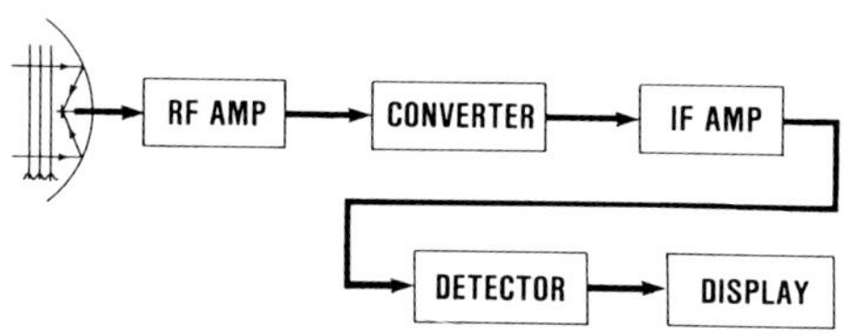

Fig. 2.13 Basic radar receiver.

frequency (IF), where substantial amplification occurs. The amplified signal is often "detected" (i.e., the pulsed nature of the signal is recovered) before the signal is inserted into a display or processing device.

The target return is not the only signal out of the receiver. The receiver always has some output because of the noise generated by thermal effects within the first few stages of the receiver. The received power from the target must exceed this thermal noise by a certain amount in order to state with a high degree of certainty, called the probability of detection, that a specific surge of output voltage is due to a target rather than noise. One can clarify this point by envisioning the receiver output connected to the vertical input of an oscilloscope: the horizontal input is triggered (begins) by a signal from the radar timer. Figure 2.14a shows the output of the receiver with no target present. This output can be related through the receiver gain to a quantity called the average noise power internally generated by the system N_{INT}, usually modeled as if it were generated at the antenna terminals.

Figure 2.14b shows the receiver output with a target present at a range such that $T = 2R/V$. As the target return decreases in amplitude, it will obviously become more difficult to discern the presence of the target. And, in the absence of a target, there is a finite chance that the receiver noise can be mistaken for a target. This possibility is called the probability of false alarm.

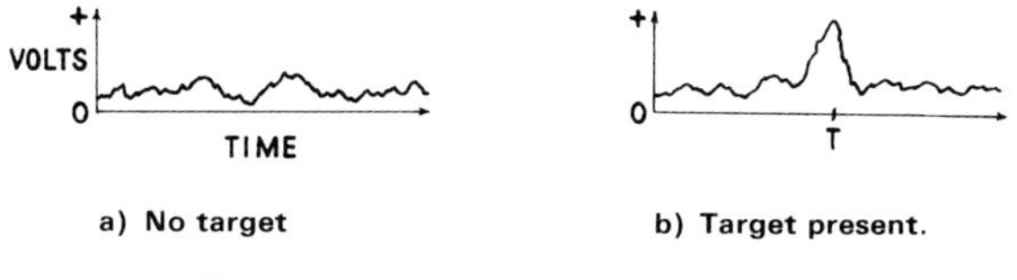

Fig. 2.14 Detected output signals.

For a radar to operate properly, the power of the target return must exceed the noise power of the system by a certain multiple called the minimum signal-to-noise ratio, $(S/N)_{\min}$. In terms of this ratio, the maximum range of the radar for specific targets and idealized propagation can be formulated, as will be shown later. The actual power received, measured at the same point at which N_{INT} is defined, is used to define the received signal-to-noise ratio per pulse as

$$\left(\frac{S}{N}\right)_p = \frac{P_r}{N_{INT}} = \frac{P_R G_R^2 \sigma_T A \lambda^2}{(4\pi)^3 R^4 N_{INT}}$$

This actual signal-to-noise ratio may or may not exceed the minimum signal-to-noise ratio required for system performance.

2.3 TYPICAL RADAR SYSTEM

A typical pulsed radar is shown in Fig. 2.15. The timing device periodically tells the transmitter when to provide power to the antenna. The time between transmitter outputs is called the pulse-repetition interval (PRI) and the inverse of the pulse-repetition interval is called the pulse-repetition frequency (PRF). The inherent assumption here is that the transmitter will shut off before it is told to transmit again. Selection of this pulse-repetition frequency is not completely random, but it is related to the desired "unambiguous" range of the radar. As shown earlier in Fig. 2.5, the reflection from an object at range R will return at a time $T = 2R/V$ (V = speed of EM wave). Thus, if the next transmit pulse occurs prior to this time, it may mask the target return or cause an erroneous interpretation of the target range (i.e., much shorter range than it really is). These effects are shown in Fig. 2.16, where the returns from target 2, at a range greater than $R_{un} = VT/2$, can overlap and be mistaken for returns from the shorter-range target.

The transmitter supplies EM power (RF power) in units of watts (typically kilowatts or megawatts) for a period typically measured in microseconds. The RF power is usually as high as possible and consists of a sinusoidally time-varying electric field. The duration of the pulse, symbolized by τ, is again chosen to attain specific objectives of range resolution. As mentioned earlier, the RF energy travels away from the antenna at $V = 3 \times 10^8$ m/s. The leading edge of a pulse of duration τ will be $V\tau$ meters

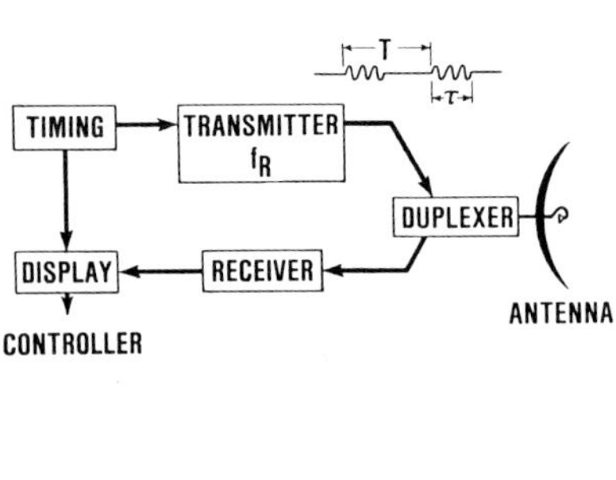

Fig. 2.15 Typical radar.

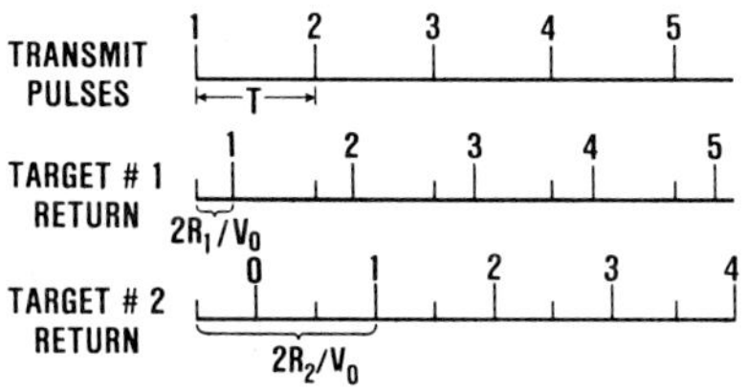

Fig. 2.16 Example of ambiguous target.

away from the antenna before the trailing edge of the pulse leaves the antenna. Thus, the pulse occupies a physical length of $V\ \tau$ meters. Other components are the duplexer, antenna, and receiver. The duplexer is a switch that connects the antenna to the transmitter or receiver, but prevents the coupling of the transmitter power into the receiver at levels that may be damaging to the receiver circuitry. As stated before, the antenna is a device to concentrate the available transmitter power in a given direction of space. The receiver amplifies the target return for display to an operator for other processing.

The earlier discussion of range resolution may have given the impression that any pulse duration is possible, but this is not the case. For the receiver to process a pulse of very short duration effectively, it must pass a continuous band of frequencies centered around the transmitter frequency. The width of this band of frequencies in hertz is called the receiver "bandwidth." For optimum operator viewing, bandwidth relates to the pulse duration through the expression [16]

$$B_R = 1.2/\tau$$

Thus, the receiver bandwidth must increase as the radar pulse becomes shorter. Typical values are shown in Table 2.5.

Increases in the receiver bandwidth results in greater internal noise power and voltage or power fluctuations at the receiver output increase because of noise. This level of noise power can be computed approximately through the following formula (a more accurate formulation involves the definition of the system's noise temperature):

$$N_{INT} = N_o B_R \,\mathrm{NF}$$

where

N_o = normal noise within a 1 MHz bandwidth given as 4×10^{-15} watts
B_R = receiver bandwidth, MHz
NF = radar system noise figure (not just receiver)

Expressed in decibel form, this equation becomes

$$N_{INT}\,(\text{in dBW}) = -144\ \text{dBW/MHz} + B_R\ (\text{in dB}) + NF_{dB}$$

In the above equations, the "noise figure" (usually expressed in decibels) for the radar system has been introduced. This quantity is a measure of the quietness of the receiver and other components.

Table 2.5 Bandwidth vs Pulse Width

Pulse width τ, μs	Bandwidth B_R, MHz
0.3	4.0
1.0	1.2
10.0	0.12

There are a large number of standardized radar displays, but only two are included in this discussion. The display could consist of an A-scope upon which the amplitude of the receiver output is displayed as a function of time. This type of display is identical to that shown in Fig. 2.14. In such a display, a new trace begins with each transmitted pulse and is then amplitude modulated. This terminology means that the deflection of the spot on the viewing screen is proportional to the amplitude of the receiver output. Another common display is the plan position indicator (PPI). The A-scope indicates only recently received signals or signals from targets in the antenna beamwidth, but the PPI can display both current and past targets. Figure 2.17 shows a sketch of a radar with a plan position indicator.

The plan position indicator presents a display as if the region under surveillance were viewed from above with the radar at the center and an arbitrary reference direction at the top. The timer originates a trace with each radar pulse; the trace is then on a radial corresponding to the current pointing direction of the antenna. Target returns appear as bright spots, the size of which is discussed in a later section. The plan position indicator is an example of an intensity-modulated display.

The ability of the plan position indicator to display old target returns is due to the persistence of phosphor on its display screen and the retina of the human eye. The light spot on the screen remains long after the electron beam forming the spot has moved on.

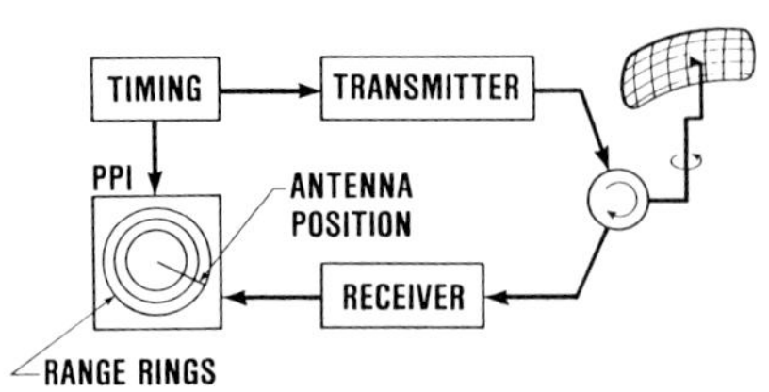

Fig. 2.17 Radar with PPI display.

2.4 RADAR DETECTION IN RECEIVER NOISE

This section uses material from earlier sections to quantify the expected performance of a pulse radar. It considers the effect of a fluctuating target cross section and pulse integration.

Single-Pulse Detection

The received signal-to-noise ratio for the situation described in Fig. 2.12 is

$$\left(\frac{S}{N}\right)_p = \frac{P_R G_R^2 \sigma_T \lambda^2}{(4\pi)^3 R^4 N_o B_R \text{ NF}}$$

where the result from Sec. 2.3, $N_{INT} = N_o B_R$ NF, is used. This signal-to-noise ratio, achieved for each received pulse, represents the target (or targets) within a single resolution cell of length $\Delta R = V\tau/2$ and width proportional to beamwidth times range. At a time τ seconds later, the content of a new resolution cell is tested. The radar system is required to test the voltage of the receiver output every τ seconds at a rate given by $B_R = 1/\tau$ or it will risk missing a return. In each of these tests, the probability of false alarm (P_{FA}) depends on the value of threshold voltage relative to the receiver noise voltage, which in turn depends on N_{INT}. A typical probability of false alarm is 10^{-8}, which means that, on average, a false alarm is declared once in 10^8 trials. Thus, the time between false alarms t_{FA} is given by

$$B_R t_{FA} = 10^8 = 1/P_{FA}$$

After the probability of false alarm is established, the probability of detection (P_D) then depends on the value of $(S/N)_p$ as shown in Table 2.6 [17] (assume that $(S/N)_p = (S/N)_{\text{min}}$).

Since radar performance is usually specified for a particular target cross section, it is necessary to find the range at which a given set of radar parameters (power, gain) provide this performance. A typical problem is shown in Example 2.1.

Table 2.6 $(S/N)_{\text{min}}$ Required for Target Detection, dB

P_D	P_{FA} 10^{-6}	10^{-8}	10^{-10}
0.3	10.4	11.7	12.8
0.5	11.2	12.6	13.6
0.8	12.6	13.7	14.6
0.9	13.2	14.2	15.0
0.99	14.2	15.1	15.8

Table 2.7 Radar Parameters

Quantity	Value	dB Value
P_R	1 MW	60 dBW
G_R	3160	35 dB
λ	0.5 m	−3 dB
σ_T	10 m^2	10 dBm^2
NF	3	5 dB
B_R	1.0 MHz	0 dB/MHz

Example 2.1

Find the range of a radar with the characteristics in Table 2.7. A $P_{FA} = 10^{-6}$ and a $P_D = 0.9$ are required.

Based on data from Table 2.6, the minimum signal-to-noise per pulse is 13.2 dB or has a numerical value 21. The radar range is

$$R^4 \leq \frac{P_R G_R{}^2 \lambda^2 \sigma_T}{(4\pi)^3 N_o B_R \mathrm{NF}(S/N)_{\min}}$$

This can be solved in decibel form as

$$4R\ \mathrm{dB} = P_R + 2G_R + 2\lambda + \sigma_T - 3(4\pi)_{\mathrm{dB}} - N_{INT} - (S/N)_{\min}$$

$$= 60 + 70 - 6 + 10 - 33 - (-139) - 13.2 \approx 227$$

$$R = 10^{(227/40)} = 473\ \mathrm{km}$$

Radar Search Rate and Multiple-Pulse Integration

The decision regarding the presence or absence of a target in a cell is usually not made on the basis of one pulse. A number of pulses, typically 10–40 are received from each cell and combined (i.e., integrated); the decision is then made on the summed signal. How many pulses are available in a specific case? In Fig. 2.18, a search beam is shown at a given location during its rotation around the radar position. The half-power beamwidth is given by θ_{az} and the rate of beam rotation is ω radians/second. Only the target returns with significant power, usually those caused by illumination from the antenna within its 3 dB beamwidth, are considered. If the radar is the same one considered in Example 2.1, the maximum radar range is 473 km (255 n.mi.). If this maximum range is set equal to the desired unambiguous range R_{un}, then, based on earlier considerations, the pulse-repetition frequency f_{PRF} must be less than 317 Hz (i.e., $f_{PRF} = V/2R_{un}$). With this pulse-repetition frequency, the antenna rotates through an angle ωT between pulses. After N pulses, the antenna has rotated through the 3 dB beamwidth, or $N\omega T = N\omega/f_{PRF} = \theta_{az}$.

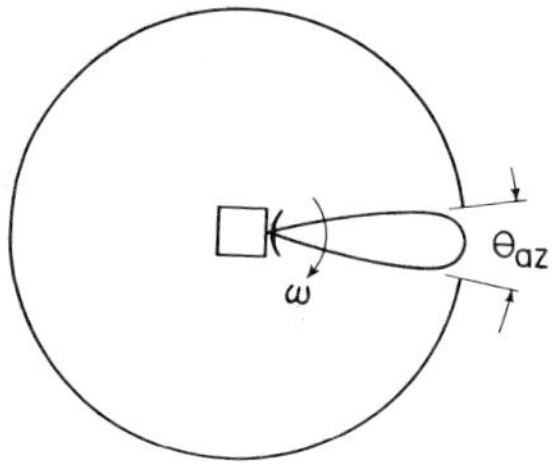

Fig. 2.18 Overhead view of search radar.

The number of pulses available for addition from one angular resolution cell is thus

$$N = \frac{f_{PRF}\theta_{az}}{\omega}$$

or, for a required number of pulses, the angular rate must be

$$\omega = \frac{f_{PRF}\theta_{az}}{N}$$

These pulses are processed (i.e., combined) in the radar receiver. In this processing (called incoherent or coherent integration depending on the design of the radar), the radar develops a signal with an integrated signal-to-noise ratio $(S/N)_{INT}$ at the point in its circuitry where a decision must be made about the existence of a target. Integration using both signal amplitude and phase information is called "coherent integration." Integration that does not use signal phase is called "noncoherent integration." Generally, coherent integration is accomplished prior to detection, resulting in the term "predetection integration" and noncoherent integration after detection for which the term is "postdetection integration." A cathode ray tube performs noncoherent integration. If the radar is coherent and it processes N pulses, the S/N ratio increases by the factor N (numeric, not dB). Then, the integrated signal-to-noise ratio is given by

$$(S/N)_{INT} = N(S/N)_P$$

A good rule of thumb, according to many sources, is that the increase in the S/N ratio is approximately proportional to $\sqrt{N}$ if the radar is incoherent. Actually, this is a pessimistic approximation for incoherent integration [17], but other loss mechanisms beyond this discourse justify its use. Thus, for incoherent pulse integration,

$$(S/N)_{INT} = \sqrt{N}(S/N)_p$$

In either case, the resulting integrated signal-to-noise ratio must equal $(S/N)_{\min}$ as required by the desired P_D and P_{FA}.

For example, the radar in the previous example might be tasked to detect a 2.5 m^2 target rather than a 10 m^2 target at 255 nm. The S/N per pulse $[(S/N_p)]$ out of the IF amplifier is now 7 dB, rather than 13 dB, because the target size has been reduced by a factor of four. To compensate for this loss of 6 dB, the radar must noncoherently integrate 16 such pulses ($\sqrt{16} = 4$, $10 \log 4 = 6$). This number of pulses requires the radar antenna to rotate at an angular velocity of $\omega = 0.45$ rad/s or 26 deg/s. The radar now scans a complete circle (360 deg) in 14 s or makes 4.3 rotations/min. Some circles state that the radar has a frame time (the time between display updates) of 14 s.

The range equation can account for the impact of pulse integration by incorporating the appropriate factor for S/N improvement. The IF signal-to-noise ratio per pulse is then $(S/N)_p$. The minimum integrated signal-to-noise ratio required for reliable target detection is related to the per pulse value through $(S/N)_{\min} = N(S/N)_p$ for coherent integration and $(S/N)_{\min} = N/L_i(S/N)_p$ for incoherent integration. The factor L_i representing the incoherent integration loss is assumed here to equal $\sqrt{N}$.

The new equation for radar range is then

$$R = \left[\frac{P_R G_R^2 \sigma_T \lambda^2 N}{(4\pi)^3 (S/N)_{\min} N_{INT} L_i}\right]^{1/4}$$

Fluctuating Targets and Other Losses

The model of the radar detection problem used thus far contains too many simplifications, the removal of which usually reduces the radar performance relative to the performance predicted by the simple model. Inclusion of "loss factors" in the range equation can often account for this reduction. These loss factors can result either from more realistic modeling of the radar system or from the environment in which the radar operates. This section considers loss factors accounting for fluctuations in the target cross section, beam scanning, and rain attenuation and then mentions some other loss factors.

As shown in Fig. 2.19, the target cross section is very dependent on the radar viewing angle. Since the exact angle at which the target is observed is unknown, the value of the cross section used in the range equation must be an expected value or the value obtained by averaging the actual cross section observed over a 5–10 deg interval. This expected value can also vary in angle. More important, the target is usually moving and the viewing angle, as well as the cross section, changes in time. If the cross section changes significantly over a time of viewing equal to the radar pulse repetition interval , it is called Swerling II. If the cross-sectional changes are significant only over times on the order of those between radar bursts (each group of pulses integrated), the target is Swerling I. The loss factor accounting for a Swerling I target is shown in Table 2.8 [18,p. 48] and it is perceived as dependent on the detection probability required. This loss factor is incorporated into the range equation by adding its decibel value to any other existing losses, such as the incoherent integration loss.

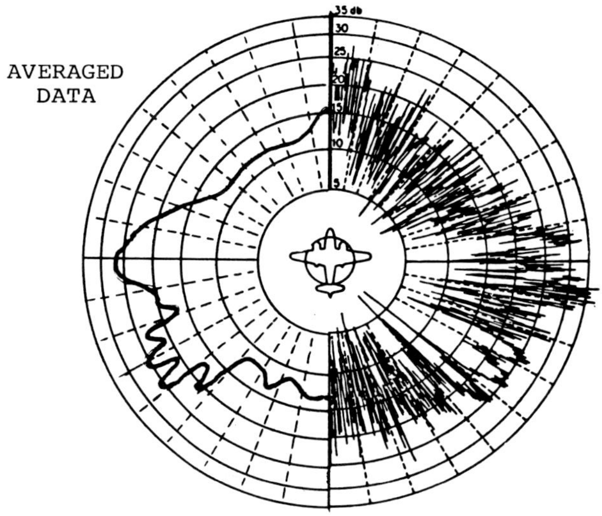

Fig. 2.19 Aircraft radar cross section [measured data from Refs. 18 (p. 40) and 2 (p. 13-10)].

The pulses received also vary in amplitude because of the varying antenna gain as the beam scans past the target (i.e., the pulses are not all received with maximum antenna gain). This variation in amplitude requires inclusion of a scanning loss L_s in the range equation. A value, $L_s = 1.6$ dB, for this scanning loss is valid in most cases for *each* direction of the scan. Thus, for a fan beam that scans in one plane, $L_s = 1.6$ dB, but it is 3.2 dB for a pencil beam scanning vertically and horizontally.

Other loss factors associated with the radar must normally be incorporated into the range equation if one desires realistic predictions of range performance. A few of these factors are listed in Table 2.9, along with typical values. Of course, the operator loss would not be included if electronic thresholding were used. Similarly, many other corrections could be required

Table 2.8 Target Fluctuation Loss

Detection probability	Loss, dB
0.5	1.4
0.8	5.6
0.9	8.1
0.95	10.8
0.99	17.2

Table 2.9 Detection Loss Factors

Reason for loss	Value, dB
Operator	1.4
Collapsing	2.0
Filter mismatch	1.0

Table 2.10 Attenuation Due to Rain, dB/km

Fall rate, mm/h	Frequency, GHz 3.0	10	30
2.5	0.000785	0.0388	0.492
12.5	0.00364	0.285	2.73
25.0	0.00728	0.656	5.47

in specific cases. For example, if the power out of the radar transmitter suffered losses on its way to the antenna, then line losses must be added. These losses and others are discussed in depth in Refs. 17 and 18. The purpose of this section is to make the reader aware of the need for, and the scope of, the corrections required in the previous simplified range equation. Consideration of other factors in system loss is not warranted at this time because they would cause undue complication of the model.

However, consideration of an environmental loss factor due to rain is essential. The loss factor accounting for rain approximates the attenuation experienced by the radar signal as it propagates to and from the target. As would be expected, this loss increases with increased rainfall rate. However, the loss increases even more rapidly as the radar frequency increases. Table 2.10 tabulates rain attenuation rates for selected frequencies and rainfall rates [69].

The loss factor is computed by multiplying the attenuation rate of Table 2.10 by twice the range to the target. Since the target range is unknown, it must be estimated and an iteration procedure must be followed to solve for the actual range value. If the computed range is smaller (larger) than the estimate, then the estimated value used to compute the rain loss is increased (decreased) until the computed and estimated values are equal.

Example 2.2

Building upon Example 2.1, what is the radar range if the radar incoherently integrates 16 pulses and scans a fan beam in azimuth, the target is a Swerling I, and other assorted losses total 4 dB?

Solution. Radar range, found from the following equation, is that at which the integrated signal-to-noise ratio $(S/N)_{INT}$ exceeds $(S/N)_{\min}$:

$$\frac{P_R G_R^2 \lambda^2 \sigma_T}{(4\pi)^3 R^4 N_{INT}} \frac{\sqrt{N}}{L_T} = \left(\frac{S}{N}\right)_{INT} \geq \left(\frac{S}{N}\right)_{\min}$$

The total losses are, in decibels,

$$L_T = L_f \text{ (fluctuations)} + L_s \text{ (scanning)} + L_o \text{ (other)}$$

or

$$L_T = 8.1 \text{ dB} + 1.6 \text{ dB} + 4 \text{ dB} = 13.7 \text{ dB}$$

Expressing the range equation in decibel form and using the previous results leads to

$$\begin{aligned} 4R_{dB} &= P_R + 2G_R + 2\lambda + \sigma_T + \tfrac{1}{2}N - 3(4\pi)_{dB} - N_{INT} - L_T - (S/N)_{\min} \\ &= 227 + 6 - 13.7 = 219.3 \end{aligned}$$

$$R = 10^{219.3/40} = 304 \text{ km}$$

Example 2.3

Continuing Example 2.2, what is the radar range if the region between the target and the radar contains rainfall at a rate of 2.5 mm/h?

Solution. The range of the radar is a solution of the equation given in Example 2.2, with the loss resulting from the attenuation due to rain included in the total loss L_T. When expressed in decibel form, this total loss is

$$L_T = 13.7 \text{ dB} + 2\alpha R_{dB}$$

where α is the loss due to the rain expressed in decibels/meter. (Normally, α is given in decibels/kilometer, but it is easily expressed per meter here because all of the other units of length are in meters.) It is assumed that $\alpha = 7.54 \times 10^{-8}$ dB/m.

After inserting the numerical values given in Examples 2.1 and 2.2, the range equation, expressed in decibels, is

$$4R_{dB} = 219.3 - 15.08 \times 10^{-8} R$$

Although both sides of this equation are expressed in decibels relative to 1 m, the quantity R on the right side is in meters. This equation may be solved iteratively, beginning with the rain-free solution of Example 2.2, i.e., $R_o = 304$ km.

The new solution R_1 is a solution of

$$4R_{1dB} = 219.3 - 15.08 \times 10^{-8} \times 3.04 \times 10^5 \cong 219.25$$

or remains essentially unchanged, demonstrating the low loss experienced by a 600 MHz signal when propagating through rainfall.

It is useful to assume a higher attenuation parameter in order to demonstrate the method of solution. To this end, it is assumed that $\alpha = 5 \times 10^{-5}$. Then,

$$4R_{1dB} = 219.3 - 10^{-5}R_o = 219.3 - 3.04$$

and

$$R_1 = 255 \text{ km}$$

Continuing this process for two more iterations results in

$$4R_{2dB} = 219.3 - 2.54, \qquad R_2 = 262 \text{ km}, \qquad R_3 = 261 \text{ km}.$$

Many more iterations are required when the rain attenuation is a greater percentage of the total loss. Then, a computer becomes very useful.

2.5 RADAR DETECTION IN CLUTTER

The most favorable situation for a radar designer is detection of a target when the return must compete only with thermal noise from the receiver. In most cases, target detection and estimation must be conducted in the presence of interfering target-like signals caused by reflections from both Earth and the atmosphere and radiation from man-made sources. This section examines these reflected signals, often called clutter, and some standard methods for eliminating them. A later section deals with man-made interference.

Sources of Clutter

Few radars operate in a homogeneous or uniform space. Most radars are required to detect targets of interest in regions containing discrete objects such as buildings that produce undesired target-like returns. Similarly, the propagation characteristics of regions containing interfaces between clear air and rain change from point to point and produce false target returns. Finally, significant reflections are created whenever the beam of the radar antenna points at the surface of Earth. (This is actually another example of a change in the medium of propagation.) All three cases produce target-like returns (the last two, of course, are extended or large targets), which are called clutter.

A ship or coastal installation, using a search radar in the presence of the smooth curved Earth, such as the one shown in Fig. 2.20, will be subjected to ground clutter returns from a large area. This area is a circle with a radius equal to the distance from the radar to the radar horizon. This horizon is somewhat greater than the optical horizon because of the nature of RF propagation in the atmosphere and is given by

$$R = \sqrt{2R_e h}$$

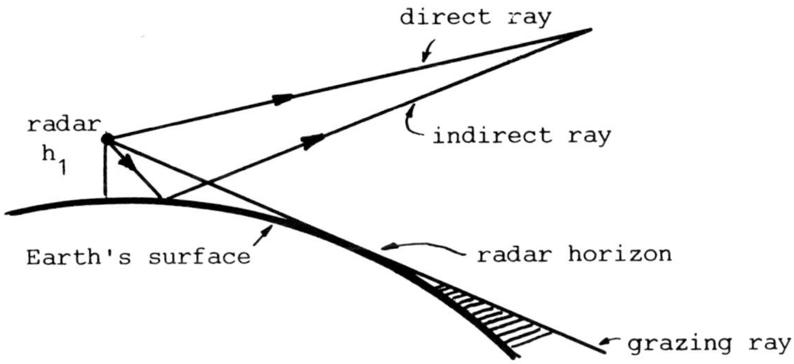

Fig. 2.20 Scattering from the Earth.

where R_e, the effective Earth radius of 8.5 Mm [15, p. 2-47], is equal to 4/3 the actual radius and h is the height of the radar. Part of the RF power that strikes the Earth is reflected away from the radar and the remaining power is scattered in all directions, including the direction back to the radar. The part reflected back to the radar is clutter that competes with any target returns.

An example of a cluttered display, shown in Fig. 2.21, is an overhead view of the radar search volume as presented on a PPI. Shown on the screen is a useless region where nearby land clutter completely masks any target return. Also shown are clutter returns from distant mountains that rise above the horizon. These mountains also shadow targets in the regions behind them if the targets are below a certain elevation. Finally, an area of rainfall presents a change in the propagation medium and thus can cause radar returns large enough to mask target returns.

The implication of Fig. 2.21 is that clutter returns are much stronger than the internal noise from the radar receiver. For a ground radar, this condition typically extends out to the radar horizon. This dominance of clutter interference over receiver noise requires reformulation of the radar range equation in terms of the signal-to-clutter ratio.

Reformulation begins with definition of a quantity called the signal-to-clutter ratio per received pulse $[(S/C)_p]$. This is simply the ratio of power received from the target during one pulse to power received from the

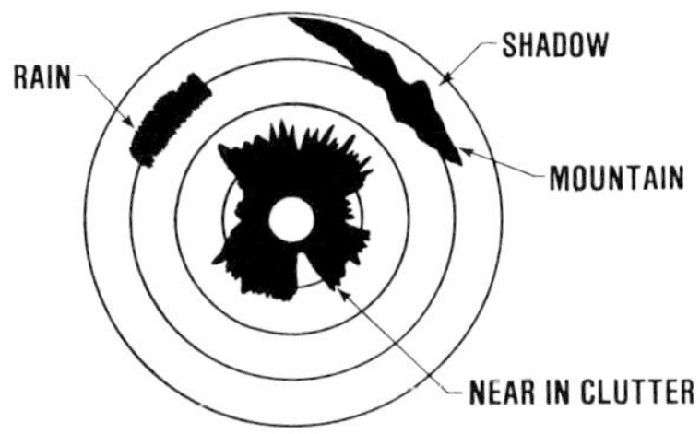

Fig. 2.21 Radar clutter.

clutter sources within a resolution cell around the target. If this clutter power is denoted by C, the ratio is

$$\left(\frac{S}{C}\right)_p = \frac{P_R G_R^2 \sigma_T \lambda^2}{(4\pi)^3 R^4 C}$$

For many ground and airborne radars, this clutter power can be expressed in terms of the ground area illuminated within the resolution cell and a clutter reflection coefficient σ^o (see Fig. 2.22). In the figure, the area illuminated by the radar is given as $R\theta_{az}(V\tau)/2$. The clutter reflection coefficient is σ^o and for surface clutter would be dimensionless. A typical value of σ^o would be 0.001 or -30 dB, which means that every 1000 m^2 of terrain would scatter power back to the radar as if it were a target of 1 m^2. This return ratio may not seem large, but, at 10 km, a radar with a 1 deg beamwidth by 1 μs pulse width illuminates an area of 26,000 m^2.

With this expression of clutter power, the signal-to-clutter per pulse becomes

$$\left(\frac{S}{C}\right)_p = \frac{\sigma_T}{R\theta_{az}\Delta R\sigma^o}$$

which can be quite small. For example, if $\sigma_T = 26$ m^2, the previously mentioned radar and clutter parameters result in a $(S/C)_p$ of unity. The actual clutter reflection coefficient is a function of many parameters, including terrain type, radar frequency, and incidence angle. Typical values are tabulated in Ref. 19 and grossly span the range of -20 to -50 dB. Thus, the expected signal-to-clutter ratio on each pulse is usually so small that target detection is virtually impossible. Under conditions of massive target cross sections and small clutter reflection coefficients, the detection of targets in clutter is possible. A logarithmic receiver (see Fig. 2.44) followed by a "fast time-constant" circuit enables the detector circuit to use a constant threshold, resulting in a fixed number of false alarms per unit of time due to returns from clutter (i.e., a constant false alarm rate or CFAR) [18, p.538]. When such a circuit is in use, an operator is more likely to see and recognize returns from targets that exceed the threshold. Reliable detection

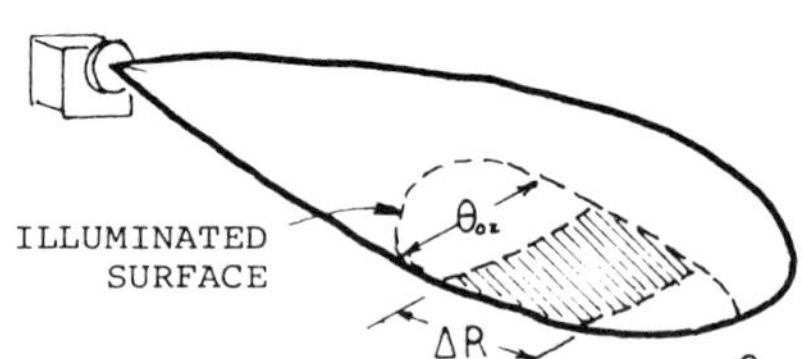

$$C = \frac{P_R G_R}{4\pi R^2}\;\frac{R\theta_{az}\Delta R\sigma^o}{4\pi R^2}\;\frac{\lambda^2 G_R}{4\pi} = \frac{P_R G_R^2 \lambda^2 \theta_{az}\Delta R\sigma^o}{(4\pi)^3 \quad R^3}$$

Fig. 2.22 Clutter power computations.

($P_D = 99\%$) with small probability of false alarm ($P_{FA} = 10^{-6}$) can be achieved with per pulse signal-to-clutter ratios on the order of 20 dB if one assumes that the clutter returns are independent for each of 30 pulses and the 30 pulses are incoherently added [20]. But this assumption is generally invalid. The clutter returns from a resolution cell are often more accurately modeled as identical during the time a cell is observed. The clutter returns are thus more "target-like." Fortunately, a clutter return appears to come from a stationary target, which provides the key to elimination of the clutter.

Clutter Processing

The key to the elimination of clutter is the Doppler principle. If one assumes that the radar transmits at a frequency f_R, the Doppler principle states that the signal reflected from an object moving toward the radar at velocity V_T will be at a frequency f_T given by

$$f_T = f_R + \frac{2V_T}{\lambda}, \qquad f_R\lambda = 3 \times 10^8 \text{ m/s}$$

If the clutter is stationary, it reflects a signal at the transmit frequency. Therefore, if the radar contains a frequency-sensitive receiver that passes the target frequency but not the frequency of the clutter, the target return at the receiver output again must compete only against receiver noise.

A radar required to detect targets embedded in clutter usually employs moving target indication (MTI) techniques based on the Doppler principle described above. It appears that such MTI techniques might better be labeled fixed-target eliminators, but clutter is not always stationary. Nevertheless, the effectiveness of MTI techniques in attenuating the competing clutter depends on many interrelated parameters, such as internal clutter motion and the radar pulse-repetition frequency. Application of MTI techniques usually imposes such additional constraints upon the radar as S/N loss and use of a constant transmit frequency during the receipt of the multiple pulses handled by the MTI processor (i.e., coherent processing). The MTI improvement factor of a radar is a measure of the receiver's ability to reduce the clutter return relative to a constant amplitude of the target return. Such improvement factors range from 0–60 dB, depending on the type of clutter (e.g., rain, chaff) and radar design. A radar with a specified MTI improvement factor I, will produce an output signal-to-clutter $(S/C)_o$ ratio given by

$$\frac{(S/C)_o}{(S/C)_p} = I$$

If this output signal-to-clutter ratio exists only after all N pulses available are coherently processed (e.g., in a fast Fourier transform or pulse Doppler processor), it must exceed the previously defined $(S/N)_{\min}$ for reliable detection. Typically, radar MTI processors are designed to make the clutter power that passes through the MTI (often called the clutter residue) less than the receiver noise power that comes out of the processor. It is further assumed that the resulting signal residue out of the processor must exceed the receiver noise or clutter residue by the value previously discussed in Table 2.6. In performing this calculation, all losses discussed in Sec. 2.4 should be applied to compute the reduction in output signal-to-clutter ratio (or signal-to-noise ratio if larger). If the clutter residue can be made less than the receiver noise with only M of the N pulses available, then we will assume $N–M+1$ pulses are available at the MTI canceller output for subsequent coherent or incoherent integration. The canceller by itself provides no reduction in broadband noise. (The actual number of pulses useful for integration is less than $N–M+1$. [11])

Example 2.4

The radar in Example 2.2 must detect the target in the presence of ground clutter at a range of 300 km. If the ground reflection coefficient equals 0.001, what must be the MTI improvement factor?

The signal-to-clutter ratio per pulse is

$$\left(\frac{S}{C}\right)_p = \frac{\sigma_T}{R\theta_{az}\Delta R\sigma^o}$$

$$= \frac{10}{(3\times 10^5)(0.0174)(180)(0.001)} = 1.06\times 10^{-2} \rightarrow -19.7\text{ dB}$$

before system losses are considered as $\tau = 1.2\ \mu$s. The beam scanning loss is still applicable as returns from clutter are always present at a constant level within the beamwidth of the antenna, but the return from the target will vary as the antenna scans past. The loss resulting from target cross-sectional variations remain and it is assumed "other" losses of Example 2.2 remain as well. Incorporating these losses, the value of $(S/C)_p$ entering the MTI processor is $(S/C)_p = -33.4$ dB.

If it is assumed the output signal-to-clutter ratio is equal to the signal-to-noise ratio specified in Example 2.2, the required improvement factor I is given by

$$\frac{(S/C)_o}{(S/C)_p} = I \qquad \left(\frac{S}{C}\right)_{odB} = I_{dB} + \left(\frac{S}{C}\right)_{pdB}$$

$$13.2 = I - 33.4 \qquad \text{or} \qquad I = 46.6\text{ dB}$$

Subclutter visibility (SCV) is also used to characterize the performance capability of a radar in clutter. If a radar can reliably detect a target in a clutter environment with a given $(S/C)_p$, then it has a subclutter visibility equal to the absolute value of the $(S/C)_p$ when it is expressed in decibels. Similarly, if the required output signal-to-clutter ratio is $(S/C)_o$, the connection between SCV and MTI improvement factor is

$$\frac{(S/C)_o}{I} = \left(\frac{S}{C}\right)_p = \frac{1}{SCV}$$

or

$$I = SCV \times (S/C)_o$$

$$I_{dB} = SCV_{dB} + (S/C)_{odB}$$

One MTI method uses delay line cancelers, a two-pulse example of which is shown in Fig. 2.23. Although operation of this clutter canceler can be described with the aid of the Doppler principle, its operation is usually described in the following manner. Radar returns from clutter and a target, produced by the first transmitted radar pulse, load the delay line. Subsequent radar returns from clutter are then canceled in the summing junction because the clutter returns are assumed to be identical for each pulse. (Note that, in Fig. 2.23, the sign of one path is changed prior to the summing junction.) Target returns are passed because they are different for each radar pulse except at specific target velocities called blind speeds, which are given by

$$V_T = \frac{n\lambda}{2} f_{PRF}$$

Example 2.5: Rain Clutter

A radar with a pulse duration of 1 μs and with equal azimuth and elevation beamwidths of 2 deg illuminates a 1 m^2 target in a region of 4 mm/h rainfall 10 km from the radar. If the rain has a clutter reflection coefficient of -70 dB m^{-1}, what is the signal-to-clutter ratio on each pulse?

Again, the signal-to-clutter ratio is given approximately by the ratio of the target cross section to the size of the resolution cell times an appropriate reflection coefficient, as

$$\left(\frac{S}{C}\right)_p = \frac{\sigma_T}{\eta_o \dfrac{V\tau}{2}\dfrac{\pi}{4} R^2 \theta_{az}\theta_{el}}$$

where η_o = reflectivity in m^2/m^3 or m^{-1}. The reflection coefficient has

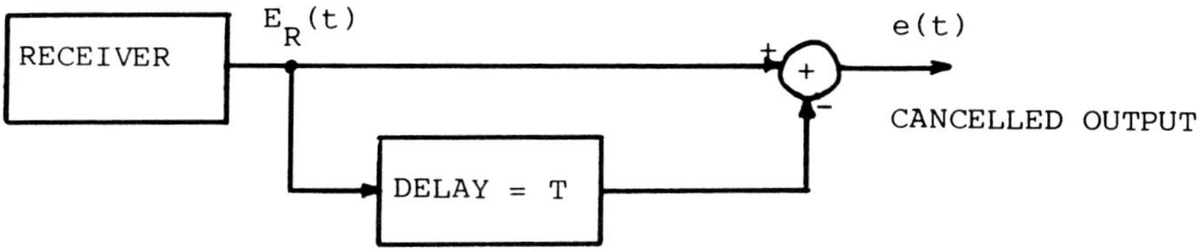

Fig. 2.23 Two-pulse MTI canceler.

dimensions because rain occupies a volume

$$\frac{\pi(R\theta_{az}/2)(R\theta_{el}/2)V\tau}{2}$$

and we desire an equivalent cross section. It must be noted that this result will be approximately 1.5 dB too high because the illumination is not uniform over the projected area of the clutter cell [19, p.61].

As $-70\ \text{dB/m}^{-1} = 10^{-7}\ \text{m}^{-1}$

$$\frac{S}{C} = \frac{1}{(10^{-7})(150)(9.6 \times 10^4)} = 0.7 \text{ or } -1.5\ \text{dB}$$

Example 2.6: Detection in Rain Clutter

If the radar in Example 2.5 searches in two dimensions and must detect a Swerling I target with a probability of detection of 90% and a probability of false alarm of 10^{-6}, what must be its MTI improvement factor? Ignore the rain attenuation and assume that, in the absence of a target, a noiselike clutter residue dominates the MTI output.

Solution. Table 2.8 shows that the fluctuation loss is 8.1 dB and other losses given in Table 2.9 add up to 4.4 dB. Scanning in two planes results in a 3.2 dB loss, for a total system loss of 15.7 dB. The required output signal-to-clutter ratio is, from Table 2.6, 13.2 dB. Thus,

$$(S/C)_o = (S/C)_p\,(I/L) = 13.2\ \text{dB}$$

the solution of which is

$$I = 13.2 + L - (S/C)_p = 13.2 + 15.7 + 1.5 = 30.4\ \text{dB}$$

Since most of the clutter enters the radar from angles at low elevations, many search radars employ multiple beams. In this way, complicated MTI processing is restricted to the lower beam (or beams) and the upper beams can be more simply processed.

2.6 RADAR SURVEILLANCE VOLUME

The three-dimensional region of space in which a radar can observe a target is called the radar's surveillance volume. A radar centered in the void of space will have a spherically shaped surveillance volume with a radius given by the equation for radar range. The surveillance volume of a radar on the Earth's surface will be approximately hemispherical. This section briefly addresses the effect of the height of the radar and the effect of large obstacles on this volume of surveillance.

A radar is located at a height h_1 above a smooth curved Earth as shown in Fig. 2.20. If the radiation pattern of the radar directs significant power toward the Earth, then there will be power scattered both away from the radar and back to the radar out to the radar horizon. This forward scattered power propagates through the region of space illuminated directly by radiation from the radar and, at some locations, combines with it to increase the total power. However, at other locations, the two powers cancel each other. The radiation coming directly from the radar is called the direct ray and radiation scattered from the ground the indirect ray. Above the line-of-sight to the radar horizon, this interaction of the two rays periodically enhances and cancels the original radiation pattern. The resulting radiation pattern is called an interference pattern that appears as a sequence of narrow lobes following the outline of the original, free-space radiation pattern.

In the direction of the radar horizon, the direct and indirect rays (now a grazing ray) combine destructively and virtually no radiation travels in this direction or below the radar horizon. Targets in the shaded region of Fig. 2.20 cannot be detected. If the exaggerated vertical dimension is removed from Fig. 2.20, then the radar appears to be on the Earth's surface and the radiation patterns before and after consideration of the reflections from the Earth's surface take the shape shown in Fig. 2.24. One effect of the very narrow nulls in the radiation pattern is the disappearance of a target as it passes through these nulls. At the usual radar frequencies (above 1 GHz), a target will be detected if it is only slightly above the grazing ray. For simplicity, then, the assumption here is that a target will be detectable even on the grazing ray or at any time that it is not obscured from the radar and is within the radar range. A target at a height h_2 is then detected at a range given by $\sqrt{2R_e h_1} + \sqrt{2R_e h_2}$, provided this sum is within the radar range.

An obstacle many wavelengths in size, such as a building or mountain between the radar and the target, will reduce the radar's surveillance volume. Given the geometry of Fig. 2.25, the range at which a target becomes visible when the intervening obstacle is of height h_3 at a distance R_{13} is given by the solution of a quadratic equation. If h_2 is greater than both h_1 and h_3, the solution for R_{12} is

$$2R_{12} = R_{13} + (h_1 - h_3)\frac{2R_e}{R_{13}} + \sqrt{\left[R_{13} + \frac{2R_e}{R_{13}}(h_1 - h_3)\right]^2 + 8R_e(h_2 - h_1)}$$

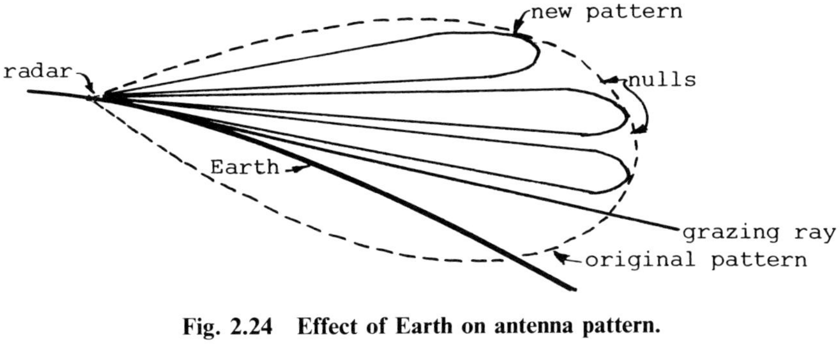

Fig. 2.24 Effect of Earth on antenna pattern.

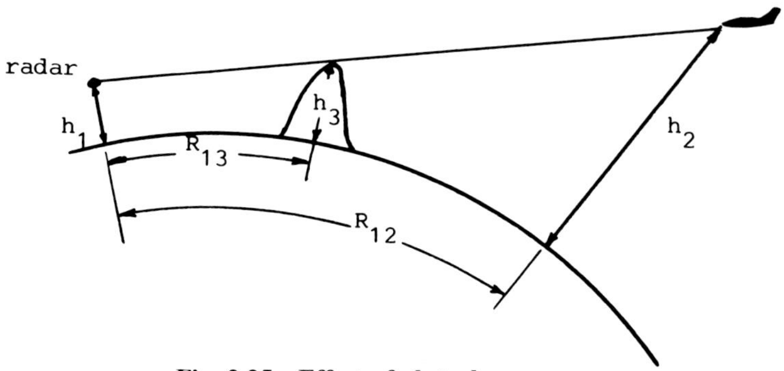

Fig. 2.25 Effect of obstacle on range.

This equation and the earlier one for the radar horizon assume that the computed range is much less than the Earth's radius and that the radar or target height is much less than the range computed. Furthermore, the slant range and distance on the Earth's surface are considered equal.

2.7 TARGET TRACKING

Following the detection of targets within specific resolution cells, subsequent estimation of a target's position within a cell is the essence of all target tracking.

Search Radar Tracking

A search radar can approximate the location of a target within its capability to resolve range and angle. Its accuracy depends on the existing signal-to-noise ratio and the number of pulses received while it scans past the target. The search radar then provides data on the position of the target to operators or computers that "track" the target by averaging past positions into estimates of future locations. In Fig. 2.26, for example, the dots represent measured target locations and the crosses are estimates of target locations based on previous measurements.

However, sufficient accuracy for weapon delivery is usually not possible because of the slow update rate (new point every 10 s, fades, dropouts, etc.). For example, the warhead on a surface-to-air missile may be lethal to an aircraft if it explodes within 100 ft of the aircraft. At a range of 50 n.mi., this miss-distance corresponds to an angular error of 0.019 deg or approxi-

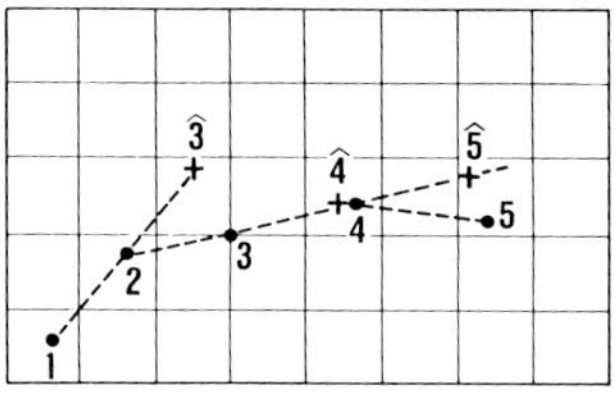

Fig. 2.26 Search radar tracking.

mately 1/50th of a 1 deg beamwidth. Although the search radar can achieve this accuracy if it has sufficient S/N ratio, it would achieve it only momentarily as it scans past the target. Errors in target position would build up before the next scan and prevent accurate weapon guidance. In addition, other considerations, such as insuring the tracking of only one of many targets or avoiding Earth reflections, usually forces designers of weapon systems to consider a separate tracking radar to insure weapons delivery.

Track-while-Scan (TWS) Radar

The first tracking radar considered in this discussion is called a track-while-scan system. Such a radar could have a block diagram similar to that of the radar shown in Fig. 2.17, with the addition of a number of new blocks to permit automatic centering of the target in the antenna beam. As shown in Fig. 2.27, the antenna beam on this new radar scans rapidly back and forth over a relatively small sector of a circle of, for example, 20 deg. Rapid scanning allows more estimates of the target position per unit time and thus reduces tracking errors.

In viewing the screen, the radar operator attempts to maintain the target return in the center of the scanned area by controlling (manually in this case) the nominal direction of the antenna. Other operators or equipment use this direction to indicate the target position.

Use of a separate radar with an antenna nodding in the vertical plane would allow estimation of target location in range, azimuth, and elevation. Such a radar system would constitute a TWS radar. Accurate target estimation usually requires different parameters for a tracking radar than for a search radar. First, since the resolutions of range and azimuth are directly related to the accuracy in estimating the target locations, the tracking radar tends to have narrower emitted pulses of RF energy and narrower antenna beamwidths caused by the use of higher RF frequencies. Second, higher radar pulse-repetition frequencies are used to get as many target returns as possible per unit time.

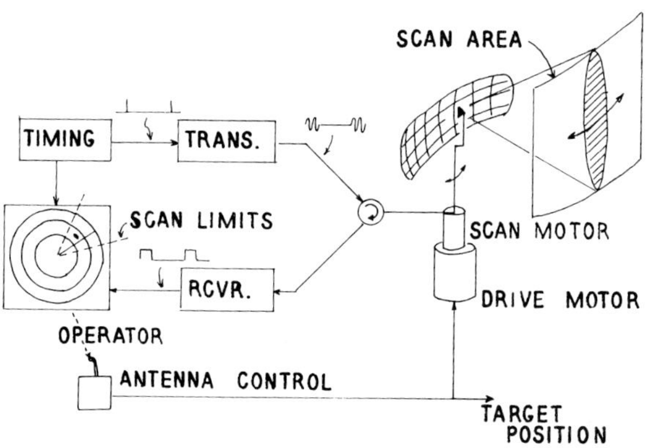

Fig. 2.27 Modified search radar.

Range and Angle Gates

In considering the automation of the radar in Fig. 2.27, one may first replace the PPI displays with a pair of "B-scopes" (see Fig. 2.28). One display is drawn for the azimuth and another could be drawn for the elevation channel. Second, a block labeled "range-gate" would be added.

Rather than automate the radar, the B-scope simply displays range vs angle, from which the operator again attempts to center the target of interest on the display. The more important addition is the range gate because the operator attempts to center the target by moving the antenna on the basis of the target location relative to the center of the screen. If other targets or clutter are present, especially in the vicinity of the selected target, the operator must ignore returns not coincident in time with the desired target. The range gate automates this target sorting process in the range coordinate by sampling the target return in such a way that it remains centered on the target return. Then, by providing a gating pulse, it allows only the target return or similarly delayed returns to reach later circuits.

A typical range gate circuit is shown in Fig. 2.29. The operator's initial estimate is used to position the early and late gates on the target return. If the target pulse is exactly centered in the gates, the product of the voltage times the time out of each gate (i.e., area) is identical and the error signal from the summing junction is zero. For the next radar pulse received, the gates will be activated at the same time relative to the transmit pulse. If the gates are no longer centered on the target return, generation of an error signal corrects the error by recentering the gate sampling time over the radar pulse received. This recentering is accomplished through use of a servo control system.

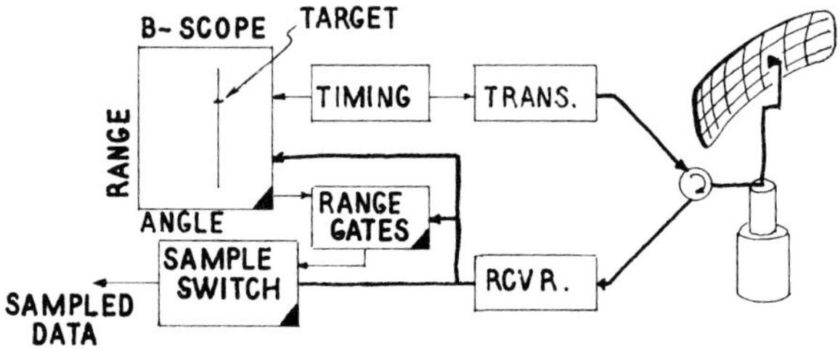

Fig. 2.28 B-scope and range-gate addition.

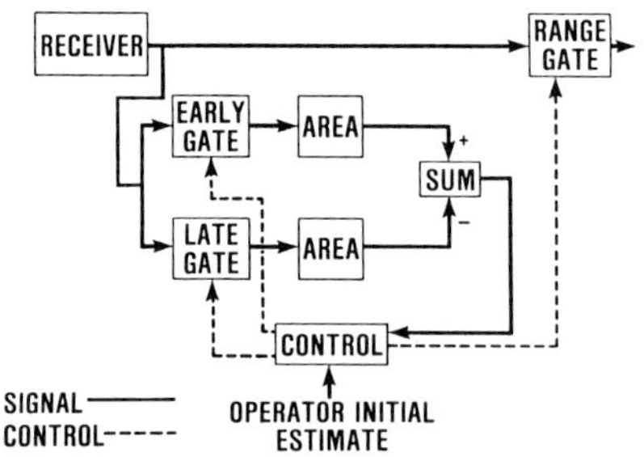

Fig. 2.29 Range-gate circuit.

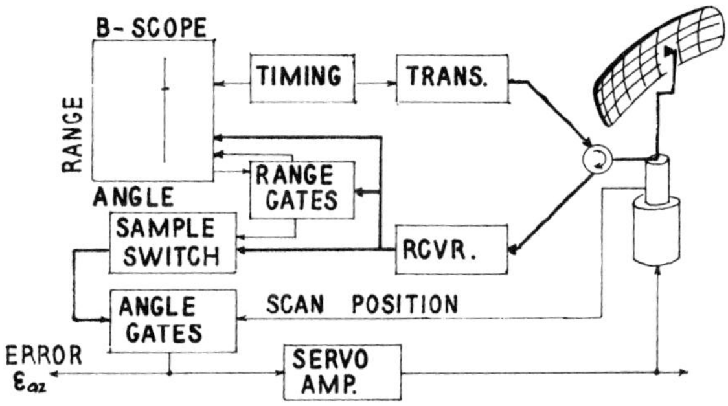

Fig. 2.30 TWS radar—azimuth channel.

The final step in automation is the addition of "angle gates" to process the sampled signals generated by the range gate (see Fig. 2.30). These angle gates sense the power received from the target on both sides of the scan center and cause the beam scan center to move toward the lower power side through a servo control network.

These angle gates operate on the same principle as the range gates. Figure 2.31 shows a sequence of pulses from the range gate. These pulses vary in amplitude in accordance with variations in the gain of the radar antenna pattern. After passing through a sample-and-hold (S&H) or "boxcar" circuit, the sequence of pulses becomes one long pulse during the scan time. This long pulse can be compared to a single pulse entering the range gates and the scan time is comparable to the radar PRI. Thus, the circuitry for angle tracking consists of early and late angle gates followed by area measurements and comparison of areas. Any bias is driven to zero by moving the center of the antenna scan over the target.

Although the block diagram appears cluttered, the basic employment concept remains the same—the operator must search for, and detect, the target.

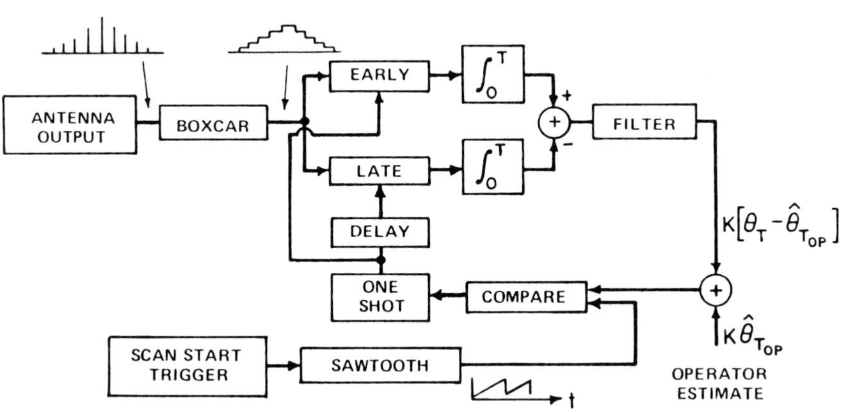

Fig. 2.31 TWS angle-gates.

(In this example, noncoherent integration of the returned pulses with the B-scope aids in detection.) The operator must then engage the automatic error-sensing circuits to maintain track. Information on the target position is transferred to other systems for weapon guidance. Such systems as the SA2 FanSong radar [12,p.312] and the AN/SPY-1A radar used with the AEGIS system [21] use TWS techniques, but most radar trackers use other forms of sequential or simultaneous lobing techniques.

Sequential Lobing Techniques

Conical-Scan (CONSCAN) Radar. This discussion of the conical-scan radar, one form of sequential lobing radar, begins at the antenna. Unlike the fan-beam pattern of the TWS radar, the radiation pattern of the conical-scan (CONSCAN) radar antenna usually has equal beamwidths for azimuth and elevation. This pattern implies an antenna aperture of roughly equal vertical and horizontal dimensions. As shown in Fig. 2.32, the antenna pattern, sometimes called a pencil beam, is scanned about a scan axis that intercepts the main beam. Even though this is a conical-scan radar, only an A-scope has been added to the TWS diagram. A range/height display is now used to acquire the target through a limited-volume search. Detection of a target on the range/height display stops the search motion of the antenna and the target return should be visible on the A-scope display. This display is identical to the display of an oscilloscope discussed earlier, but the time axis is calibrated in range. In viewing the A-scope, the operator will see the target return "modulated" by the conical-scanning pencil beam. This means that, as a new target return appears on the screen, its amplitude will differ from the preceding return, depending on the variation

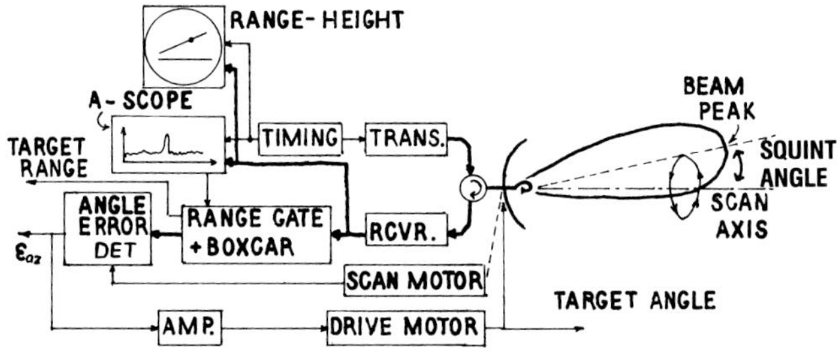

Fig. 2.32 Conical-scan radar.

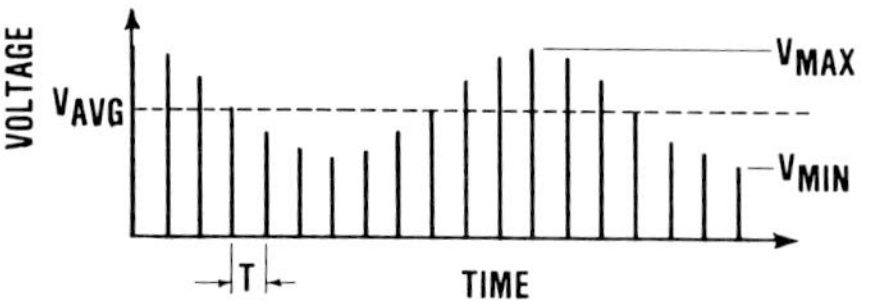

Fig. 2.33 Range-gated receiver output.

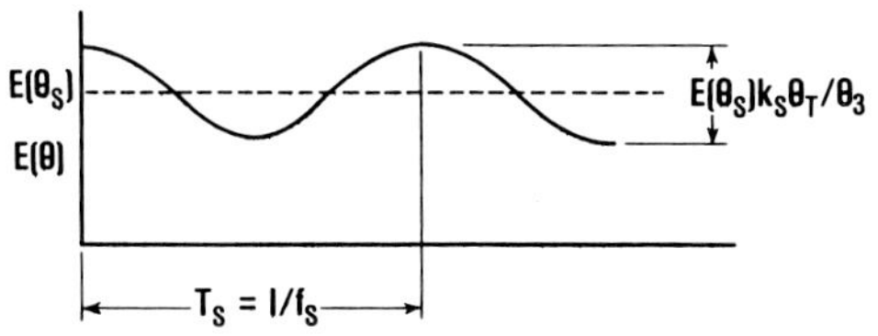

Fig. 2.34 Smoothed CONSCAN receiver output E (θ).

of gain in the antenna pattern toward the target. (Actual modulation rates may be too high for visual observation.) A sequence of such target returns is shown in Fig. 2.33.

As with the TWS radar, this pulse train is extracted from the detected IF output through the use of a "range gate." In viewing the A-scope (or the A-scope on an expanded scale), the operator engages circuitry that samples the receiver output only at the instant when the target return is present. Analysis of this pulse train shows that the variation in the voltage level is proportional to the angular distance of the target from the scan axis, defined as θ_T. After "boxcar" processing extends each pulse to the point that it equals the pulse-repetition interval, this gated receiver output (called the error voltage) finally appears as shown in Fig. 2.34. One important thing about this error voltage is that the peak-to-peak variation (the modulation depth), given by $\varepsilon = E(\theta_s)(k_s\theta_T)/(\theta_s)$, is a function of both the average received voltage $E(\theta_s)$ and the tracking error θ_T. Because of this dependency, every tracking radar employs automatic gain control (AGC) in an attempt to maintain a constant average voltage out of its receiver. To the extent that this control is successful, the error voltage gives a true indication of the angular error.

The variation in time of the curve in Fig. 2.34 can be explained as follows. As for any other radar, the power received by the conical-scan radar is

$$P_r = \frac{P_R G_R^2 \sigma_T \lambda^2}{(4\pi)^3 R^4}$$

This power is sensed only during the reception of a pulse at the peak of the antenna gain pattern G_R and is a strong function of range. However, the peak gain does not point at the target as the antenna is scanning. The gain of the antenna as a function of the angle θ away from the peak gain direction can be formulated as

$$G_R(\theta) = G_R \exp\left[(-\ell n\ 2)\left(\frac{2\theta}{\theta_3}\right)^2\right]$$

This formulation is reasonable only within one-half beamwidth from the maximum gain.

The conical-scan radar scans the antenna beam shown in Fig. 2.35 around a line called the scanning axis. As shown in Fig. 2.36, the angle between the beam center and this scanning axis is called the squint angle. Also shown in

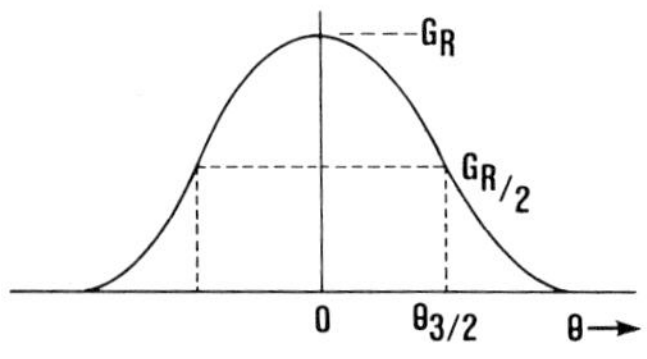

Fig. 2.35 Gaussian shaped antenna pattern.

Fig. 2.36 is the difference in received voltage when the antenna is pointed toward and away from the target. For a pencil beam scanning in a circular pattern and for small target errors ($\theta_T/\theta_3 \ll \frac{1}{2}$), the amplitude of the received voltage can be shown to be

$$E(\theta_T) = E(\theta_s)\left[1 + \frac{k_s}{\theta_3}\theta_T \cos(2\pi f_s t)\right]$$

which is the curve shown in Fig. 2.34. $E^2(\theta_s)$ equals the power received on the scan axis and is a function of the target range. The quantity k_s is called the error slope coefficient. This coefficient depends on the antenna pattern and squint angle, but here it is assumed to be constant and equal to approximately 1.5. The antenna scanning frequency is $f_s(f_s = 1/T_s)$.

After the angle error detector, the modulated voltage shown in Fig. 2.34 is converted to a constant voltage proportional to the physical angular tracking error. This constant voltage is then applied to a servo system that repositions the antenna scan axis onto the target. In principle, the angle error detectors operate the same as the TWS angle gates, but they usually take the form of phase detectors.

The CONSCAN tracking radar must perform in the same environment as the search radar. However, such characteristics as radar frequency or antenna beamwidth may cause various environmental factors to have a different order of importance relative to the radar's performance. For example, rain produces greater clutter return because of the higher radar frequency. Ground clutter may not be as important for targets at high elevation angles because the narrow-beam antennas used in tracking radars may not have significant sidelobes intercepting the ground. Again, the radar should have an MTI capability if it must track targets in rain or chaff or at low elevation angles.

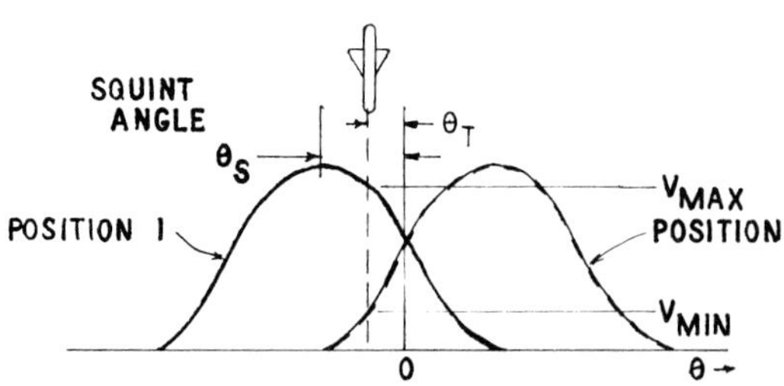

Fig. 2.36 CONSCAN antenna beam positions.

Lobing Variants. The radar shown in Fig. 2.32 can be modified by incorporating two antennas as shown in Fig. 2.37. The transmit antenna is mechanically coupled to the receive antenna so that both still point at the target. Called "conical scan on receive only" (COSRO), this system generates the necessary information for angle and range tracking even though only the antenna beam on the receiver scans. An advantage of this method is that the scanning frequency is not available to the target. The significance of this method will be apparent later. COSRO systems can also be constructed using one antenna.

If the receive antenna is properly constructed, various electronic switches could instantly place the antenna pattern at selected positions instead of employing mechanical scanning. Called lobe on receive only (LORO), the received pulse train in this variation would appear, after range gating, as in Fig. 2.38. Four receive beam positions are used: two in azimuth and two in elevation. Again, the information on angular error is contained in the received signal level variation between the antenna positions (designated as EL-1 and EL-2 or AZ-1 and AZ-2) and the scanning speeds are now limited by other than mechanical considerations.

Many other implementations of a tracking radar employing a single receive beam exist. Some use variable scan rates, random scan rates, or variation of the transmitter power. Additional information on these variants is contained in a paper by Johnston [95].

Sequential Tracking Accuracy. At this point, it is appropriate to provide an estimate of the tracking accuracy of a sequential lobing system. With proper consideration given to signal integration, the range equation derived for the search radar can be used to calculate the detection capability of a tracking radar. After determining the presence of the target, one must then estimate the target position in range, azimuth, and elevation with some

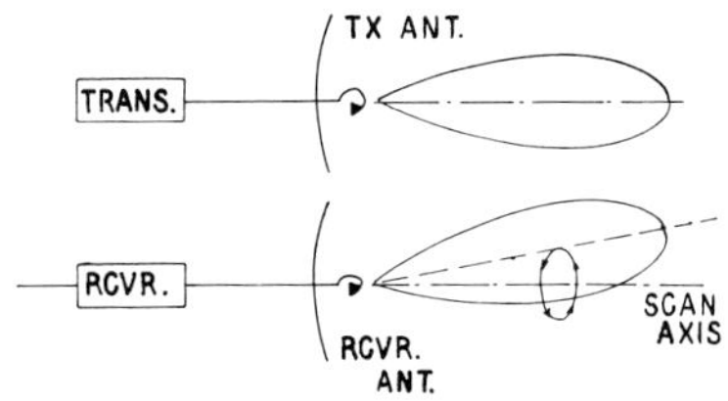

Fig. 2.37 COSRO tracker.

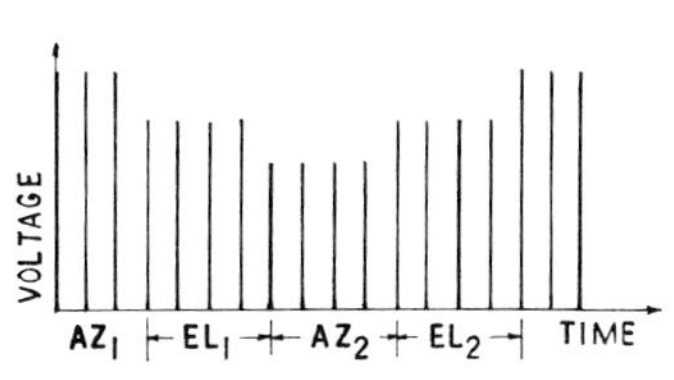

Fig. 2.38 LORO output.

degree of accuracy. In making these estimates, a number of authors [22–24] have achieved the following results when only receiver noise is present. Tracking accuracy is always formulated in terms of the single pulse signal-to-noise ratio. The root-mean-square (rms) error in range, normalized to the range resolution ΔR, is

$$\frac{\sigma_R}{\Delta R} = \frac{1}{2\sqrt{(S/N)_p(f_{PRF}/B_{N_1})}}$$

The rms error in angle σ_θ, normalized to the antenna beamwidth (sometimes specified as the one-way half-power beamwidth), is

$$\frac{\sigma_\theta}{\theta_3} = \frac{1}{1.5\sqrt{(S/N)_p(f_{PRF}/2B_{N_2})}}$$

Both of these equations contain certain usually valid assumptions, and they are for single-pulse IF S/N ratios greater than 10 dB. The quantities B_{N1} and B_{N2} are the closed-loop bandwidths of the range- and angle-tracking servo loops, respectively. These bandwidths are typically a few hertz wide.

These equations apply when the size of the target is extremely small compared to the ability of the radar to resolve in range and angle. As the distance between the target and radar decreases, the dimensions of the target become an appreciable fraction of the resolution in angle of the radar. Most of the time, the radar will track the power centroid of the many scatterers on the target and the track axis remains on the target. At times, glint effects will force the track axis off the target. In either case, the above equations no longer apply.

Usually, the range-tracking servo has no difficulty in keeping up with a target. But, because of their inertia, mechanical angle-tracking servos can introduce dynamic lag errors that may be more significant than the noise-limited errors indicated above.

Simultaneous Lobing Techniques

Scanning with Compensation (SWC). In the conical-scan radar receiver shown in Fig. 2.39, a sinusoidally modulated output voltage is generated whenever an angle error exists between the tracking (scanning) axis and

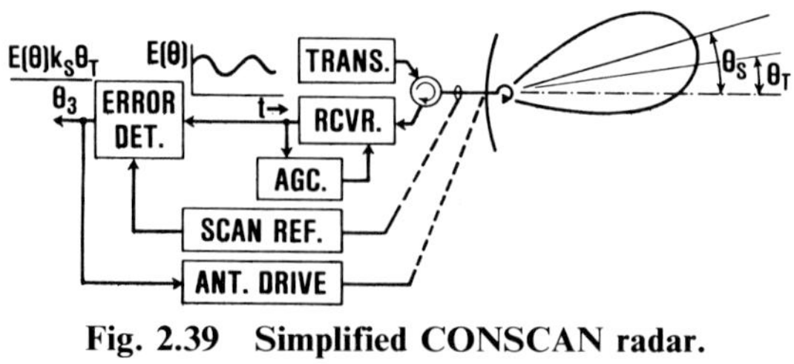

Fig. 2.39 Simplified CONSCAN radar.

the target line-of-sight. This modulated signal is generated under the assumption of a constant target cross section. It is well known that such a conical-scan radar is extremely susceptible to jammers that modulate their transmit power at the scan frequency. Scan-with-compensation (SWC) techniques can be incorporated into the radar to defeat such jammers and to reduce the tracking errors normally caused by signal fluctuations.

The simplified SWC radar shown in Fig. 2.40 can be analyzed to show how this sensitivity to amplitude modulation is reduced [88,p. 540;94]. The SWC radar is very similar to the CONSCAN radar. The only difference is the addition of a separate receive channel, including a receive-only antenna beam and a subtractor circuit. The voltages out of the main and auxiliary receivers for a constant target cross section and for a constant jammer of extremely high power are shown in Figs. 2.41a and 2.41b. The constant voltage of the auxiliary channel for the constant target results from the balancing nature of the two antenna patterns at an assumed small angular tracking error (i.e., when the main or transmit pattern is high and the auxiliary pattern low and vice versa). The respective output voltages can be derived from a study of Fig. 2.41d.

Of most importance is the error voltage generated in the presence of a strong jammer employing inverse-gain modulation. (Inverse-gain modulation means that the jammer transmits maximum power with minimum radar illumination and vice versa.) As shown in Fig. 2.41c, an appropriately phased error voltage is still generated (this is more fully explained in Sec. 3.2). Thus, inverse-gain jamming is ineffective against an *ideal* SWC radar.

The essential element in the above result is the simultaneous measurement of the target return on both sides of the scan axis. In this way, amplitude variations common to both channels can be eliminated in the subtraction circuitry.

Monopulse Radar. Monopulse is another radar technique that employs the concept of simultaneously measuring the target return on both sides of the tracking axis. This section first addresses the principles of angle tracking used by the most common form of monopulse and then briefly describes the many variations of monopulse.

The most common form of monopulse is called amplitude sum-difference.

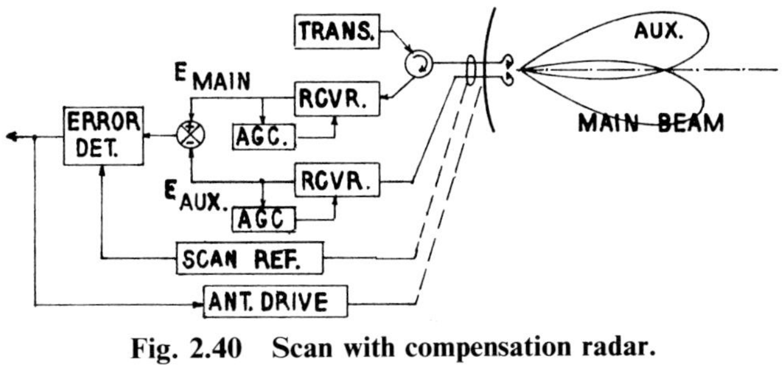

Fig. 2.40 Scan with compensation radar.

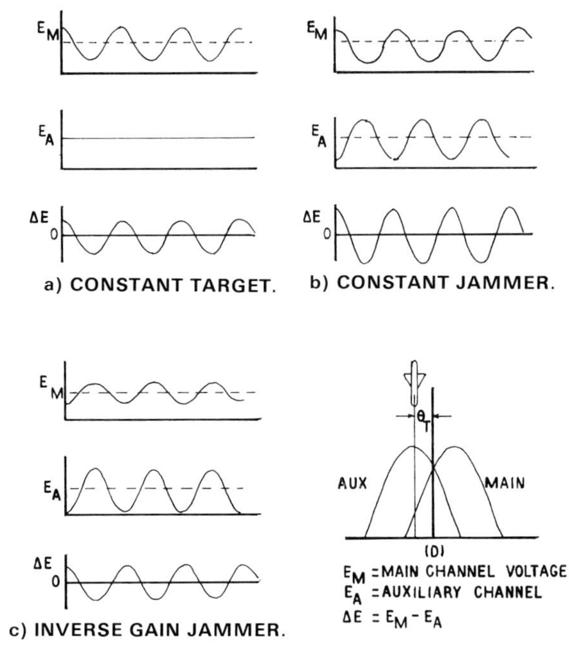

Fig. 2.41 Scan with compensation error voltage generation.

The radar antenna for this form usually employs five beams, in contrast to the two beams used by the SWC radar. The full monopulse has one transmit and four receive beams. The angle tracking principles of the monopulse radar can be addressed by first considering the SWC antenna, which transmits the radar signal on one beam but receives it on both (see Fig. 2.40). However, the radar signal can be transmitted on both beams if a microwave hybrid is installed between the transmitter and the SWC antenna. (A microwave hybrid is a four-port device capable of controlling the direction of flow of electromagnetic power.) The hybrid used in this application is usually a folded magic-T shown in Fig. 2.42. If a signal Σ is applied as shown, it is split into two in-phase parts, and, if a signal Δ is applied at the remaining port, it is split into two equal out-of-phase parts.

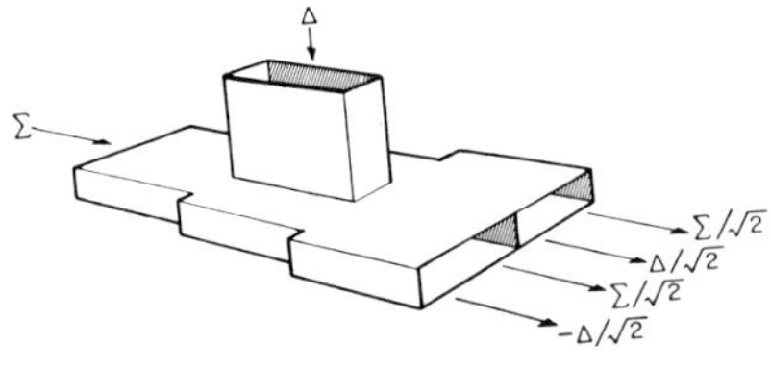

Fig. 2.42 Folded Magic-T.

If this hybrid is inserted in the nonscanning SWC radar of Fig. 2.43, monopulse angle tracking in one plane can be explained. The antenna feeds have been relabeled A and B rather than main and auxiliary. The receiver outputs are relabeled sum (Σ) and difference (Δ) to correspond to their connection with the hybrid. The hybrid causes a signal inserted at the sum port to become two in-phase signals at A and B. The signals A and B are radiated through their respective feeds to create antenna patterns A and B. These patterns combine in free space to form a single pattern called the sum pattern. The hybrid would also take a signal inserted at the difference port and produce two out-of-phase signals at A and B and a difference antenna pattern in space. (Of course, no signal is inserted at the difference port.) Signals flowing in the opposite direction are added and subtracted at the sum and difference ports. These sum and difference patterns are shown in Fig. 2.43b.

If the target of the monopulse radar is on the tracking axis, the peak of the sum pattern points at it, but, for the SWC or CONSCAN antenna, the beam peak was squinted away. The resulting greater power radiated toward the target yields slightly greater range and/or accuracy in angle tracking. This sum pattern, of course, also determines the power received from a target vs the angle of the target from the pattern peak.

Likewise, the difference-channel receiver amplifies the signal derived from the difference port of the hybrid. For tracking errors that are small in relation to the beamwidth of the sum antenna, one can show that this difference-port signal is linearly related to the tracking error. For a tracking error θ_T, the magnitudes of the voltages from the A and B patterns are

$$V_A = V(\theta_s)(1 - k\theta_T)$$

$$V_B = V(\theta_s)(1 + k\theta_T)$$

Thus, the sum, difference, and error detector voltage amplitudes are

$$V_{\Sigma_i} = V_A + V_B = 2V(\theta_s)$$

$$V_{\Delta_i} = V_A - V_B = -2V(\theta_s)k\theta_T$$

$$\varepsilon = \frac{V_{\Sigma_o}}{|V_{\Sigma_o}|} \frac{V_{\Delta_o}}{|V_{\Sigma_o}|} = -k\theta_T$$

As shown in Fig. 2.43, the sum channel usually normalizes the difference channel. The resulting voltage in the difference channel relates directly to the tracking error and is independent of the amplitude of the received signal.

As opposed to the CONSCAN and SWC radars, the monopulse radar is not required to wait until the feed has scanned halfway around (180 deg) to arrive at an estimate of angle error. After proper normalization by the sum

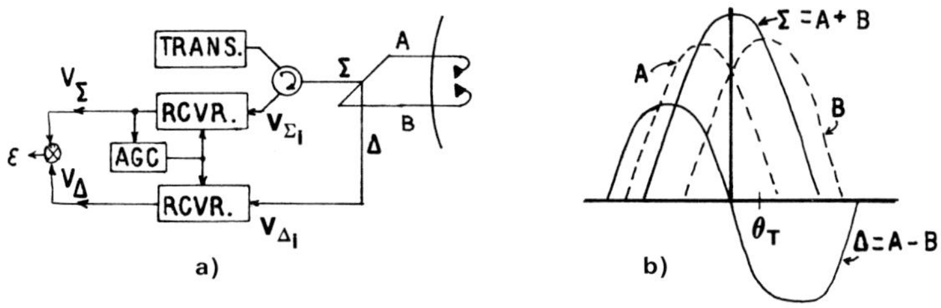

Fig. 2.43 Monopulse radar and antenna patterns.

channel, the radar ideally yields an error voltage out of the difference-channel receiver on a single pulse—hence the term monopulse. This error voltage is related in a known way to the angle tracking error. As shown in Fig. 2.43a, the antenna feeds do not rotate (or scan). Instead, there are usually two sets of feeds—one for errors in azimuth angle and one for errors in elevation. This, of course, requires a more complicated antenna feed and another receiver channel, resulting in a total of three receive channels.

The various forms of monopulse radars have been categorized [25,26] according to the method used for sensing and detecting the angle error. The previous example showed amplitude-sensing sum-difference detection, which is the most commonly used form of monopulse because the tracking axis will not be biased by mismatches in the IF channels. It is the equivalent of phase-sensing sum-difference detection [25]. The forms of angle sensing are amplitude and phase-sensing. Since amplitude and phase detection can be used with each of these sensing methods, six forms of monopulse are possible.

An example of an amplitude-amplitude monopulse is shown in Fig. 2.44. For this monopulse, the output voltage of the log-IF amplifier is proportional to the logarithm of the amplitude of the input voltage at the carrier frequency. Subtraction of the voltages of the log-IF outputs from each other

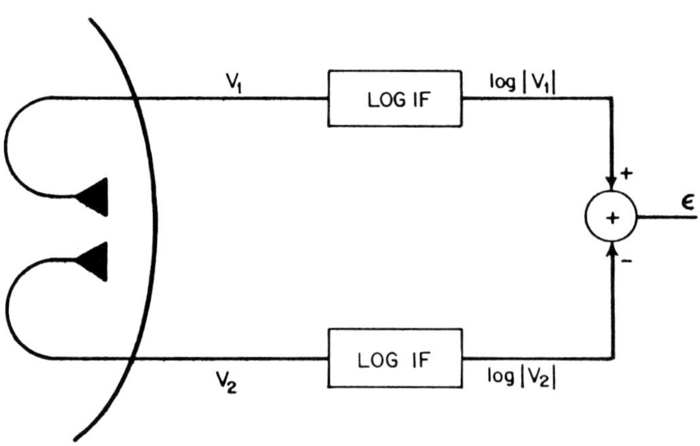

Fig. 2.44 Amplitude-amplitude monopulse.

results in a normalization of the signals equivalent to normalization of the AGC circuit in Fig. 2.43. Again, the error voltage is independent of the amplitude of the target return. This signal manipulation proceeds as follows. If the received signals are

$$V_1 = A(1 + K\theta_T) \cos 2\pi f_R t$$

$$V_2 = A(1 - K\theta_T) \cos 2\pi f_R t$$

Then the error voltage is

$$\varepsilon = \log |V_1| - \log |V_2| = \log \left|\frac{V_1}{V_2}\right|$$

$$= \log \left(\frac{1 + k\theta_T}{1 - k\theta_T}\right) \approx \log (1 + 2k\theta_T) \approx \frac{2k}{2.3} \theta_T$$

where the following approximations have been made:

$$\log_{10} x = \frac{1}{2.3} \ell n\, x \qquad \ell n(1 + x) \approx x$$

A phase-phase monopulse is shown in Fig. 2.45. In this monopulse, the hard limiter produces the required normalization of signals. If a point between the two antennas is used for a phase reference, the original signals are

$$V_1 = A_1 \cos 2\pi f_R \left(t - \frac{\mathrm{d} \sin \theta_T}{2V}\right) \approx A \cos \left[2\pi f_R \left(t - \frac{\mathrm{d}\, \theta_T}{2V}\right)\right]$$

$$V_2 = A_1 \cos 2\pi f_R \left(t + \frac{\mathrm{d} \sin \theta_T}{2V}\right) \approx A \cos \left[2\pi f_R \left(t + \frac{\mathrm{d}\, \theta_T}{2V}\right)\right]$$

The limiter output voltage is fixed at a peak value, arbitrarily chosen to be unity in Fig. 2.45. The error voltage after low-pass filtering is then

$$\varepsilon = \left\{\sin \left[2\pi f_R \left(t + \frac{\mathrm{d}\, \theta_T}{2V}\right)\right] \cos \left[2\pi f_R \left(t - \frac{\mathrm{d}\, \theta T}{2V}\right)\right]\right\}_{LPF}$$

$$= \tfrac{1}{2} \sin \left(2\pi f_R \frac{\mathrm{d}\, \theta_T}{V}\right) = \tfrac{1}{2} \sin \left(\frac{2\pi}{\lambda} \mathrm{d}\, \theta_T\right) \approx \mathrm{d} \frac{\pi}{\lambda} \theta_T$$

The tracking error has been assumed to be small in the above approximations (i.e., $2\pi \mathrm{d}\, \theta_T/\lambda \ll 1$).

As stated earlier, tracking in two planes requires formulation of error signals in both planes by inserting more antenna feeds and hybrids, as shown in Fig. 2.46. The cassegranian feed discussed in Chapter 1 is used in the

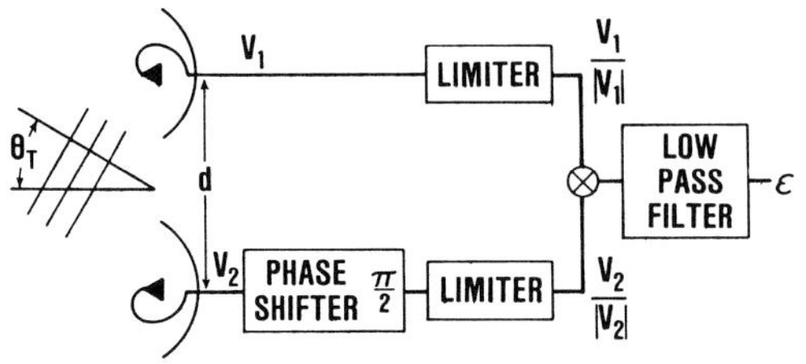

Fig. 2.45 Phase-phase monopulse.

four-horn monopulse radar. This structure permits a large focal-length to diameter ratio (f/D) in a smaller space because of feed imaging in the subreflector, and a larger f/D ratio reduces cross polarization. The four-horn feed structure is inefficient and methods to increase the antenna gain usually center on multimode feeds. These methods include use of higher-order modes in the feed structure to form apparent feed horns that maximize the efficiency of the available aperture.

Many monopulse systems use two rather than three IF amplifiers to reduce weight and cost. In fact, one form of monopulse uses only one IF amplifier [70]. A configuration of two IF amplifiers uses one amplifier or channel for alternate sampling of the two difference channels, while the second channel is dedicated to the sum antenna output to maintain normalization of angle error. This sampling can be slow or fast relative to the PRF and the necessary information can be encoded into phase or amplitude variations or delayed in time. In other words, there seems to be an endless number of innovative methods to construct a two-channel system. One approach is shown in Fig. 2.47 where the difference channels are combined in "phase quadrature" before the difference IF channel. The circuitry after

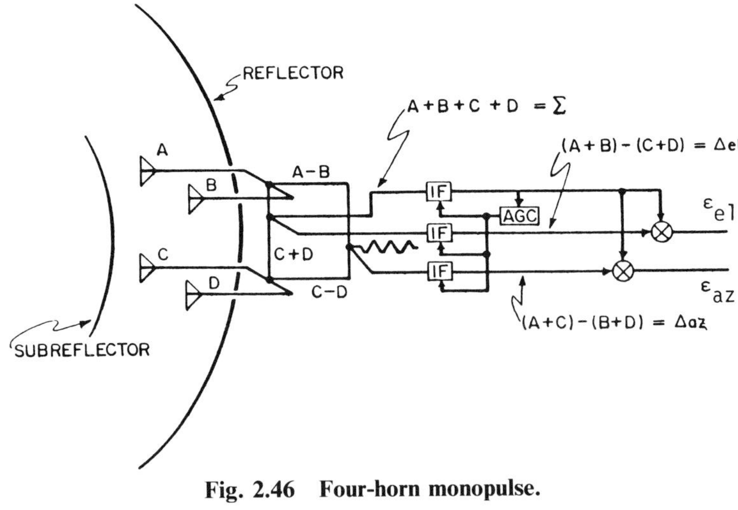

Fig. 2.46 Four-horn monopulse.

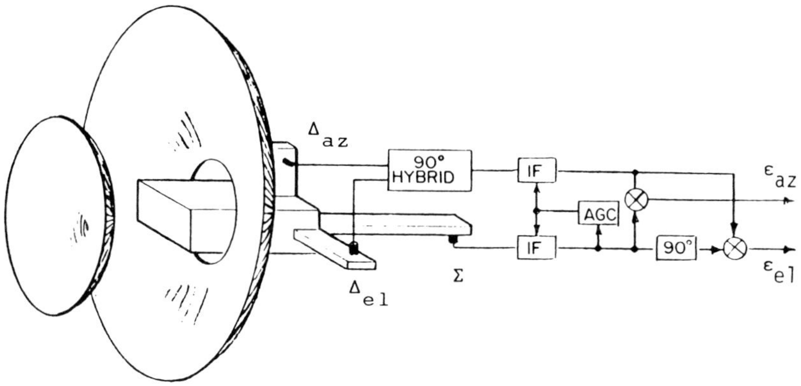

Fig. 2.47 Two-channel monopulse.

the IF amplifiers can easily recover the separate errors in azimuth and elevation.

This two-channel approach is not without penalty. Amplitude or phase errors in the IF circuitry will now cause cross coupling of the azimuth and elevation errors that, if severe enough, can result in loss of angle tracking.

During the acquisition of a target, only the monopulse sum-channel signal is processed. The range at which a target will be acquired (detected) is computed in the manner outlined in Secs. 2.4 and 2.5. When clutter limits detection, computation of the clutter power must usually consider the differing antenna gains toward target and Earth. Tracking of a target in elevation is impossible, without implementing special techniques, when the angle of target elevation is less than approximately 0.7 of the beamwidth of the radar [71].

2.8 DOPPLER PROCESSING

All of the radars discussed thus far have emitted simple, low-duty-cycle repetitive bursts of radio frequency power. Except for the discussion of moving target indicators in Sec. 2.5, each pulse could have had a starting phase unrelated to the other pulses in the burst. Such a burst of pulses is called incoherent. This section reviews the distinction between coherent and incoherent pulse trains and points out the advantages of a very low or very high PRF. The waveform has no effect on the methods of angle tracking discussed above except for consideration of interfering sources.

Sections of typical coherent and incoherent pulse trains are shown in Fig. 2.48. The pulses in the coherent train are merely samples of the continuous sinusoidal waveform. The samples of the incoherent train have the same starting phase, but the phase of each pulse relative to a continuous sine wave (as shown by the dashed pattern) is random in nature. Either pulsed radar requires a receiver bandwidth given by $B_R = 1/\tau$ to process the target returns

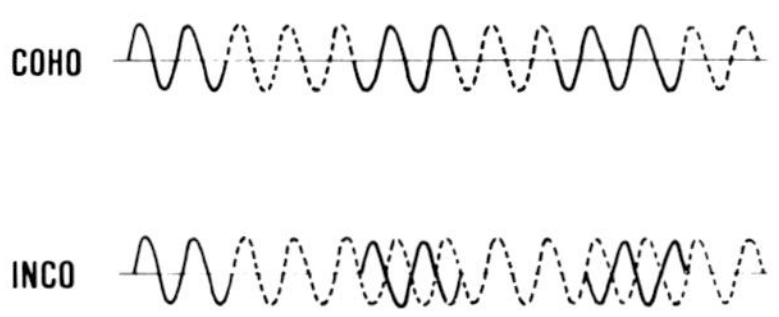

Fig. 2.48 Coherent and incoherent pulse trains.

effectively. If the radar is coherent, the radar transmission and target return is not uniformly distributed over the receiver bandwidth, but has a frequency distribution as sketched in Fig. 2.49. Such a figure is called a frequency spectrum, and it indicates frequencies at which signal power is present. Apparently, both the transmitted and received spectra consist of "spectral" lines separated by the PRF. Furthermore, the target lines are offset from the transmitted spectral lines by the Doppler frequency f_d, which is related to the "radial" velocity between the radar and the target. Again, this relationship is $f_d = (2V_r)/(\lambda_o)$, where $f_o \lambda_o = V =$ velocity of light.

This "line" spectrum enhances the ability of the radar to see the target by allowing the radar to place a narrow filter around a target line. That is, the narrow filter eliminates noise or jamming that could compete with the signal. The frequency of the target return uniquely specifies the radial velocity of the target, provided that the target Doppler shift is less than one-half the PRF. A high-PRF pulse-Doppler (PD) radar operates in this mode because it allows filtering of competing ground clutter. Effectiveness in target detection depends on the relative signal-to-noise power present in the filter used to enhance the target in the noise background. This filter bandwidth is restricted by the radar antenna search/track scan rates, any target-induced modulations, and the amount of tracking error allowed. If the filter bandwidth is greater than the bandwidth induced by the target, a correction for filter mismatch must also be included in any computations of detection.

In the continuous-wave (CW) radar shown in Fig. 2.50, no pulses are transmitted and the transmit and receive spectrum consists of single lines (with scanning sidebands) separated by the Doppler frequency. The receiver consists of either a tunable narrow-band filter or a band of such filters as shown in the diagram. The radar operator places a narrow-band filter over the target return and the target is tracked in angle by follow-on circuits much the same as the pulsed radars.

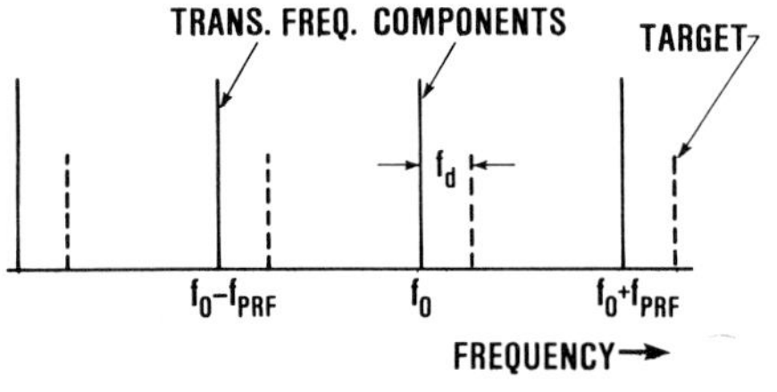

Fig. 2.49 Coherent radar spectrum.

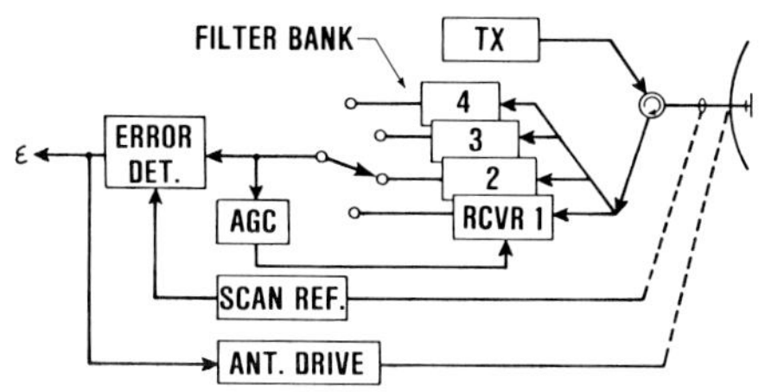

Fig. 2.50 CW radar—CONSCAN

In both the PD and CW radar systems, this narrow-band filter must remain on the target return to insure uninterrupted angle tracking. The operation is usually accomplished with the use of a frequency discriminator and associated circuitry (the whole assembly is termed a "velocity gate"). A typical velocity gate is shown in Fig. 2.51. The discriminator generates an output e_o whenever the output of the IF filter differs from the frequency of the IF center, f_{IF}. The discriminator output is positive (negative) when the actual IF frequency is greater (less) than the frequency of the IF center. The voltage-controlled oscillator (VCO) has an output frequency differing from its quiescent operating frequency, $f_R - f_{IF}$, by an amount proportional to the output of the discriminator. Thus, a discriminator output e_o results in a VCO offset of $f_d - \Delta f = K_v e_o$, where $f_d - \Delta f$ is the oscillator offset from $f_R - f_{IF}$ and K_v is assumed to be the gain of the amplifier [also called the velocity loop direct-current (DC) gain]. If the discriminator output and IF offset are related by

$$e_o = 2\pi \Delta f$$

then the resulting error in frequency tracking is

$$\Delta f = \frac{f_d}{1 + 2\pi K_v}$$

Thus, there is a finite-frequency tracking error Δf for finite velocity-loop gains. Obviously, the frequency tracking error must be smaller than one-half the bandwidth of the IF filter or the target will be lost.

The above frequency tracking circuit is also used in a pulse Doppler radar. As stated before, the PRF in such a radar must be large enough to allow

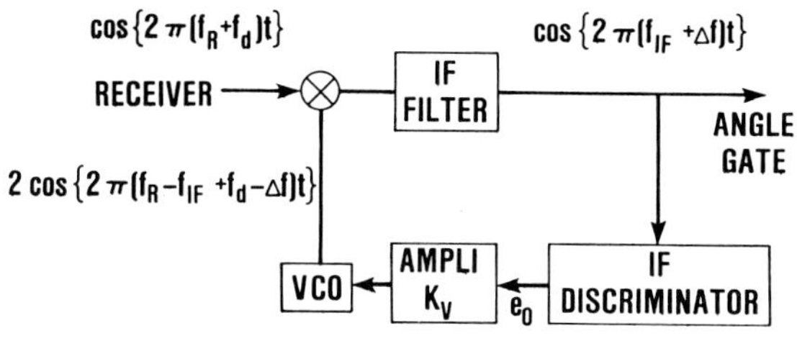

Fig. 2.51 Velocity gate.

unambiguous determination of target velocity. The transmit and target spectrum then appear as shown in Fig. 2.49, and the velocity gate is centered on one of the many target spectral lines. Normally, all of the transmit and receive spectral lines but the central one are suppressed. In either the CW or PD case, the output of the narrow-band receiver is then detected and angle-tracking circuitry similar to the circuitry of a pulsed radar performs angle tracking.

Measurement of range in the CW radar can be accomplished by varying the RF at a selected rate—for example, α Hz/s, as shown in Fig. 2.52. For a stationary target, the returned frequency as a function of time is shown in

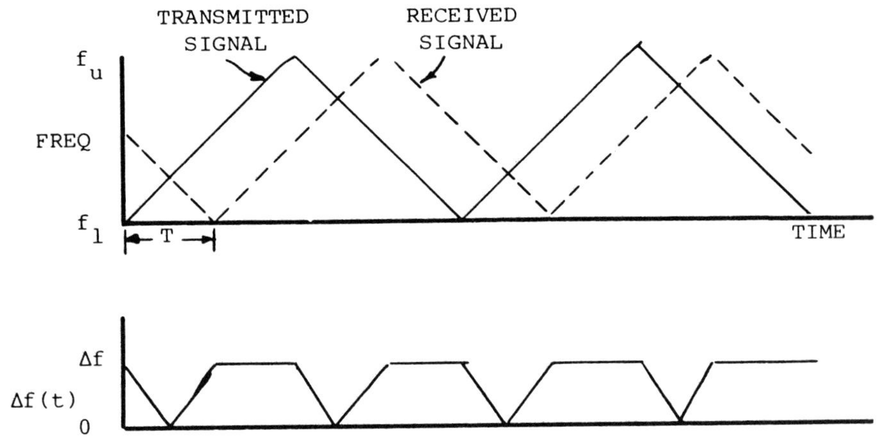

a) FIXED TARGET.

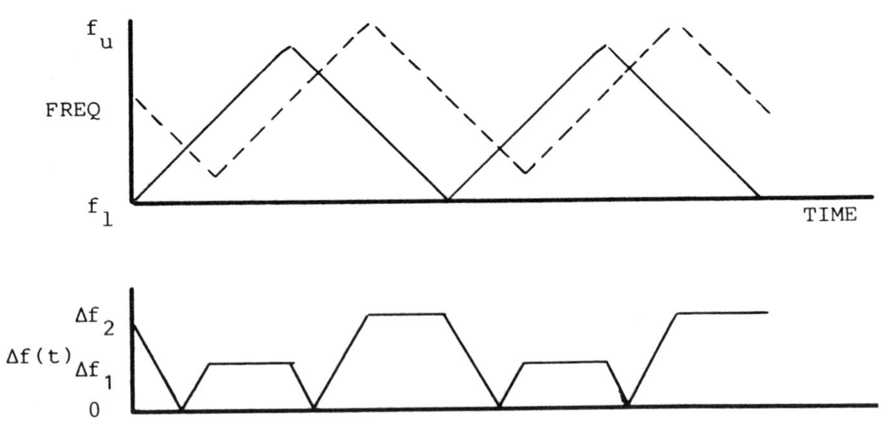

b) MOVING TARGET.

Fig. 2.52 FM/CW radar range determination.

Fig. 2.52a. The transmitted frequency during the increasing frequency portion of the cycle is

$$f_R(t) = f_l + \alpha t$$

The difference in frequency between the frequency of the transmitter and the frequency received is

$$\begin{aligned}\Delta f(t) &= f_R(t) - f_R(t - T)\\ &= [f_l + \alpha t] - [f_l + \alpha(t - T)] = \alpha T\end{aligned}$$

As T is the delay associated with a signal traveling to and from a target at a range R, $T = 2R/V$. Also, as α and Δf are known, $R = \Delta f V/2\alpha$.

A moving target merely offsets the return frequency by the Doppler frequency, as shown in Fig. 2.52b. Now, the average difference frequency yields range and the absolute frequency difference yields velocity. With the magnitude of the larger offset defined as $|\Delta f_2|$,

$$|\Delta f_2| = \alpha T + f_d, \qquad \text{if} \quad \alpha T > f_d$$

and that of the smaller offset $|\Delta f_1|$

$$|\Delta f_1| = \alpha T - f_d$$

Then the average effect is

$$\overline{\Delta f} = \frac{|\Delta f_1| + |\Delta f_2|}{2} = \alpha T$$

and the Doppler shift is

$$f_d = \frac{|\Delta f_2| - |\Delta f_1|}{2}$$

Example 2.7: Doppler Shift

A target is moving at 600 m/s on a radial path toward a radar operating at 9 GHz. What will be the Doppler offset? If both approaching and receding targets must be tracked, what is the minimum PRF that will allow unambiguous determination of target velocity?

Answer:

$$f_d = \frac{2V_r}{\lambda} = \frac{2x600}{1/30} = 36\ \text{KHz}$$

The PRF must be slightly greater than twice the Doppler or 72 kHz. This is apparent in Fig. 2.49. With any lower PRF, a receding target could be mistaken for an approaching target.

2.9 RADAR APPLICATIONS

The radar systems included in this discussion have many uses in weapon systems, but they have an obvious role in target detection and tracking (estimation). They are also used for guidance of terminal weapons, warhead fuzing, target identification, ground mapping, and warning of incoming missiles. They can be applied to all these missions, although the same radar cannot be applied to all tasks with equal success. For example, target identification favors higher transmit frequencies than does target detection at long ranges in rain. Beyond target detection and tracking, their principal tasks are weapons guidance and fuzing. Methods of carrying out these tasks may vary from weapon system to weapon system, but the most common possibilities are listed below.

1) Weapon guidance

Active homing—The missile is guided to the target by its own radar.

Semiactive homing—The missile contains a radar receiver only. This receiver guides the missile by detecting signals reflected from the target. An illuminating source provides the signal incident on the target.

Command guided—The missile is guided to the target by steering commands transmitted over a communications link. Some means must be provided for determining missile location.

Beam rider—Same as command guided, but the missile is tracked by the same radar that tracks the target.

2) Weapon fuzing

Homing radar—The warhead detonates on the basis of signals received in its on-board radar.

Active fuzing—The missile contains a separate radar solely for fuzing (proximity fuze).

An excellent article by Walsh [27] discusses the past, present, and potential role of radar systems in many types of weapon systems.

Problems

2.1 Name and briefly describe the major subdivisions of electronic warfare.

2.2 **(a)** What is the range resolution of a 0.25 μs radar pulse?

(a) 25 m (b) 37.5 m (c) 30 m

(b) A 2 kHz PRF allows what unambiguous radar range?

(a) 40.5 nm (b) 45 nm (c) 37.5 nm

(c) What is the wavelength in air of an electromagnetic wave at a frequency of 9 GHz?

(a) 0.15 m (b) 0.033 cm (c) 3.3 cm

2.3 **(a)** A transmitter power of 150,000 W equals how many decibels above 1 W?

(a) 45 dBW (b) 50 dBW (c) 52 dBW

(b) If this power is radiated through an antenna having a gain of 30 dB, what is the power density at 20 km?

(a) 10^{-6} W/m^2 (b) 5 W/cm^2 (c) 0.03 W/m^2

(c) If 150,000 W is the average power, the pulse repetition frequency is 1 kHz, and the pulse duration is 1 μs, what is the peak power?

(a) 1.5 MW (b) 150 MW (c) 150 W

2.4 Given the radar parameters in Table 2.7:

(a) What would be the radar range if the target cross section were 2.5 m^2?

(a) 180 nm (b) 200 nm (c) 225 nm

(b) What would be the power level of the internal noise of the receiver if the bandwidth changes to 5 MHz?

(a) 0.05 V (b) −132 dBW (c) −90 dBm

2.5 **(a)** How many pulses are available for integration if the pulse-repetition frequency is 200 kHz, beamwidth is 2 deg, and scan rate is 60 deg/s?

(a) 6667 (b) 100 (c) 66

(b) If a probability of detection of 0.8 and a false alarm probability of 10^{-8} are required, what signal-to-noise ratio is necessary from a target of constant cross section?

(a) 13.2 dB (b) 12.6 dB (c) 13.7 dB

2.6 Given an antenna with a circular aperture 3 ft in diameter operating at a frequency of 5 GHz, what will be the "approximate" antenna beamwidth and antenna gain if a 60% aperture efficiency is assumed?

2.7 During acquisition, a radar with the antenna in problem 2.6, a 1800 Hz PRF, a 0.5 μs pulse width, and the additional parameters listed below scans vertically at the rate of 200 deg/s. What must be the transmitter power for a detection range of 45 nm against a 1 m^2 target if a 13 dB signal-to-noise ratio is required? The system noise figure is 8 dB and the radar noncoherently integrates the returns on a cathode ray tube.

2.8 Which type of monopulse exhibits a bore site tracking error least affected by active IF component stability?

(a) Phase-Phase (b) Phase Sum-Difference (c) Amplitude-Amplitude and (d) a and b

2.9 A search radar with the parameters listed below must detect a 1 m^2 target in a region of clutter where the clutter reflectivity is -20 dB. If an output signal-to-clutter ratio of 20 dB is required for the desired level of detectability at 100 nm, what must be the MTI improvement factor?

P_R = 100 kW radar peak power
G_R = 30 dB antenna gain
RF = 3 GHz frequency
NF = 8 dB noise figure
rpm = 12
PRF = 300 Hz
θ_3 = 1.5 deg azimuth beamwidth
τ = 2 μs pulse width

2.10 A new SAM radar has been photographed. The measured parameters are:

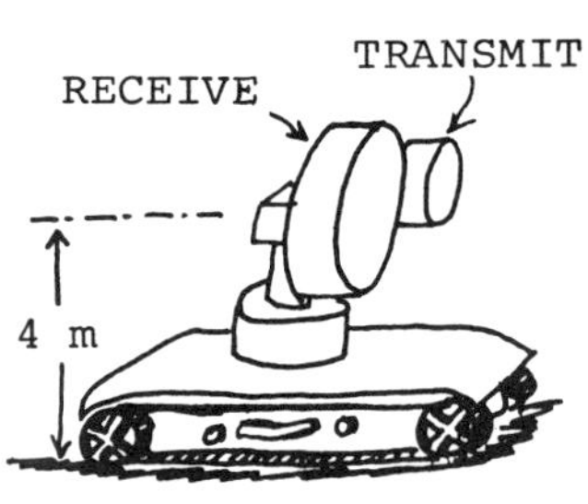

RF	= 9.5 GHz
Antenna diameter	
Transmit	= 1.0 m
Receive	= 2.0 m
Pulse-repetition frequency	= 5 KHz
Pulse duration	= 0.4 μs

(a) What is the radar horizon?

(b) Assuming a 70% aperture efficiency and a target must be on the scan axis of the receive beam to be tracked, what must be the height of the target at 7 km if the lower -3 dB point of the antenna cannot go below the radar horizon? Assume a squint angle.

(c) For the radar to track a 1 m^2 target at lower altitudes at 7 km, what must be its MTI improvement factor? Assume that $\sigma^o = -20$ dB(m^2/m^2) and that a $(S/C)_{min}$ of 20 dB is required.

3 RADAR ELECTRONIC COUNTERMEASURES

Electronic countermeasures have been defined as actions that deny the enemy use of the electromagnetic spectrum. Since total denial for all time is impossible, the primary reasons for employing ECM is to "buy time" to delay an adversary's response [35]. This chapter describes many of these actions (ECM techniques) and their effectiveness against various classes of radars. The techniques are organized according to the pattern outlined in Sec. 3.1 and are followed by a review of the critical microwave components used in implementing the techniques. The chapter concludes with a discussion of the operation of idealized ECM systems. Also included are descriptions of various electronic counter-countermeasures (ECCM) when such descriptions are necessary for explaining variations in the basic ECM techniques.

3.1 THE HIERARCHY OF ECM TECHNIQUES

A number of approaches can be used to classify the seemingly infinite variety of ECM techniques [28]. One way is to classify them from the perspective of the intended threat, for example, search radar and track radar ECMs. The official Air Force approach [1] is to classify them according to the anticipated effect, such as masking or deception jamming. Each approach has unique advantages and disadvantages. For example, in the masking/deception approach, technique A may deny angle estimation by radar B and is thus called deceptive. Against radar C, however, technique A may act as a beacon. Where should it be listed? A third approach is to designate techniques as active or passive. Subclassification depends on the actual signal radiated or action taken as opposed to the desired effect of those signals or actions on a radar.

Electronic countermeasures can be classified as follows:

Active ECM—radiation of electromagnetic energy to deny the enemy full use of his radar or weapon system.

Passive ECM—no electromagnetic energy radiated, but some positive action taken to degrade the performance of the enemy's weapon system.

Active ECM techniques can be further classified as noise-like or target-like according to the type of emission [96]. Sometimes, these two classifications are considered synonymous with masking and deception. However, some noise techniques are deceptive and target-like signals can mask the targets. Perhaps some of the confusion can be avoided by distinguishing between these techniques according to the nature of their emissions, rather than the intended impact of the emissions on a radar.

What is meant by target-like or noise-like jamming? The target-like jammer radiates a signal much like that from the victim's radar. To be effective in creating erroneous data, however, this signal must possess properties sufficiently different from those of the actual target return. Such differences might include a time delay from the real pulse, a frequency offset, or a time variation in the amplitude pattern. On the whole, however, the output of the target-like jammer is just as deterministic as the output of the radar signal scattered by the target.

Noise jamming, on the other hand, has none of the deterministic qualities usually found in target-like jamming. Measured in terms of amplitude or frequency, the output of the noise jammer at any point in time must be independent of the output at any other point in time. Presumably, enough of this randomness will prevail at the receiver output to prevent detection of the target or accurate estimation of the target parameters. Of course, various statistical descriptors associated with noise jamming, such as average frequency or average power, may vary in a controlled manner with time.

Passive ECM techniques can be further classified as either observable reduction or clutter enhancement. Any tendency to regard these classifications as goals rather than physical actions should be avoided. The classifications represent actions taken to achieve the objectives of target masking and/or radar deception. Viewing observable reduction as a goal would be similar to viewing noise jamming as a goal and asking whether radiated noise is "good" according to some measure of goodness. This subject is treated in a later discussion; but, in organizing a hierarchy of techniques, one must view both observable reduction and noise jamming as possible actions. Thus, the main thrust of this chapter is the effect of each category of actions on radars.

A hierarchy of ECM techniques is shown in Fig. 3.1. Further classifications are based on the parameters used in implementing a particular ECM. Many techniques simultaneously employ more than one of the variables shown in Fig. 3.1. This point becomes clearer in a detailed description of the techniques.

3.2 NOISE JAMMING

A typical noise jammer and its emission is depicted in Fig. 3.2. The figure shows only a short time segment of continuous radiated power from the transmitter. The actual emission can be viewed as a series of pulses (like the pulses of the radar transmitter), each of which has an amplitude unrelated to the amplitude and phase of any other pulse. The jammer power P_J is

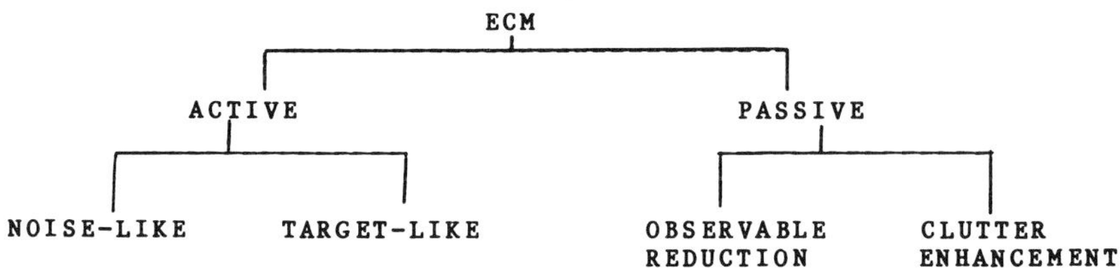

Fig. 3.1 Hierarchy of ECM techniques.

defined as the average power radiated. The averaging is performed in time and the averaging time is much larger than the duration of the individual pulses τ_J. For example, if P_{J_1} is the radiated power of the first pulse and P_{J_i} the radiated power on the ith pulse emitted from the jammer, the jammer power is defined as

$$P_J = \frac{1}{N} \sum_{i=1}^{N} P_{J_i}$$

If the jammer pulses have the same frequency as the frequency of the radar being jammed, some of the jammer power must be received by the radar. The bandwidth of the jammer emission B_J must be given approximately as

$$B_J \approx \frac{1.0}{\tau_j}$$

which is exactly the IF bandwidth of the radar if $\tau_j = \tau_R$. Then, the average jamming power received by the radar P_{r_J} at a distance R from the jammer is (in this idealized case)

$$P_{r_J} = \frac{P_J G_J}{4\pi R^2} A_e = \frac{P_J G_J}{4\pi R^2} \frac{\lambda^2}{4\pi} G_R$$

where G_J is the jammer antenna gain toward the radar.

Both theoretical and experimental studies have found that the amplitude of the output of the radar receiver should have a certain characteristic for maximum effectiveness in masking jamming [4,p.4; 29]. The amplitude of the output measured at many points in time will result in a range of measured values. When these measured values are plotted on a graph showing their frequency of occurrence, the resulting curve should follow a Rayleigh probability density distribution. Such a receiver output will have an amplitude

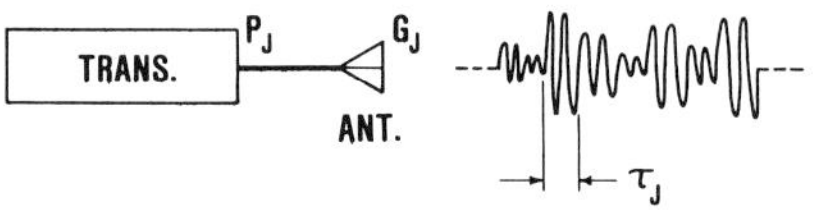

Fig. 3.2 Noise jammer and output.

that varies with time. The exact amplitude at any instant cannot be predicted, but the percentage of time that the output will lie between a range of amplitudes can be determined. Table 3.1 shows the average percentage of time that a quantity following a Rayleigh distribution will be bounded by specific values expressed as some multiple of the average value. For such an ideal jammer, the receiver output appears exactly as it would if it were caused only by receiver thermal noise, except for perhaps the power level.

In this instance, maximum effectiveness is synonymous with the inability of the radar operator to determine the correct target range cell at a given signal-to-noise ratio. Practical limitations in jammer equipment, such as saturation effects and dynamic range, require increased bandwidth for the noise jammer relative to the bandwidth of the radar receiver in order to create this Rayleigh noise out of the receiver. A probability theorem called the "central limit theorem" proves that this can happen and we can rationalize this Rayleigh noise generation with the aid of Fig. 3.3. Figure 3.3a shows the idealized receiver output when the jammer and radar have identical bandwidths. The peak voltage of the radar output is fixed by an assumed maximum jammer output. As shown in Table 3.1, the desired output voltage for the receiver does not have such a limiting value and thus Fig. 3.3a does not represent optimum jamming. But, in Fig. 3.3b, the jammer pulse width has been decreased (its bandwidth increased) and the output of the radar receiver at any instant is now due to a large number of independent jammer inputs. The actual number of independent input signals contributing to any given output signal is approximately τ_R/τ_J. Thus, the input of the radar filter still suffers from a fixed maximum value, but the receiver output could be

Table 3.1 Rayleigh Distribution

Range of values (X = average value)	Time spent in range, %
0–0.2X	3.1
0.2X–0.5X	14.7
0.5X–X	36.6
X–2X	41.3
2X–3X	4.2
3X–∞	1.0

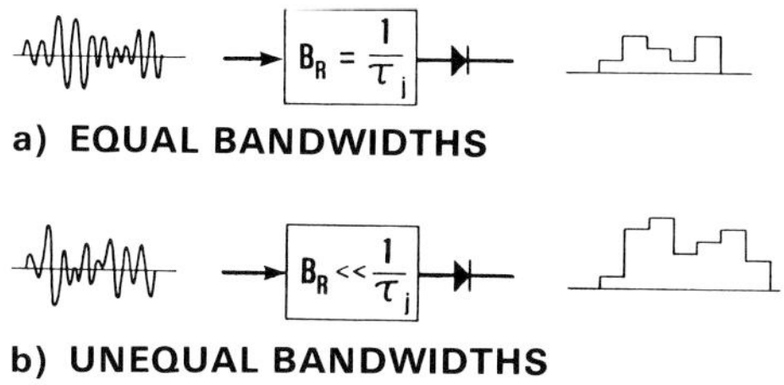

Fig. 3.3 Effect of jammer bandwidth on video signal.

much larger if all inputs result in outputs that combine constructively at some time. Of course, this line of reasoning does not prove that the receiver output has the distribution desired; more sophisticated mathematics are required for that. This argument does outline the principles involved in creating effective noise by a realistic, saturated-signal source. This increase in jammer bandwidth causes the above equation for received jamming power to become

$$P_{r_J} = \frac{P_J G_J}{4\pi R^2} \frac{\lambda^2 G_R}{4\pi} \frac{B_R}{B_J} = \frac{P_J G_J G_R \lambda^2}{(4\pi R)^2 B_J} B_R$$

indicating the presence of jamming energy outside of the IF bandwidth B_R of the radar. Such jamming is called continuous-wave (CW) spot noise jamming, since the jammer radiates a constant average power over a relatively narrow bandwidth.

Continuous-Wave, Self-Screening Noise Jammer

Such a CW noise jammer can now be used to define the radar "burn-through" range and jamming-to-signal ratio. Based on the preceding discussion, the power received from a target at a range R is

$$P_{r_T} = \frac{P_R G_R^2 \sigma_T \lambda^2}{(4\pi)^3 R^4}$$

If the jammer is carried on the target aircraft and if the jamming power received is the dominant noise source, the operator or automatic detector must attempt to operate in a per pulse signal-to-noise ratio of

$$\frac{P_{r_T}}{P_{r_J}} = \left(\frac{S}{N}\right)_p = \frac{P_R G_R^2 \sigma_T \lambda^2}{(4\pi)^3 R^4} \frac{(4\pi)^2 R^2 B_J}{P_J G_J \lambda^2 G_R B_R} = \frac{P_R G_R \sigma_T B_J}{4\pi R^2 P_J G_J B_R}$$

The inverse of this equation is called the jamming-to-signal ratio J/S and can be written

$$\frac{J}{S} = \frac{P_J G_J B_R 4\pi R^2}{P_R G_R B_j \sigma_T}$$

This is actually an expected J/S ratio, since the jammer power is the average power.

What does this equation imply about detection performance? First, if the jammer can produce Rayleigh noise out of the receiver and if the radar uses circuitry for automatic detection, this equation allows use of the procedures discussed in Chapter 2 to predict radar performance at any given range. If neither of the above is true (for example, an operator is used rather than an electronic threshold), then the jamming-to-signal ratio providing the desired level of masking or screening must be empirically evaluated. The above equation can then be used to evaluate the range at which a target becomes visible in the presence of the jamming if this required J/S ratio is inserted on the left-hand side. This condition is defined as "burn through" [28, p. 41].

Others define "burn through" as the range at which the signal-to-jamming ratio equals the initially specified signal-to-noise ratio for a desired level of radar performance [30, 31]. For example, specifications for the radar system may call for a detection probability of 0.5 and a false alarm probability of 10^{-6}. These performance quantities then require an IF signal-to-noise ratio that depends on the assumed target model, number of pulses integrated, etc. But, with the exception of simple automated detection systems, it seems unreasonable to require the same probability of false alarm in the presence as in the absence of jamming. The low probabilities specified for false alarm prevent unnecessary operator fatigue or overloading of computers during day-to-day operation of the radar. The obvious presence of jamming will usually result in a change from the day-to-day routine (unless jamming is continually present over periods of days or weeks). After all, something has changed. The presence of jamming normally means a high probability that a target is present. One such reasonable change would be to allow for a higher probability of false alarm per scan and to compare the scan-to-scan detections visually for patterns meaningful to a skilled operator. The operator's job (or that of a sophisticated computer algorithm) is now to determine the target range through an ill-defined interaction between man and machine.

Experimental efforts have attempted to determine the J/S ratio at which a target burns through on a PPI display [29]. These experiments involved a single target and a laboratory observer and showed that, at a J/S ratio of 3 dB, there is only a 50% probability of correctly determining the true location of the target. This forced-choice experiment implied a 50% probability of selecting an incorrect target position. The same experiment showed a decrease in the required J/S ratio when the jamming produced a voltage distribution that was not Rayleigh, in disagreement with a previous statement. The reason for this enhancement of jamming is that the non-Rayleigh jamming created target-like outputs that confused the operators. However, simple ECCMs easily removed this confusion and again resulted in a much higher J/S ratio for non-Rayleigh jamming that was as effective as Rayleigh jamming. Other experiments [2, p. 7] have shown that, for multiple targets, the J/S ratio required to reduce correct selection of target position to 50% is less than 3 dB (−1 dB for two targets). Certainly, it is impossible to estimate operator performance in the heat of battle.

An operator will usually have alternative displays, such as an A- or J-scope to extract information on the amplitude of signal returns. For such displays, the J/S ratio required to reduce the probability of correct range selection to 50% is typically 5 dB higher than the J/S ratio required for PPI displays [32].

Continuous-wave noise jamming can deny information about target range to any one radar, but denial of angular information is another question. If the operator has sufficient time in a one jammer on one radar encounter, he should be able to determine the jammer angle by reducing the PPI drive signal. At some point, jamming signals received in the sidelobes of the radar antenna will no longer be visible or the pattern on the PPI display will be symmetric enough to determine the jammer's angular position.

Figure 3.4 shows the effect of range on the J/S ratio. Obviously, the J/S ratio increases 20 dB for every tenfold increase in range because the reflected target power decreases more rapidly with increased range than the received jamming power decreases.

The J/S ratio computed so far is a per-pulse J/S ratio. The signal-to-noise ratio in a radar that integrates N pulses is enhanced in the presence of noise jamming, just as it is enhanced for receiver noise. The previous J/S equation must be modified to account for this integration if the decision regarding the presence of a target is based on this average receiver output. In a radar employing noncoherent integration of N pulses, the J/S ratio is reduced by $1/\sqrt{N}$. The J/S reduction for a radar employing an MTI canceller is computed as follows. If N pulses are available, but the canceller processes M at a time followed by incoherent integration of the $N - M + 1$ pulses out of the MTI canceller, the J/S reduction is $1/\sqrt{N - M + 1}$. In a coherent radar employing fast Fourier transform processing or narrow-band filters, either the final coherent radar bandwidth must be used or the jamming-to-signal ratio must be reduced by $1/N$ if the IF bandwidth is used. A second factor p accounting for differences in jammer and radar polarization should also be added. Thus, the final equation for the jamming-to-signal ratio caused by a self-screening noise jammer is

$$\frac{J}{S} = \frac{P_J G_J B_R L_i}{P_R G_R B_J N} p \frac{4\pi R^2}{\sigma_T}$$

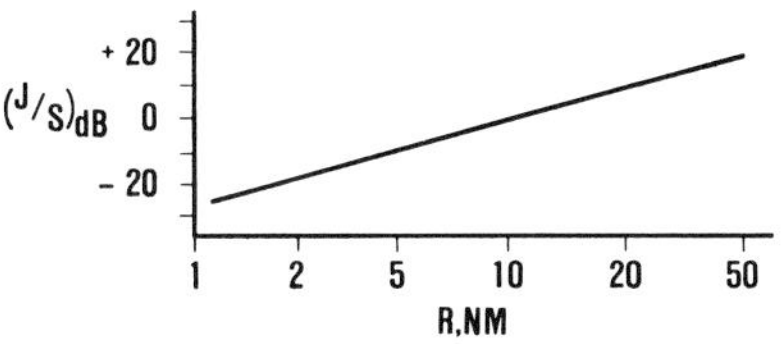

Fig. 3.4 Variation of J/S with range.

Example 3.1

The preceding discussion can be summarized by computing the J/S ratio for a typical problem. The jammer is described in Table 3.2 and the radar in Table 3.3. The target is at 20 n.mi. and has a 6 m^2 cross section.

Answer: Inserting the pertinent parameters into the J/S equation and noting that the radar noncoherently integrates 16 pulses, the J/S ratio is computed as follows:

$$(J/S)_{dB} = P_J + G_J + B_R + 4\pi + R^2 - P_R - G_R - B_J - \sqrt{N} - \sigma_T$$

$$= 20 + 3 + 0 + 11 + 91.4 - 60 - 35 - 10 - 6 - 4 = 10.4 \text{ dB}$$

Note that this is the J/S ratio plotted in Fig. 3.4. If the polarization of the jammer had not been identical to the polarization of the radar, the J/S ratio would be reduced to account for the mismatch in polarization. For linear polarization, this reduction factor equals $\cos^2\phi$ if the respective electric field vectors for the radar and jammer subtend an angle of ϕ degrees. If one antenna is linearly polarized and the second antenna circularly polarized, the factor p is $\frac{1}{2}$. In Example 3.1, the J/S ratio would be reduced to 7.4 dB ($\cos^2 45 \text{ deg} = \frac{1}{2}$) if the jammer's polarization were 45 deg away from the vertical.

The noise jammer in Table 3.2 had a bandwidth only as large as required to obtain high-quality noise. In many instances, one jammer must operate against many radars, each at a different frequency. With one CW noise jammer, effective jamming is possible only by increasing the bandwidth of the jammer until it covers all threat-radar frequencies. This, of course, dilutes the radiated power density of the jammer. In Example 3.1, the jammer radiated 100 W over a 10 MHz bandwidth or 10 W/MHz (see Fig. 3.5a).

Table 3.2 Jammer Parameters

Power	$P_J = 100$ W, 100% duty cycle (i.e., CW)
Antenna gain	$G_J = 3$ dB, horizontally polarized
Bandwidth	$B_J = 10$ MHz

Table 3.3 Incoherent Search Radar Parameters

$P_R = 1$ MW	$B_R = 1.0$ MHz
$G_R = 35$ dB	rpm = 4.3
RF = 600 MHz	$\theta_{az} = 1.3$ deg
PRF = 317 Hz	

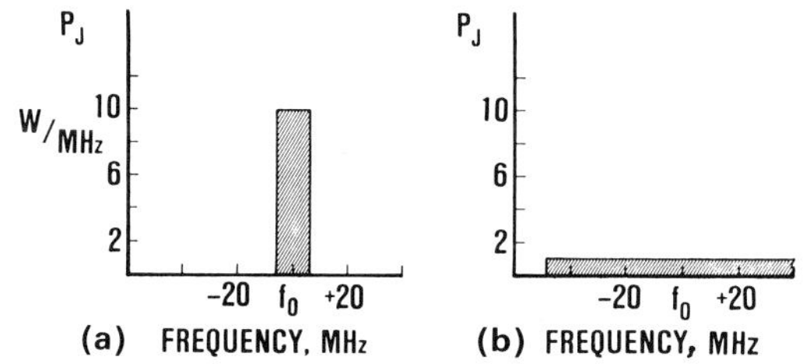

Fig. 3.5 Jammer power density.

If the jammer bandwidth is increased to 100 MHz, then the jammer will still radiate 100 W but only at 1 W/MHz, as shown in Fig. 3.5b. When the gain of the jammer antenna is included in these descriptions, the jammer is said to produce effective radiated power (ERP), where ERP $= P_J G_J$. In Example 3.1, the ERP is 200 W, or a density of effective radiated power of 2 W/MHz is produced. When the noise-jamming bandwidth is much greater than the IF bandwidth of a typical radar, the jamming is termed barrage jamming.

Amplitude-Modulated Noise Jammer

As previously stated, a CW noise jammer does not by itself create significant angular errors. In fact, a CW noise-jamming signal can provide a very good angle-tracking signal for all tracking radars. Against sequential lobing or TWS radars, however, the average noise-jamming power can be modulated in amplitude at the lobing rate to induce large angular tracking errors or even breaklocks. Description of the jammer emission and the voltages induced in a conical scan and an SWC radar should clarify this phenomenon.

Figure 3.6 shows the signal that might be emitted from an amplitude-modulated jammer. The jammer radiates power continuously, but only a short period in relation to the radar-scan period T_s is shown. The average power is computed over a much larger period of time than the individual pulses τ_J, but much smaller than the scan period T_s. This average power varies over time in a manner determined by the ECM designer to maximize the angular error in the jammed radar. The diagram of error voltage from the conical-scan receiver in Fig. 3.7 shows how this angular error is generated.

Figure. 3.7a shows the two extreme angular positions of the radiation pattern for a conical-scan radar antenna. Figure 3.7b shows the error voltage from a constant target that can be used to reposition the scan axis on the target, and Fig. 3.7c shows the effects of a constant-power jammer on output error voltage. The fluctuations of the jammer power have been averaged, since the radar servo system will track only the average error. The peak-to-peak variation caused by the constant jammer is only one-half

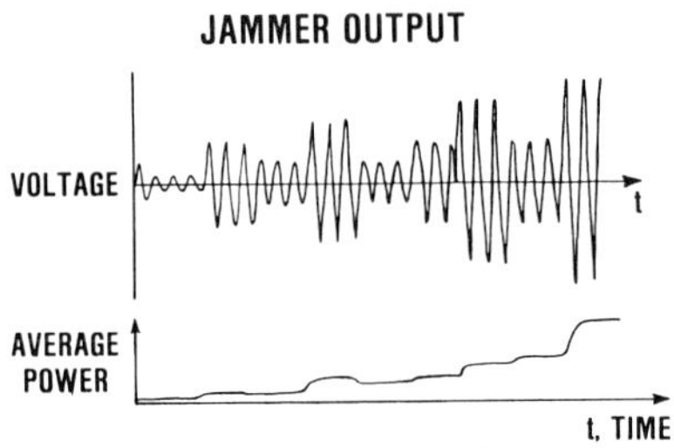

Fig. 3.6 Amplitude-modulated noise-jammer output.

the variation of the constant target, since only one radar antenna pattern (receive) is involved in forming the error signal. However, the similarity between the phase of this error voltage and the phase of the constant target error voltage implies the eventual, although somewhat more sluggish, nulling of tracking errors.

The effect of the relative phase of jammer modulation is an important consideration in analyzing the modulated jammer. Figure 3.8a shows a simplified CONSCAN radar and a target above the tracking axis. The phase of the indicated error voltage will move the boresight up. The error voltage shown in Fig. 3.8a could be written as

$$\varepsilon_y(t) = \text{const} + \Delta\cos(2\pi f_s t + 0^\circ)$$

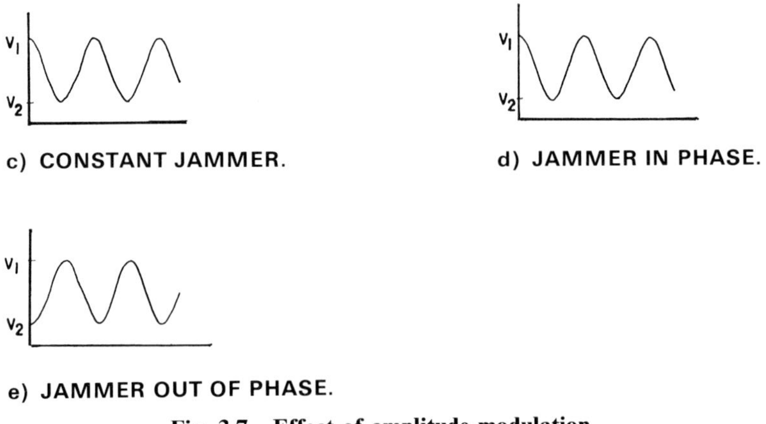

Fig. 3.7 Effect of amplitude modulation.

This is a cosine wave with a phase of 0 deg. A cosine wave with a phase of 180 deg will cause the error voltage at zero time to be less than the constant value and will correspond to a target below the scan axis. Therefore, if the target had been below the tracking axis, the error voltage would have had an opposite phase and the tracking axis would move down.

If the jammer is on the tracking axis and if the jammer power is constant, the error voltage is also constant and the boresight axis does not move. If the jammer is modulated in-phase with the antenna scanner, the error voltage will be as indicated and the seeker will move up until it reaches an angle at which the competing target returns, which will tend to drive the seeker down, balancing the jamming. If the jammer is modulated out of phase, the seeker moves down until the target balances the jamming. It is thus apparent that the jamming phase determines the direction of the seeker error and is otherwise unimportant. For example, a jammer in quadrature with the seeker (at a relative phase of 90 deg) will move the seeker horizontally.

Thus, Fig. 3.7d presents the error voltage when the jammer power is modulated "in phase" with the antenna scanning waveform. Again, in phase means that the jammer achieves peak power at the time that the maximum antenna gain is directed at the target. Figure 3.7e shows the error voltage when the jammer modulation is out of phase with the radar-scan pattern. In both Figs. 3.7d and 3.7e, the error voltage tends to increase rather than minimize tracking errors. For the in-phase case, the track axis moves beyond the target and again results in Fig. 3.7e. Figures 3.7d and 3.7e correspond to an infinite J/S ratio, but it seems that effective jamming should be possible above some J/S ratio.

What constitutes effective jamming? A large tracking error may be sufficient for effective ECM against AAA. But this same angular error induced in a SAM system employing semi-actively guided missiles may be ineffective because the missile will still reach the target if sufficient power is scattered by the target. If the error in angle can be made large enough, possibly through inertia of the radar servo system or additional target maneuvers, the radar seeker cannot continue following the target automatically and the result will be a breaklock. At this point, reacquisition is necessary, probably with operator assistance, before tracking can be restored. During this period, AAA weapons may not be effective and irrecoverable errors may develop in a missile during flight.

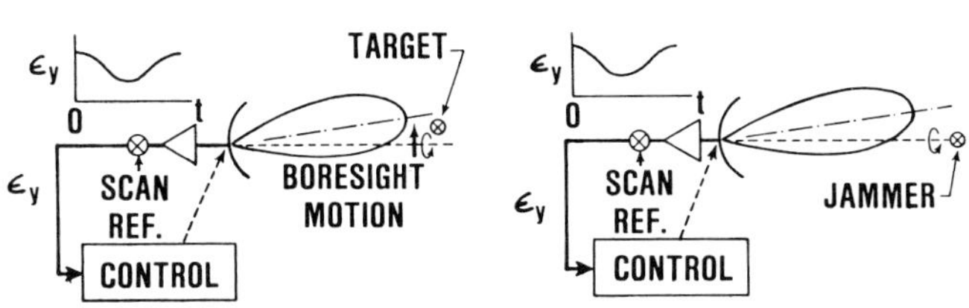

(a) TARGET. (b) JAMMER.

Fig. 3.8 CONSCAN error correction.

This J/S ratio must be on the order of 20–30 dB for a target-like jammer (or repeater) to induce sufficient angular error for breaklock [2, p. 15-16; 33]. This J/S ratio is based on the assumption that, when the squint angle is 0.5 beamwidth, the jammer must create a tracking error of 1.12 beamwidth before breaklock occurs. A computer simulation by Cikalo and Greenbaum [41] demonstrated that radar breaklocks begin when the induced tracking error exceeds 0.85 beamwidth and require a J/S ratio of 17.5 dB. For the CW noise jammer, range tracking on the target may be lost when the J/S ratio exceeds 10–13 dB [2, p. 15-14]. When this happens, there is no longer a competing skin return and it is necessary that only sufficient depth of modulation is present to induce the desired magnitude of angular error and that the average J/S ratio over the scan period prevents burn-through.

There are two ways to modulate jammer amplitude. One, called inverse gain, derives phase information from the scan pattern of the radar antenna to modulate the jammer power out of phase. As shown in Fig. 3.9, such phase information is available at the output of a receiver tuned to the radar. Analysis of this jammer/radar combination in Part II shows that an unstable servo control loop results. In-phase modulation of such a jammer will result in a radar with rapid response and overshoots, but with a stable tracking point on the target. As shown in Fig. 3.9, the generation of inverse-gain modulation requires only one receiver to handle all of the radar-scan frequencies.

The second approach is simply to modulate the amplitude of the jammer output, usually in the form of a square wave. This approach requires knowledge of the radar-scan frequency, from which the analysis proceeds as shown in Figs. 3.7 and 3.8. In this case, the rate selected for jammer amplitude modulation is that believed to be best for defeating the specific radar. Selection of the rate may be automatic or manual, based on the output of a radar warning receiver, or it may be preset before a specific mission in anticipation of a specific radar threat.

Figures 3.7 and 3.8 do not reflect the requirement that modulation of the jammer amplitude must be at or near the scan frequency of the radar antenna (or at or near a multiple or subharmonic of the scan frequency). This modulation rate is necessary because the radar receiver includes filters that are tuned to the known scan frequency. These filters are typically positioned after the boxcar detector. Signals that vary at other than the scan rate will

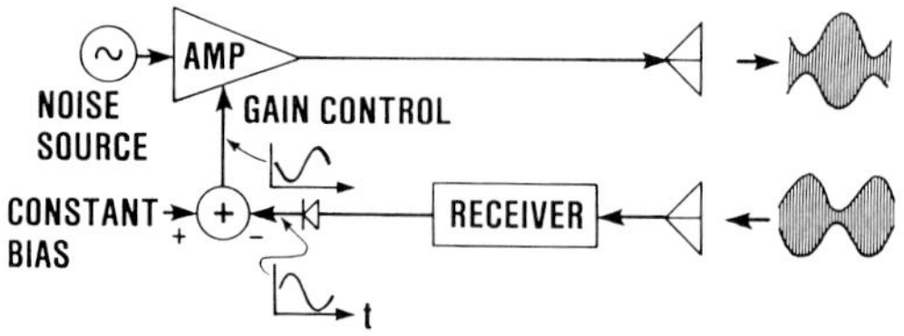

Fig. 3.9 Inverse-gain noise jammer.

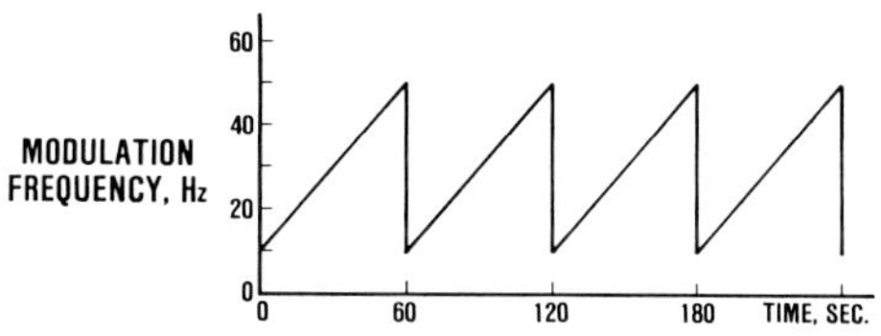

Fig. 3.10 Variation in amplitude modulation frequency.

not pass through these filters. This latter factor is especially important in jamming COSRO or LORO radars because these radars do not reveal their scan rates. Inverse gain is ineffective against these radars, but square-wave modulation can still be used if the period of the jammer modulation is correct or if it is varied over the suspected radar-scan frequency as shown in Fig. 3.10 [95].

Although a linear variation in frequency is shown in Fig. 3.10, nonlinear variations and more complicated combinations of variations are possible. The rate of change in the frequency of modulation must be slow enough to allow the radar to respond during the time that the frequency of modulation is near the scan frequency. The exact rate of change chosen is typically verified through experimental simulations. High rates are possible against short-range radars and slower rates must be used against long-range radars. This difference in the modulation rate change is reasonable because the short-range radar must respond to targets with higher angular rates than long-range radars.

Against an ideal scan-with-compensation radar, amplitude modulation ofa noise jammer's radiated power will have no effect. A look at Fig. 3.11 shows a succession of signals that would exist in an SWC radar if amplitude-modulated (AM) noise were used against it. The main channel phase has been reversed; the root cause of an error is CONSCAN. However, the auxiliary channel signal has a greater depth of modulation than the main channel, since it results from jammer and auxiliary antenna gains that go up and down together. If the separate main and auxiliary AGCs adjust

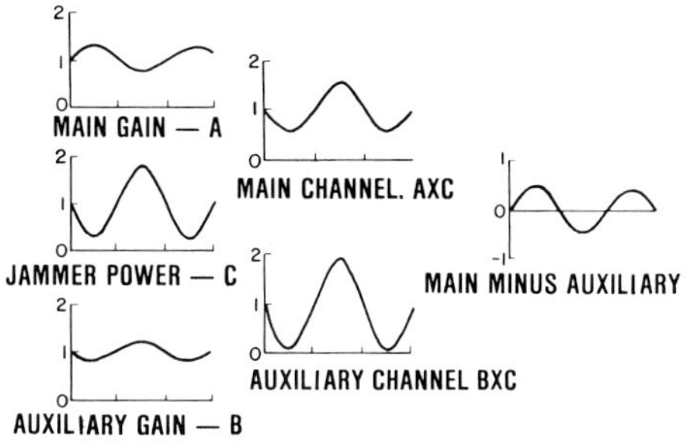

Fig. 3.11 SWC signals in AM noise.

the average levels to the same value, the difference voltage is again in the correct phase.

Average Center-Frequency Variation of a Noise Jammer

Efficient conversion of aircraft prime power into RF power sometimes requires amplitude modulation in the form of on-off modulation. As opposed to turning off the jammer, varying the power spectrum emitted by the jammer may be more advantageous. Such emissions produce a number of effects on both search and track radars.

Against an improperly designed search radar, sweeping of the high-level jamming signal through the IF passband can saturate later stages of amplification. In tube-type receivers and, presumably, in types employing field-effect transistors, this saturation effect can take many PRIs to rectify itself and thus nullify the usefulness of the radar. The addition of limiters to the front end of the radar have all but eliminated the effectiveness of this jammer modulation shown in Fig. 3.12. The Dicke-fix, an example of such a circuit, is discussed in Chapter 4.

Against a scanning radar, CONSCAN or TWS, this frequency-modulated wave will again produce a variation in the signal level out of the error detector and will thus cause a tracking error. However, the frequency deviation Δf must be larger than twice the sum of the radar and jammer bandwidth $[2(B_R + B_J)]$ to produce a sufficient depth of modulation for large angle errors. As shown in Fig. 3.13, any uncertainty in the IF bandwidth of the radar must result in larger deviations in jammer frequency and a lowering of the duty cycle of the jamming signal from the radar. Since wider jammer bandwidths than the bandwidth used in Fig. 3.13 are necessary, larger frequency deviations are also required. The result is that as much or more prime power may be required to defeat the same threat as would be required with simple amplitude modulation. The radiation of a jamming signal over a broader bandwidth than that required for spot noise may or may not be to the jammer's advantage. For example, another CONSCAN radar at a different scan frequency might be able to track the jammer.

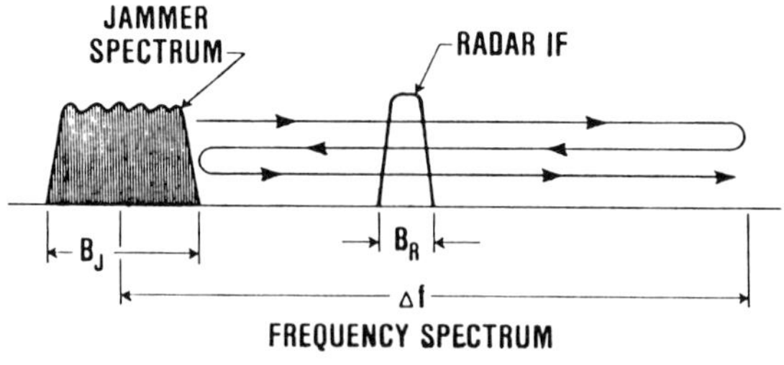

Fig. 3.12 Frequency-modulated noise spectrum.

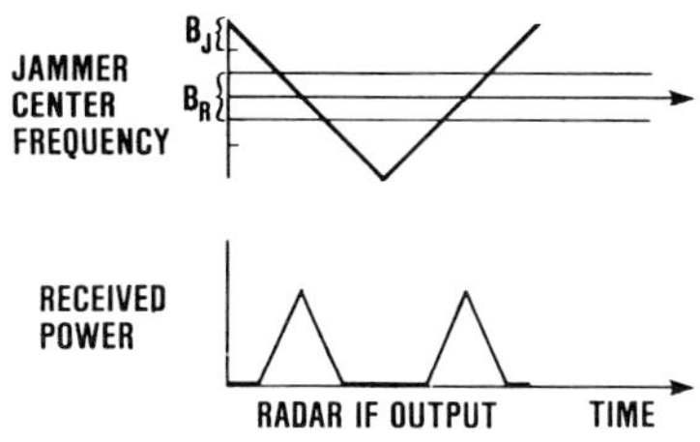

Fig. 3.13 IF output resulting from swept jammer.

Modulation of Noise Jammer Polarization

A CW noise jammer can produce an amplitude-modulated signal in a scanning radar if the polarization of the jammer relative to that of the radar antenna varies over time. In fact, if a coherent signal is modulated in polarization rapidly enough, it can generate a noise-like signal that will couple to a radar antenna of arbitrary polarization to within 3 dB of the maximum possible coupling [34]. Polarization can also be used to introduce angular errors into monopulse and SWC radars, but this effect will be explored in the target-like jamming section and in Part II. This discussion is concerned only with the effect on a lobing radar of a noise source that is modulated in polarization.

For linear polarization, the radar electric field can be considered to lie in a particular plane. Then, without loss of generality, the radar signal incident on the jammer can be called vertical. If the time-varying jammer polarization (electric field orientation) makes an angle ϕ with respect to the vertical, then the signal coupled into the radar will vary as the cosine of ϕ^2. Figure 3.14 shows this coupled signal for a variety of ϕ variations with time.

Figures 3.14a and 3.14b indicate that the curves of received power would be reversed if the radar polarization had been 45 deg from the vertical. Thus, blindly switching between states of "orthogonal" polarization may create only a beacon rather than a jammer because the signal received in Fig. 3.14b is the signal of a CW noise source. Of course, a pure match or mismatch in states of polarization is unlikely and some modulation of the received signal is likely. Part II shows that the expected depth of modulation is only 2 dB, hardly enough for large angular errors.

Figure 3.14c corresponds to the jammer polarization vector rotating around the line-of-sight direction every T_s seconds. The resulting modulation is twice this rate or $2/T_s$ hertz. The large depth of modulation should result in appreciable tracking error.

Multiple Noise/Target Sources

The discussion thus far has been limited to one noise jammer vs one radar, but many-on-many cases introduce new possibilities. This section addresses two unresolved jammers, stand-off jammers, and ghosts.

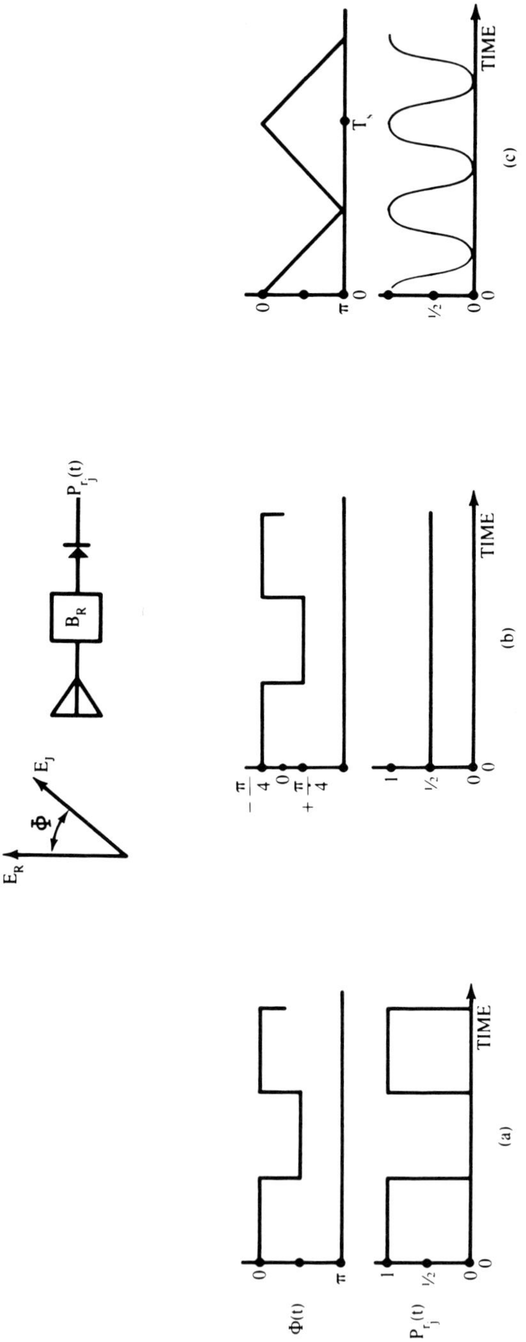

Fig. 3.14 Noise jammer polarization modulation.

If two noise jammers operate simultaneously within the main beam of a radar to the extent that they cannot be resolved as separate targets, the radar will track the power centroid between the two jammers [4, p. 190]. Figure 3.15 shows the geometry considered in this discussion. Jammer 1 radiates an average power of P_1 and is separated by an angle of $\Delta\theta$, as seen by the radar, from jammer 2 radiating P_2. The tracking error measured as an error in the tracking of jammer 1 is $\bar{\theta}$ and is given by

$$\bar{\theta} = \frac{P_2}{P_1 + P_2} \Delta\theta$$

The track axis simply points to the power centroid, the position of which is found in much the same manner as the center of mass of an object. This concept of tracking the power centroid extends to any number of unresolved noise-like sources (or uncorrelated targets). Although tracking of the power centroid follows from an assumption that the angular separation of the targets is much less than the half-power beamwidth of the radar, more exact derivations in Part II show that it is still a reasonable result for angular separations of targets that are an appreciable fraction of the beamwidth.

The two jammers have complete control of the radar tracking axis until the radar resolves the jammers as separate sources. This ability to resolve and then track one of two equally powered sources typically occurs when $\Delta\theta \approx 0.9\ \theta_3$. If the two jammer sources have equal power, the radar initially tracks the power centroid that lies halfway between the targets. The radar seeker's response to the movement of the power centroid slows as the source separation increases as long as the targets remain unresolved. Once resolution occurs, the seeker quickly aligns itself with one of the two targets.

Obviously, two aircraft could fly in formation and attempt to manipulate the radar track axis at a rate that induces the maximum system error. This error might be derived from some particular susceptibility of the missile to the modulation frequency or it might be due to reducing the missile velocity because of excessive maneuvering. Two aircraft are not required, since the reflection of jamming from the terrain can simulate a second jammer. (For an analysis of this configuration, see the subsection on multiple repeater

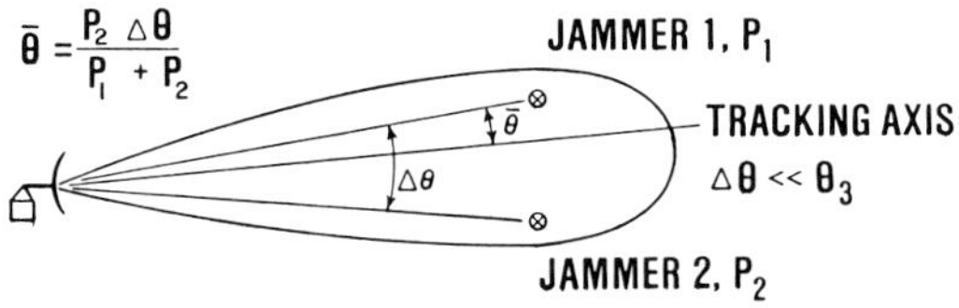

Fig. 3.15 Power centroid tracking.

sources in Sec. 3.3.) Other possibilities for this second source include remotely piloted vehicles or active expendable noise sources.

An expendable source is simply a noise source (or a target-like source) meant to operate for a short period and not to be recovered. Since even monopulse radars will track the power centroid, and thus track the desired target with some error, it seems that expendables should be very popular. But there are many conceptual and practical employment problems. For example: How can it be ejected? Is the next expendable in the ejector at the right frequency? Is it really cheap enough to dump three? Many have asked these questions, but there does not appear to be agreement about the answers [72; 73].

If the noise jammers are not contained on the penetrating aircraft but are carried on vehicles beyond the lethal range of the defensive forces, the jamming is called standoff jamming. Figure 3.16 depicts a typical standoff jammer (SOJ) geometry.

In this case, one standoff jammer is in line with, and is attempting to mask, the penetrator's position. Its effectiveness depends on the resulting J/S ratio in the radar. At the radar, the received jamming and target powers are

$$P_{r_J} = \frac{P_J G_J}{4\pi R_J^2} A_e \frac{B_R}{B_J}, \qquad R_J = \text{range jammer}$$

$$P_{r_T} = \frac{P_R G_R \sigma_T}{(4\pi R_T^2)^2} A_e, \qquad R_T = \text{range target}$$

As $R_T \neq R_J$, the J/S ratio per pulse is

$$\left(\frac{J}{S}\right)_p = \frac{P_{r_J}}{P_{r_T}} = \frac{P_J G_J B_R 4\pi R_T^4}{P_R G_R B_J \sigma_T R_J^2}$$

If the range to the SOJ is constant, the J/S ratio decreases by 40 dB for every tenfold reduction in target range or by 12 dB for every halving in the target range. If Example 3.1 is used with the SOJ and penetrator at 20 n.mi., the J/S ratio can be read from Fig. 3.4. The J/S ratio vs range can then be replotted on a similar figure (see Fig. 3.17). In such calculations, the SOJ must

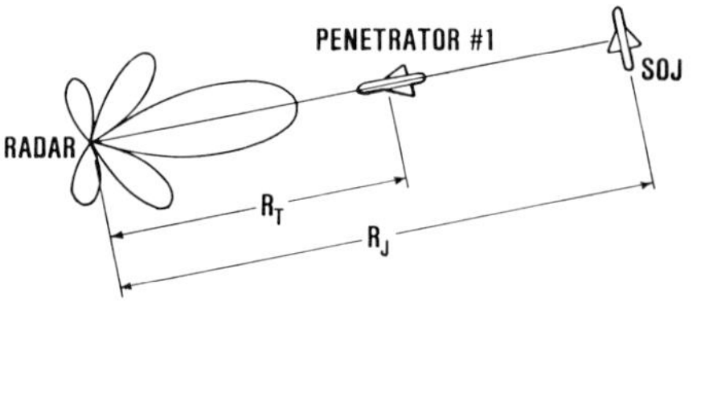

Fig. 3.16 Mainlobe standoff jamming.

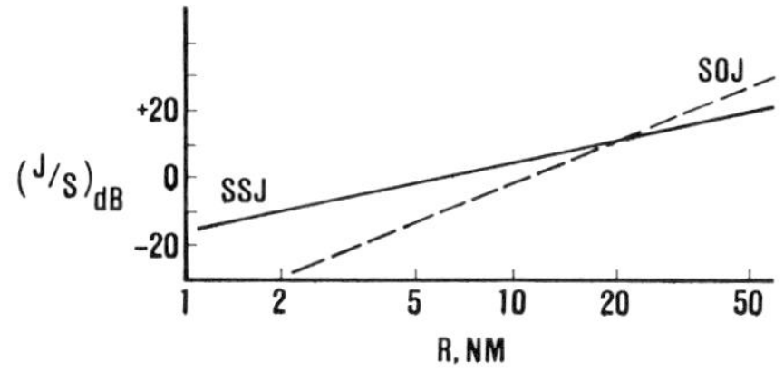

Fig. 3.17 Mainlobe noise jammer *J/S* vs range.

be kept at an elevation sufficient to be subject to the same radar antenna gain as the penetrator. If it is too low, the SOJ can be in the null near the grazing angle and its jamming would thus be ineffective.

Another geometry that subjects the SOJ and penetrator to different radar antenna gains is depicted by the penetrator in Fig. 3.18. In this instance, the penetrator is subject to the maximum antenna gains and the jammer is radiating only into the radar's sidelobes. Under these conditions, the standoff jamming equation is modified by reducing the jamming-to-signal ratio by the sidelobe level relative to the peak gain. If G_R is the peak radar gain and G_{RJ} is that gain in the direction of the jammer, then the J/S ratio becomes

$$\left(\frac{J}{S}\right)_p = \frac{P_J G_J G_{RJ} B_R 4\pi R_T^4}{P_R G_R^2 B_J \sigma_T R_J^2}$$

Figure 3.19 provides a comparison of all three cases. It is assumed that the radar has a -30 dB sidelobe in the jammer's direction. Both the sidelobe level and the J/S ratio will actually vary with the angle from the main beam. Targets separated by a small angle from the jammer should be screened better, since higher sidelobes generally exist near the main beam. If more than one SOJ is used and if each jammer is spaced on the order of a radar beamwidth from the adjacent jammers, a whole sector of angular width equal to the sector covered by the physical locations of the jammers should be masked. The mismatch in jammer-radar polarization must be considered in any computation.

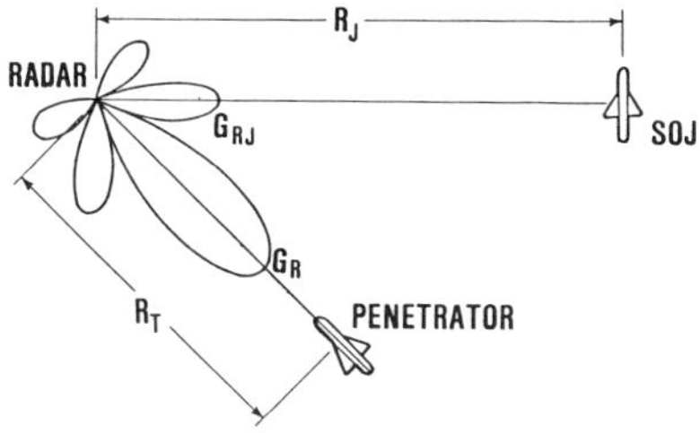

Fig. 3.18 Sidelobe standoff jamming.

Example 3.2: SOJ Problem

Given the radar described in Table 2.6 and the information provided below, what must be the effective radiated power of a jammer with a 30 MHz bandwidth if a 10 dB J/S ratio is required for screening? The jammer is at twice the target range of 100 km and its polarization is 45 deg from the vertical.

Additional Radar Parameters

Sidelobe level toward jammer relative to peak gain	−20 dB
Polarization	Vertical
Number of pulses incoherently integrated	30

Answer: The appropriate equation is

$$\frac{J}{S}=\frac{P_JG_JG_{RJ}B_R4\pi L_iR_T^4}{P_RG_RG_RB_J\sigma_TNR_J^2}p$$

The ratio G_{RJ}/G_R is equal to the sidelobe level of −20 dB, and, for incoherent integration, the integration loss L_i is equal to $\sqrt{N}$. The 45 deg angle between jammer and radar radiated fields results in a polarization mismatch of $p = 1/2$.

Insertion of all quantities into the above equation gives

$$\frac{J}{S}=P_JG_J+\frac{G_{RJ}}{G_R}+4\pi+\frac{R_T^4}{R_J^2}-P_R-G_R-B_J-\sigma_T-\sqrt{N}+p$$

$$10=J_{ERP}-20+11+94-60-35-14.8-10-7.4-3$$

$$J_{ERP}=55.2\ \text{dBW}$$

If many CW noise jammers are simultaneously jamming many radars, the jamming effects are enhanced by a factor greater than the mere increase in the number of jammers. Since the two radars shown in Fig. 3.20 cannot resolve either of the jammers in range (denoted by the dots), the radar sites may rely on triangulation in an attempt to determine range. This process

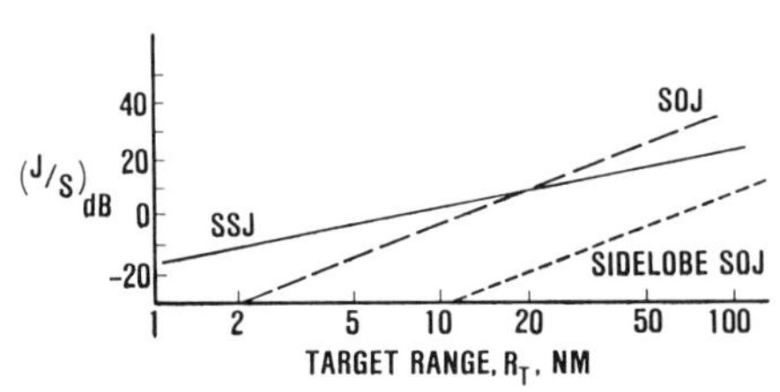

Fig. 3.19 Sidelobe noise jammer J/S vs range.

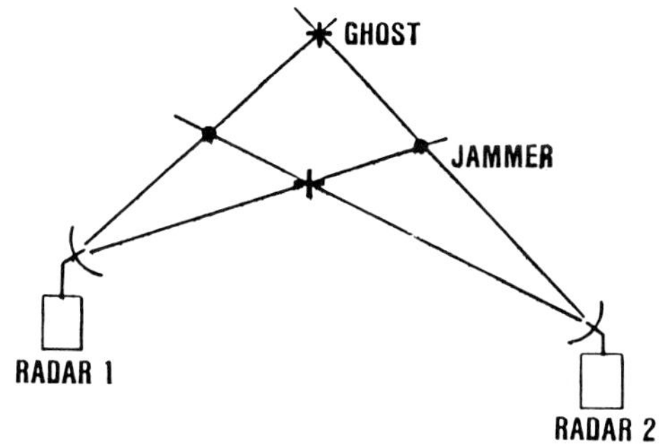

Fig. 3.20 Example of ghosts.

results in four possible jammer locations, two of which are correct and two "ghosts" (designated by the Xs.). Determination of the intersections that are true targets requires effort and time. Approaches used to make this determination are visual observation by an interceptor, observation of potential targets for unrealistic velocity, excessive range, or use of a separate radar at an unjammed frequency or polarization to reduce the number of unknowns.

For three jammers, there are nine intersections and six ghosts. If another radar is added as shown in Fig. 3.21, the jammers can be sorted from the ghosts since radars 1 and 2 locate at most N^2 possible jammer locations, of which $N^2 - N$ are ghosts. Some of these positions may be ignored since the associated intersection occurs at great range (point A in Fig. 3.21). Of the remaining points (N^2 at most if none are eliminated), the third radar will observe jamming from only N actual jammer positions. This figure is an oversimplification because the target angles are available at different times from the radar sites; also, target motion, fades, and inaccuracies in measurement will prohibit an exact crossover of all three beams. Thus, some errors in jammer location should result and the offense should gain some time.

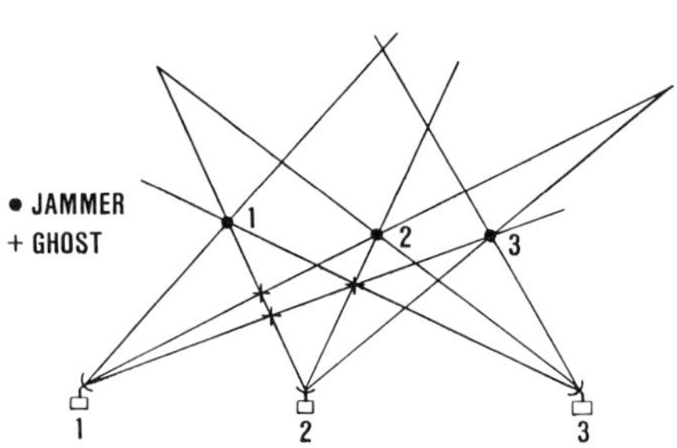

Fig. 3.21 Jammer deghosting.

Noise Summary

With sufficient J/S ratios, all noise emissions deny the range of the target to the radar. Likewise, if the average jammer power is constant, angle information is not denied. In many cases, some average parameter associated with the noise has been varied in a specific manner to create an error in angle. Varied parameters include power, center frequency, and polarization. For single jammers, only polarization has much use against ideal monopulse or SWC radars. If multiple unresolved jammers are used, all radars will normally track the power centroid rather than the individual jammers. Finally, multiple resolved jammers can generate confusion in a radar net.

3.3 TARGET-LIKE JAMMING

If the radiated jamming signals are similar to the signal scattered by the target, the jamming is called target-like. Normally, one or more characteristics of the radiated jamming signals are different from the signals scattered by the target in order to deceive or confuse a radar operator. This section describes some desirable and undesirable features of target-like jamming, typical jamming signals, and the effects of these signals on various radars.

Since target-like jamming signals can be identical to target returns, a radar gains no jamming-to-signal ratio reduction through the integration of many target pulses. A jammer signal can thus easily overpower the target signal and become the dominant signal tracked by range- and angle-tracking circuits. This beacon mode is not desirable, however, and the jammer signal must be modified (or modulated) to control the range- and/or angle-tracking point. The modifications available are limited to modulations in the time, amplitude, frequency, polarization, and number of jamming sources.

Quite often, target-like signals are generated in a repeater jammer. A simplified repeater jammer, as shown in Fig. 3.22, receives, amplifies, and retransmits the illuminating radar signal. This repeated signal is then appropriately modulated in time, frequency, amplitude, and polarization to create the desired jamming effect. Such a repeater system will generate a J/S ratio at the threat radar as derived below. In Fig. 3.22, G_{JR} is the jammer receive antenna gain in the direction of the radar, G the power gain

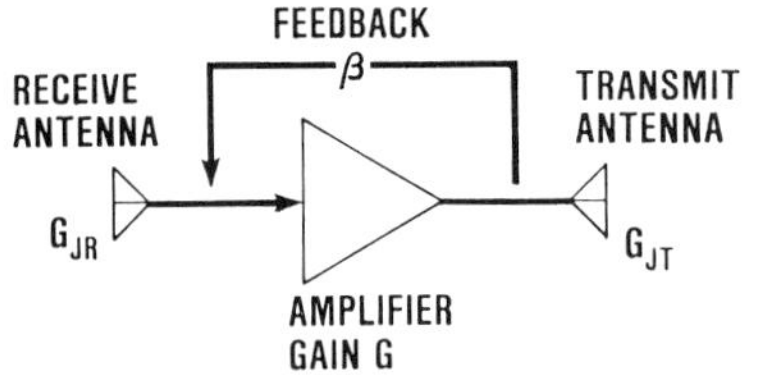

Fig. 3.22 Simplified linear repeater.

of the repeater, and G_{JT} the jammer transmit antenna gain in the radar direction. The jammer and target power for a self-screening repeater are then

$$P_{r_J} = \frac{P_R G_R}{4\pi R^2} \frac{\lambda^2 G_{JR}}{4\pi} \frac{GG_{JT}}{4\pi R^2} \frac{\lambda^2}{4\pi} G_R$$

$$P_{r_T} = \frac{P_R G_R}{4\pi R^2} \frac{\sigma_T}{4\pi R^2} \frac{\lambda^2}{4\pi} G_R$$

A linear repeater has a constant power gain G and radiates a signal that is a fixed multiple of the input signal. For example, the output signal is 10 times the input signal. A repeater is linear only if this relationship holds for all levels of input signal. The jamming-to-signal ratio produced by a linear self-screening jammer without a mismatch in polarization is then

$$\frac{J}{S} = \frac{P_{r_J}}{P_{r_T}} = \frac{\lambda^2}{4\pi} \frac{G_{JR} G G_{JT}}{\sigma_T}$$

The assumption of a linear repeater results in a J/S ratio independent of the target range. It may appear that an arbitrarily high J/S ratio can be achieved by increasing the jammer gain G, but this is not so. In Fig. 3.22, an arrow labeled β represents the jammer-radiated power that is coupled to the jammer input antenna. Such power, coupled from output to input, is called feedback. Since a repeater receives and retransmits almost simultaneously, this feedback limits the maximum usable gain of the system. For example, the time delay for propagation of a typical signal through a repeater is 100 ns [37]. If the radar pulse is 1 μs long, then the receive antenna will sense a fraction of the jammer-transmitted power at the same time that it receives the last 900 ns of the radar pulse. If the power measured at the output port of the jammer's receive antenna is S_1, then the power coupled to the input port of the transmit antenna of the jammer is GS_1. As defined here, the fraction of the jammer-transmitted power sensed at the output of the receive antenna is $\beta G S_1$. Thus, if $\beta G > 1$, the total received signal can be dominated by the jammer's own emission. In fact, the jammer system may go into "oscillation" and not be able to amplify any radar signals. For these reasons,

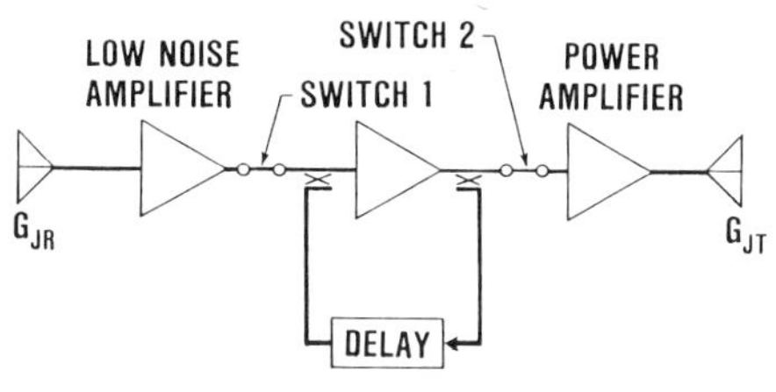

Fig. 3.23 Simplified delay line transponder.

it is necessary that the product of the jammer gain and feedback, as measured at the antenna ports, be less than one.

Many target-like jammers store all or part of the received radar signal in a memory device and use this stored information to reconstruct the jamming signal. Called transponders, these systems do not suffer a gain limitation determined by available antenna isolation (isolation is the inverse of feedback, i.e., $1/\beta = \text{isolation}$), but they are usually more complex than repeaters. A typical delay-line transponder, shown in Fig. 3.23, consists of at least three amplifiers, usually traveling wave tubes (TWT). Initially, the closing of switch 1 and opening of switch 2 allows the radar signal to enter the storage medium shown here as a delay line. When the delay line is full, both switches 1 and 2 are open and the signal is "kept alive" in the recirculating loop and center TWT. Finally, switch 2 closes at the desired time and a portion of the radar pulse is extracted from the loop, amplified, and transmitted. This transmission can go on for much longer periods than the time delay of the delay line and very powerful target-like signals can be radiated. However, the timing of the control signals is both critical and complex in a modern radar environment.

Another approach to transponder construction uses a keyed oscillator shown in Fig. 3.24. The frequency of the received radar signal is compared on the first pulse with an oscillator, possibly a voltage-controlled oscillator. The frequency of the oscillator is varied until it equals the frequency of the radar. This oscillator is the "keyed oscillator" since it is keyed by the radar pulse. With the radar frequency precisely measured, the onboard oscillator can be used as a replacement for the incident radar pulse in the generation of jamming waveforms. Chapter 4 shows that both of these approaches to transponder construction experience difficulty in covering the true target return against a radar that has agility in the RF frequency or pulse-repetition interval.

In both repeaters and transponders, the amplifier actually consists of a number of low-power amplifiers followed by a high-power amplifier. This high-power amplifier can be either a traveling wave tube or a cross-field amplifier. In most cases, the jammer is designed for saturation of the high-power amplifier (driven to produce its maximum output) because it is most efficient in this mode. Efficiency is defined as the percentage of prime aircraft power actually converted to RF power. Under such conditions, the jammer is called a saturated repeater with constant output power and variable gain.

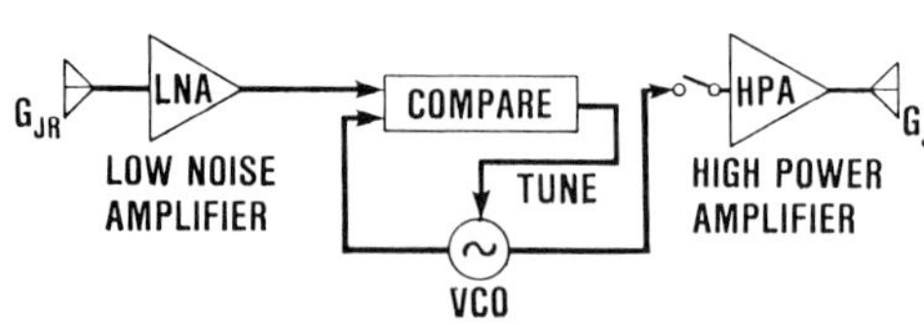

Fig. 3.24 Keyed oscillator transponder.

The gain must be variable because an increase in the received power cannot result in an increase in the radiated power.

The J/S ratio produced by such a saturated repeater or transponder in the self-screening mode is then

$$\frac{J}{S} = \frac{P_J}{P_R}\frac{G_J}{G_R}\frac{4\pi R^2}{\sigma_T}$$

where G_J is the gain of the jammer transmit antenna toward the radar.

This will be recognized as the J/S ratio of the noise jammer minus the reduction factors due to the ratio of the radar bandwidth to the jammer bandwidth and any radar integration. These factors are missing because emissions from the jammer resemble the radar emissions and, in many cases, suffer only the same losses suffered by the radar signal itself. This is one of the benefits of repeater jamming. It results in a lower average power requirement than that needed for noise jamming and leads to reductions in the size and weight of the jammer system as well. As stated earlier, the output of the repeater must be different from the reflection of the target in order to be effective. Determining which modifications to make and which to avoid requires more information about the threat radar than is required to set up a simple CW noise jammer. This is usually considered a drawback of repeater jammers. However, if the noise jammer provides no protection, what is the value of its apparent simplicity?

Repeater Modulations

Whether target-like jammers operate in the repeater or transponder mode, they are usually known collectively as repeaters. Up to this point, the repeater has enhanced the radar's ability to detect or track the target. Thus, to degrade radar performance, the jammer must obviously do something beyond repeating the radar emissions.

What are the tasks attempted by the radar? First, the radar must detect the target. Once a radar experiences jamming or once it recognizes that its increased false alarm rate is not due to system malfunction, target detection means assessing the range and azimuth cell of the target. As opposed to normal detection by a search radar where the presence of the target is unknown, detection of this type is probably attempted with a much larger allowed probability of false alarm on each scan and with an operator observing multiple scans for recognizable patterns. Typically, a tracking radar is brought into action only after the target is detected.

Both noise and repeater jammers defeat the normal detection mode of search radars and, at the same time, enhance this mode since the radar shows an increased level of false alarms. The detection, or rough estimation process, utilized after the presence of a target has been assumed is countered differently by the two jammers. The noise jammer was described previously as masking the target by burying it in a powerful noise signal. The repeater jammer can also mask the target if it generates false targets. However, when presented with one real and many false targets, the search or track radar operator must decide which return is due to the real target. This distinction is necessary because the tracking resources and search performance of the tracking radars are limited and many tracking radars can follow only one target at a time.

Once the operator has selected a target, the task of the tracking radar is to estimate the position of the target within the selected range and azimuth resolution cell. The repeater jammer again modifies the repeated signal to degrade this function or to destroy it.

The jamming signal variables available for modification to create false and deceptive signals are limited. As shown in Fig. 3.25, transmission of the radar and thus the output of the repeater can be described as a pulse of RF power. The most obvious variables describing the radar signal are the time of occurrence, amplitude, and frequency. Each of these variables can be controlled in the jamming signal to induce the effects listed in Table 3.4. Other variables not obvious in Fig. 3.25 are the polarization of the returned signal and the number and positions of the source of jamming power. Many jammers control more than one variable, either simultaneously or sequentially. For example, errors in range tracking induced through time delay may be followed by amplitude modulation to generate angular errors.

A repeater capable of implementing most of these techniques is shown in Fig. 3.26. Added to the simplified repeater shown in Fig. 3.23 are methods for modulating the jammer's signal in frequency, amplitude, and polarization; switches 1 and 2 implement the time delay/advance. Of course, the modulations can be implemented in other ways [37]. Any specific mechanization will depend on the available technology, the priority being placed

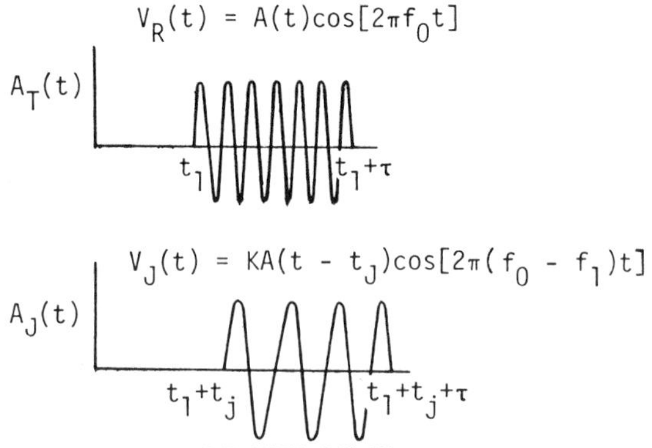

Fig. 3.25 Modified repeater return.

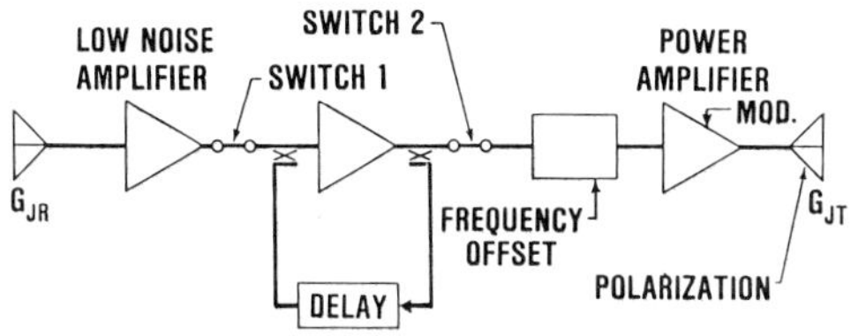

Fig. 3.26 Typical repeater jammer.

on the implementation of various techniques and the compatibility of the hardware with the various techniques desired. Not shown in Fig. 3.26 are the all-important ESM receiver and control circuits required for jammer operation.

Repeater Time Delay/Advance Modulation

If the jammer merely repeats an amplified version of the radar signal, the detected IF amplifier output of the radar should appear as shown in Fig. 3.27. The output signal is caused primarily by the jammer after an initial delay due to jammer placement on the airframe, signal propagation through the jammer, or simply the time while the jammer is turned on. (Note the leading and trailing edges of the radar pulse.)

As discussed in Chapter 2, this video signal is normally sampled by early and late gates. The energy sensed by each gate is measured and the gate positions readjusted until the energy is the same in each. In this manner, the early/late gates are kept centered on the radar signal and another gate (switch), called the range gate, can then pass the target video to the boxcar and angle-tracking circuits.

In the presence of the jamming signal shown in Fig. 3.27, the early/late gates will center themselves on the jammer signal if it exceeds the target by 10 dB or more [2, p. 15-14]. This action is called "range gate capture." Once the range gate has been captured, a variety of ECM tactics may be employed.

Increasing the time delay even more causes an apparent movement of the target in range, which is called a "range gate walk-off." This increased time

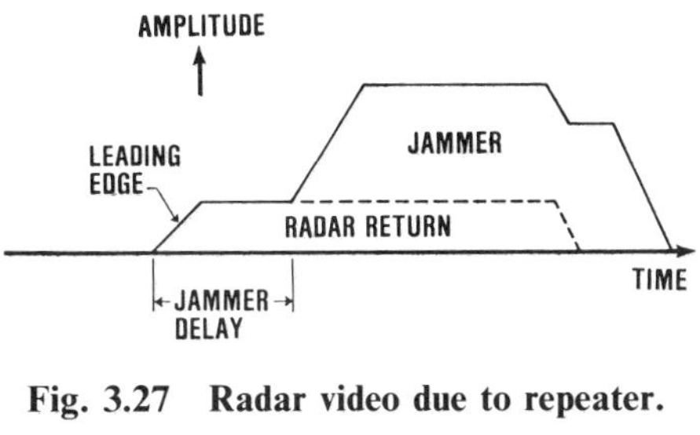

Fig. 3.27 Radar video due to repeater.

delay can occur at a linear or parabolic rate. When the jammer pulse is delayed by more than a pulse width from the target, the radar angle tracks on the jammer alone. Once the radar is tracking away from the target, the jammer can simply stop repeating. The radar then has no information on angle or range and operator intervention may be necessary to reacquire the target. Rather than simply stop repeating, the jammer may begin another form of modulation to generate angle, as well as range, errors in the radar. All the modulations listed in Table 3.4 are possible. The advantage gained by the jammer in applying these modulations after stealing the range gate is that the jamming signal does not compete with the target signal or skin return. Later sections address the question of whether these additional modulations will generate angle errors.

Another reason for the jammer to attempt inducing angle errors is that most radar angle trackers must be capable of operating under conditions of signal fading in the nonjamming environment. Such operation usually implies continued tracking with fades of a specified duration. How is this done? The target angular rate at the time of signal loss is stored and used to keep the antenna moving in the direction that it was moving prior to the loss of track information. Unless the target maneuvers violently, the antenna will point in the direction of the target for a short period. The range gate is often automatically swept out in range, possibly following a brief search at the last known target range. Thus, if the movement of the antenna is correct, the target will be quickly reacquired. Weapon guidance will be briefly interrupted and, if the jammer continually walks off the range gate and drops it, the percentage of time without tracking information may significantly degrade the effectiveness of the weapon system.

Since the repeater response is normally later in time than the target return, the radar can employ "leading edge tracking" to defeat the repeater. The radar video is passed through a fast-time-constant circuit whose output is

Table 3.4 ECM Modulations and Effects

Variable	Primary error: Created in	Primary error: Through
Time delay/advance duration	Range	Range gate walk off, false targets
Frequency/phase	Velocity	Velocity gate walk off, false targets
Amplitude	Angle	Sequential lobing, AGC effects
Polarization	Angle	Cross polarization of radar antenna
Multiple sources	Angle	Angle of arrival

proportional to the slope of its input. The output of such an idealized circuit when the input takes the form of the jammed signal in Fig. 3.27 is shown in Fig. 3.28. (Note that the unjammed leading edge of the target return becomes a pulse free of jamming.) If an early/late gate is centered on the output of this leading edge, the target can be tracked in the presence of time-delay jamming.

To counter the leading edge tracker, the jammer must reduce its time delay—not an easy task for either a repeater or a transponder. It is undesirable to amplify all incident signals in the repeater, since a dense (or abundant) signal environment may cause the average power rating of the final amplifier to be exceeded. Thus, the repeater must recognize the desired signal and gate the power amplifier on for each pulse. Even if the time to gate on the power amplifier can be effectively reduced to zero, there are other sources of time delay. In the transponder, the jammer must measure the PRI and store the initial radar pulse. Then, if it is assumed that the RF frequency or PRI does not change, the jammer can generate a pulse that will lead, be coincident with, or lag the next radar pulse reflected from the target.

Usually, the transponder pulses are intentionally longer than the pulses of the radar to compensate for error in the PRI measurement and to account for the random variation (jitter) of the PRI on a pulse-to-pulse basis. The transponder pulses can now be positioned anywhere in time and, if their time delay is continually reduced, the radar will conclude that the target range is decreasing. No longer can a radar operator conclude that the first target on his PPI scope is the real target. However, if the range information is extremely important to weapon system operation, a separate A-scope and range-track operator is an electromagnetic counter-countermeasure that can still be employed. Now, the jammer pulse must "look" like a target return to this operator. If the target return normally varies in amplitude, a constant-amplitude jammer pulse probably will not fool the trained operator. To deceive this operator, the jammer output must be controlled to appear as a true false target. Similarly, the true target must be masked to prevent its discrimination by the operator on the basis of power alone. After all, how many jamming signals are weaker than the target return?

In many weapon systems, for example, the single-seat aircraft, the workload of the operator is such that he cannot dedicate himself solely to range tracking. If range is essential in such cases, it may be acquired from an alternate sensor, such as a different frequency radar or laser. Alternatively,

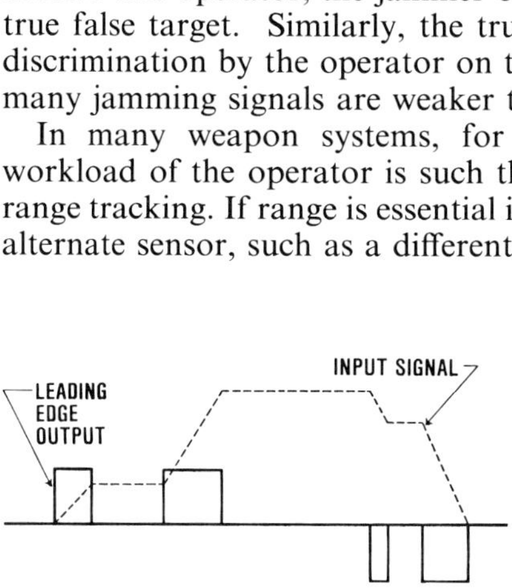

Fig. 3.28 Input of leading edge tracker.

range may be given secondary importance by employing semi-active, active, beam-rider, or home-on-jam (HOJ) missiles. With such missiles, the exact range of the target is not required, but knowledge that the target is within some interval is still necessary.

When jammer emissions are not coincident with target returns, they may be interpreted as false targets. Whether they are so interpreted depends on the discriminative capability of the jammed radar.

Repeater Frequency Modulations

Repeater frequency modulation is useful against coherent radars employing narrow-band filters to detect and velocity gates to track the Doppler shift of the target. Figure 3.29 shows the frequency spectra of a CW radar and of a target and jammer return. The radar emission is centered at f_o. The radar has located its narrow-band receiver over the target return located at $f_o + f_d$. The jammer return is also close enough to the target return to pass through the receiver filter. This spectrum is also applicable to FM/CW and pulse Doppler trackers.

A jammer employing frequency modulation will now cause the frequency of the jammer emission to change from the frequency of the target in some prescribed manner. For example, it may increase or decrease linearly relative to the target Doppler to some fixed offset and then stop. What would be the result if the velocity gate in Chapter 2 were used to track the target? If the signal from the jammer is at least 10 dB [2, p. 15-17] larger than the signal from the target, the velocity gate will initially be centered on the jammer Doppler. The velocity gate will then follow the jammer as its frequency changes if the rate of change in the jammer's frequency is not excessive. Similar to range gate pull-off, this sequence is known as velocity gate capture and velocity gate walk-off (VGWO). Of course, the walk-off rate need not be linear and, as opposed to range gate advance, there is no problem in shifting the jammer response up or down in frequency relative to the target return.

Once the radar is tracking the jammer, other modulations may be used to generate angle errors. Since there is no skin return under this condition, these modulations will function at their maximum advantage. The generation of the jammer Doppler offset, initially only a few hundred hertz or less at a frequency of possibly 10 GHz, is not simple. One approach is to "serrodyne the helix" of a traveling wave tube (see Fig. 3.30 for a simplified jammer using this approach). As described in Sec. 3.5, this traveling wave tube contains a delay line structure, often called a helix because it sometimes has

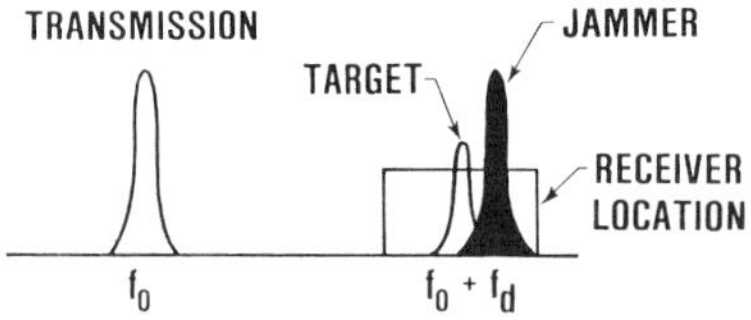

Fig. 3.29 Typical CW radar and target spectrum.

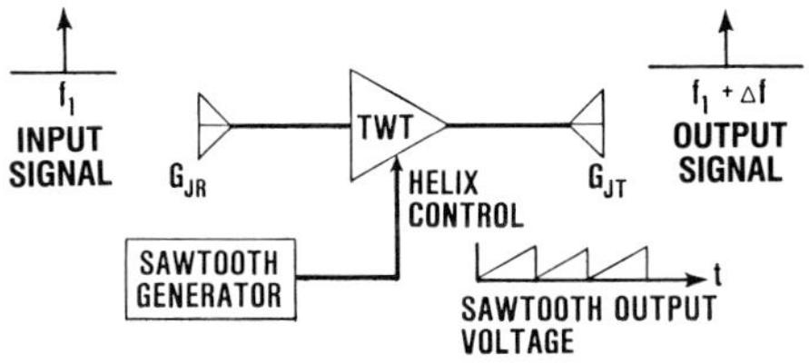

Fig. 3.30 Doppler offset by serrodyne method.

the form of a helical spring. If the helix voltage, normally at a few thousand volts, is varied linearly, the signal amplified by the traveling wave tube is also shifted in Doppler. Since a linear variation in voltage cannot continue indefinitely, the variation in helix voltage usually takes the form of the sawtooth shown in Fig. 3.30. Normally, a specifically designed traveling wave tube is required for serrodyning and it is followed by a power amplifier.

The jammer emission will not appear simply as an offset target because of the finite isolation β between jammer input and output. If it is assumed that the jammer radiates a power P_J at a frequency given by $f_1 + \Delta f$, then, in addition to the target signal received at f_1, the jammer also receives a portion of the transmitted signal, given by βP_J at a frequency $f_1 + \Delta f$, which it reradiates at a frequency $f_1 + 2\Delta f$. If the target signal received is P_{in}, then the power radiated is $P_J = GP_{in}$. The feedback signal that competes with the target is $\beta P_J = \beta GP_{in}$. As $\beta G < 1$, this ring around signal is smaller than the original target signal by the factor βG. This process of harmonic generation continues indefinitely, but, practically, it is terminated when the signals received are at the level of the front-end noise of the jammer. The resulting transmitted spectrum is shown in Fig. 3.31. Such a signal may cause problems for the jammer by taking power away from the signal present in the radar velocity gate. It also gives away the usage of velocity gate walk-off to an operator viewing the environment with a spectrum analyzer.

Of course, frequency modulation can be employed for reasons other than the simulation of target velocity. If the jammer signal is swept in frequency through the bandwidth of the radar receiver at the scan rate of the radar antenna (assuming a CONSCAN radar), then an amplitude modulation generated in the radar video can induce errors in angle tracking. As with

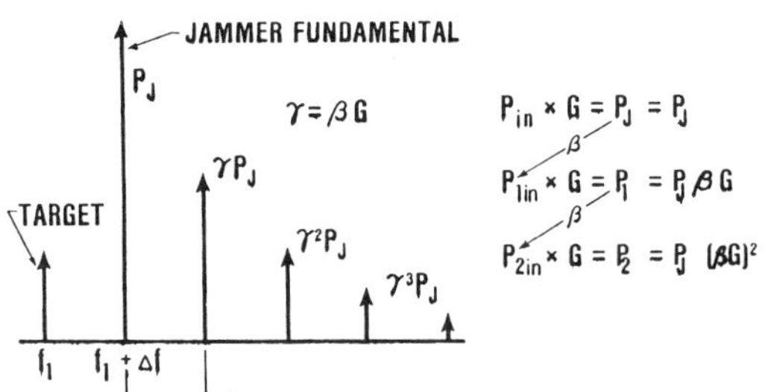

Fig. 3.31 Coherent jammer emission spectrum.

modulating the polarization of noise sources, the rate of frequency sweep and the frequency deviation are extremely important in determining the level of signal that competes with the target return.

Repeater Amplitude Modulation

Once a range or velocity gate has been stolen, amplitude modulation (AM) is the form of modulation usually employed to generate errors in angle. Again, for efficient operation of the power amplifier, the amplitude modulation is usually applied in an on-off form. This discussion has already treated two forms of amplitude modulation: inverse gain and square-wave modulation. These modulations are just as useful for repeater jammers as for noise jammers. Of course, the AM rate need not be directed at the scan rate of the radar antenna. Lower rates of amplitude modulation may excite "resonances" in the radar servo or in the radar-servo-missile combination, which, in turn, may amplify any existing dynamic tracking error.

Another form of amplitude modulation, called AGC deception [2, p. 171], consists of transmitting a strong jamming signal, but not on every radar pulse. The strong jamming pulses control the AGC voltage in the radar receiver, which is thus improperly adjusted for the intervening target returns. In Fig. 3.32, for example, a sequence of pulses from a target slightly off the tracking axis is displayed for both unjammed and jammed signals. For the jammed signal, the average received power is much larger than the power in the unjammed signal because of the strong jamming spikes. As mentioned earlier, the output of the radar receiver prior to the scan filter has the form $E = E_o(1 + k_s\theta_T/\theta_3 \cos\omega_s t)$. In this formulation, the average signal E_o multiplies the angular error and the voltage that drives the antenna servo is proportional to it. The actual error voltage driving the antenna servo is $E_o k_s \theta_T/\theta_3$.

Thus, if the average signal level measured by the radar is not correctly related to the average signal of the target return, the error voltage developed will be incorrect. The high average power level in Fig. 3.32 will cause the AGC voltage to reduce the receiver gain and the peak-to-peak voltage variation from the sample-and-hold (S&H), as shown in Fig. 3.33.

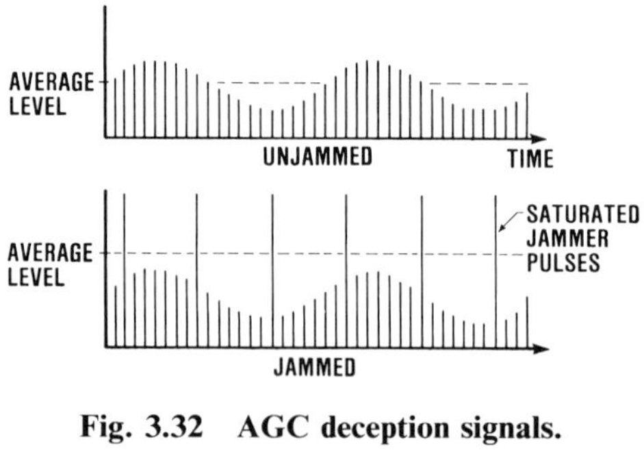

Fig. 3.32 AGC deception signals.

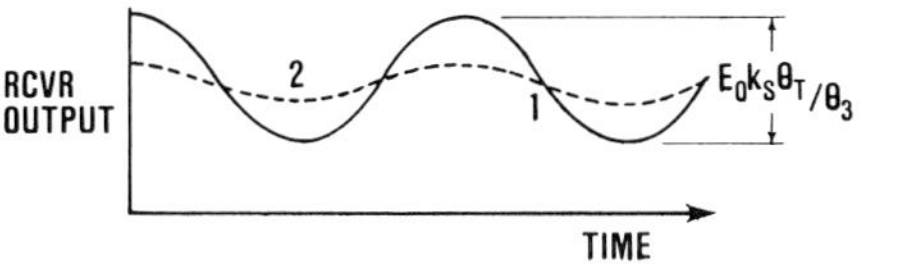

Fig. 3.33 Effect of AGC deception.

If curve 1 in Fig. 3.33 represents the correct output for a target tracked with an angular error θ_T, curve 2 represents the error voltage resulting from AGC deception where the incorrect AGC voltage has suppressed E_o. The radar is now driven by a smaller error and will respond more slowly. This sluggishness can cause excessive dynamic tracking errors and an unstable servo control loop in some radar systems.

Repeater Polarization Modulation

As stated earlier, the electric field created by an antenna has an orientation in space, possibly a time-changing orientation known as its polarization. For example, the dipole antennas in Fig. 3.34a and 3.34b create, respectively, a vertically and horizontally polarized field at the position indicated along the direction of the y axis. A fundamental antenna theorem is reciprocity, which states that a receiving antenna is most responsive (matched) to an incident signal with a polarization identical to the polarization created by the antenna itself when it is used as a transmitting antenna. Thus, the antenna in Fig. 3.34a is matched to an incoming field that has its electric field in the y-z plane; the dipole of Fig. 3.34b receives best an incident signal oriented in the x-y plane. For example, of all electromagnetic fields of equal power per unit area incident on the vertical dipole, the field that produces the largest signal at the antenna terminals is the field with its electric vector oriented vertically at all times.

If vertical and horizontal dipoles are driven in phase with equal power, the result is a combined field that is simply the "vector" sum of the individual fields. The total field is then described as linear, with a polarization 45 deg from the vertical. This situation is shown in Fig. 3.35a.

If only the radiated field in one direction is desired, the two antennas can be combined to yield a new antenna (see Fig. 3.35b) oriented at 45 deg from the vertical. This antenna will radiate a field identical to the field of the

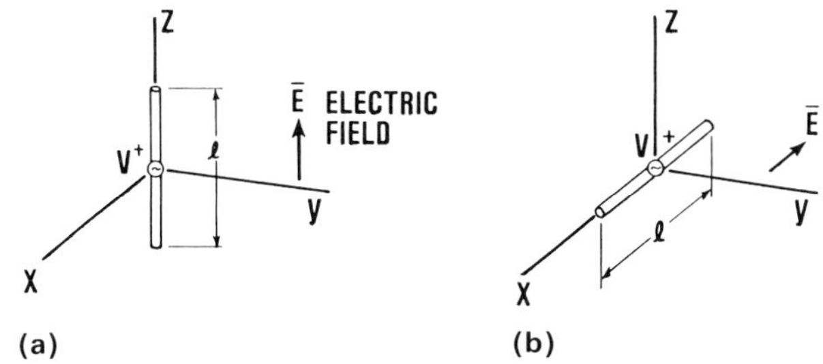

Fig. 3.34 Dipole antenna polarization.

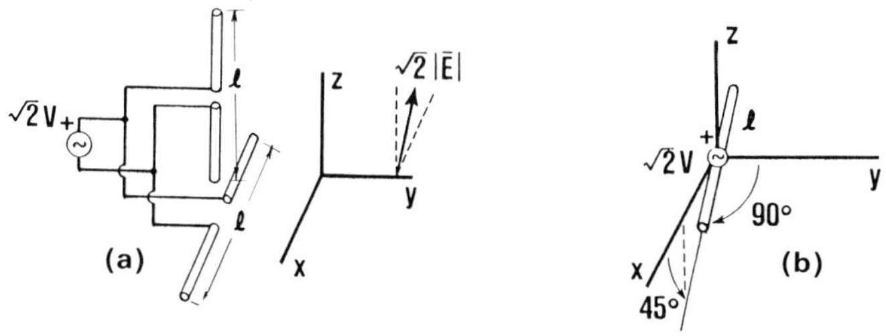

Fig. 3.35 Equivalent antennas.

original antennas in the specified direction. In general, the field radiated in other directions will not be the same.

Of course, the power distribution to the two dipoles in Fig. 3.35 can be unequal and the driving voltages need not be in phase. With unequal power, the result is a linearly polarized radiated field, but the electric field is oriented in a direction other than 45 deg from the vertical. The orientation of the electric field will become more vertical as the vertical dipole power is increased. If the dipoles are driven by generators that have any phase relationship other than 0 or 180 deg, some form of elliptical polarization will result. The polarization is described as elliptical since the magnitude of the electric field vector, which now rotates about the y axis, outlines the shape of an ellipse. If the generators have a 90 deg phase relationship and equal power, the result is circular polarization which can be considered an extreme form of elliptical polarization. See Table 3.5.

Of course, few radar antennas radiate a field with a single type of polarization in all directions. As mentioned earlier, the power density, or intensity, varies with direction. The intensity in a given direction may be divided in no more than two different polarizations with the proportion in each polarization varying with angle. For example, the dipole in Fig. 3.34b radiates a horizontal electric field that is x-directed toward the y-axis. However, if the radiated field is observed along a line still in the x-y plane but 45 deg from the x axis, the radiated field must have both x- and y-directed components.

Table 3.5 Polarization Relationships

Radiated field and matched polarization	Orthogonal polarization
Vertical	Horizontal
Horizontal	Vertical
Circular, right hand*	Circular, left hand
Circular, left hand	Circular, right hand

*Right hand circular polarization has an electric field rotating clockwise when observed while looking in the direction of propagation.

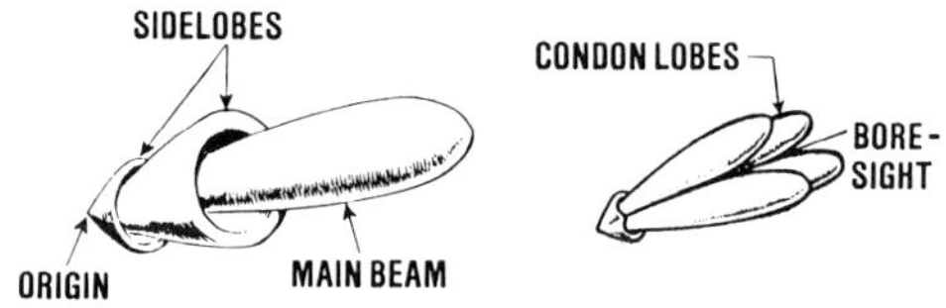

Fig. 3.36 Paraboloid radiation patterns.

Actual fields from typical radar antennas, such as the paraboloid, are much more complicated than the fields discussed thus far. Therefore, the power density radiated in each polarization is usually described separately. Since radar designers usually consider the radiation pattern in one polarization as desirable, this polarization is called the copolarized field or working polarization. Such a vertical, copolarized field is shown in Fig. 3.36a for a paraboloid. The horizontal radiation pattern or the cross-polarization field for the paraboloid is shown in Fig. 3.36b and is characterized by the formation of four lobes (sometimes called Condon lobes)[39]. The gain of these Condon lobes relative to the main beam can be controlled by varying the diameter of the aperture relative to its focal length. The cross-polarization lobes decrease as the focal length increases for an aperture with a fixed diameter. Without special precautions, these cross-polarized sidelobes are higher than the first copolarized sidelobes. An example in Ref. 39 has a value -16 dB relative to the main beam in an antenna designed for maximum efficiency of the aperture.

As discussed in Chapter 2, a monopulse antenna would form a sum pattern by combining four of the lobes in Fig. 3.36a. Since the difference patterns are different combinations of the same four lobes, the polarization of the difference patterns should be different from the polarization of the sum pattern.

What polarization should a jammer radiate to induce an error in a radar? As might be expected, the polarization radiated depends on the type of error desired and the radar polarization. The discussion of noise jammers indicated that amplitude modulation can be induced by polarization modulation. Although such an approach is also useful in this instance, emphasis is given to an approach commonly known as cross polarization.

As shown in Fig. 3.37, some scanning radars use a nutating feed and others have rotating feeds. The polarization of the nutating feed is essentially fixed; in Fig. 3.37, for example, it is fixed in a vertical position. In this case, the

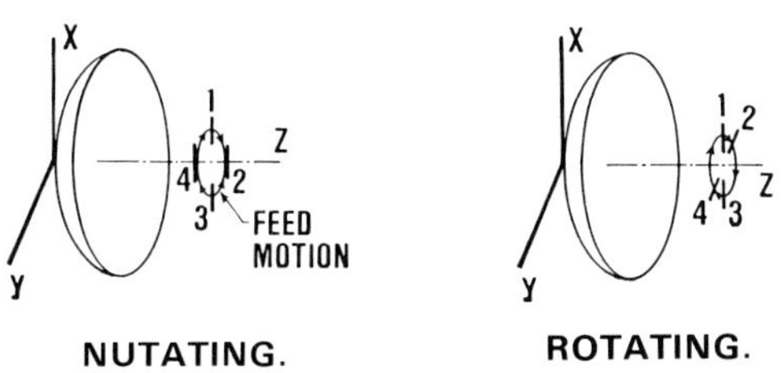

Fig. 3.37 Scanning feeds.

radar should lock up and track the jammer on one of the four cross-polarized lobes if the jammer radiates horizontal polarization. The lobe selected would be a matter of chance and initial conditions.

Figure 3.38 shows the tracking sequence on a cross-polarized lobe. Figure 3.38a shows the original, unscanned positions of the vertically polarized maximum and the horizontally polarized maximums. In all parts of Fig. 3.38, the term "CO" represents the copolarized maximum. The circle represents the locus of points of the copolarized pattern with −3 dB gain. The sequence of Figs. 3.38b–3.38h represents successive beam positions as scanning occurs around the scan axis, and Fig. 3.38i summarizes the sequence of positions taken up by the upper right cross-polarized lobe. Given a purely horizontal incident signal, the output voltage of the radar depends on the position of the horizontally sensitive beams; thus, tracking about a new axis is possible. Development of the same effect in the rotating feed requires the jammer polarization to rotate physically ahead of the radar polarization by 90 deg.

For a monopulse radar, the polarization of the difference-receive pattern will differ from the polarization of the sum-transmit pattern. Therefore, even though an incident orthogonally polarized signal will produce a zero sum-channel signal, it will generate a nonzero difference-channel signal that will degrade the tracking capability of the radar.

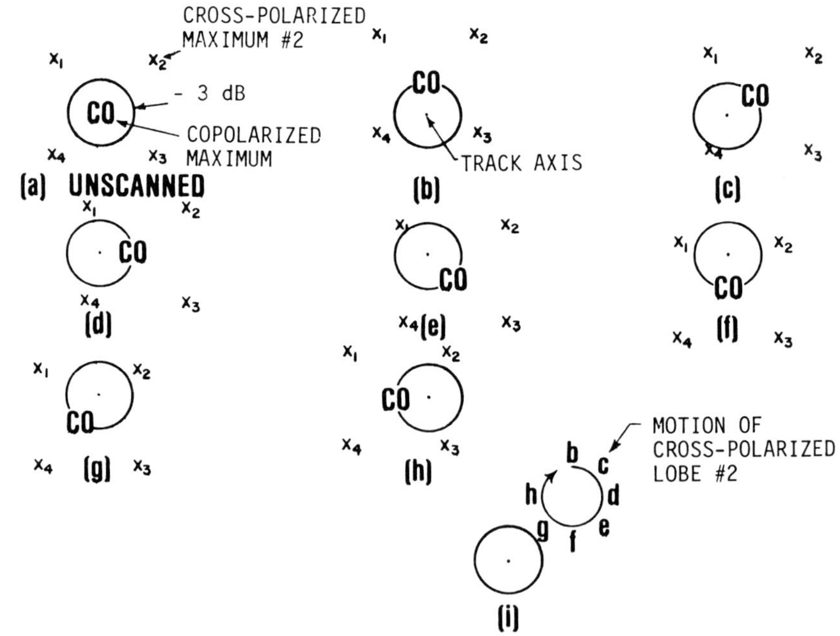

Fig. 3.38 Nutating feed beam positions.

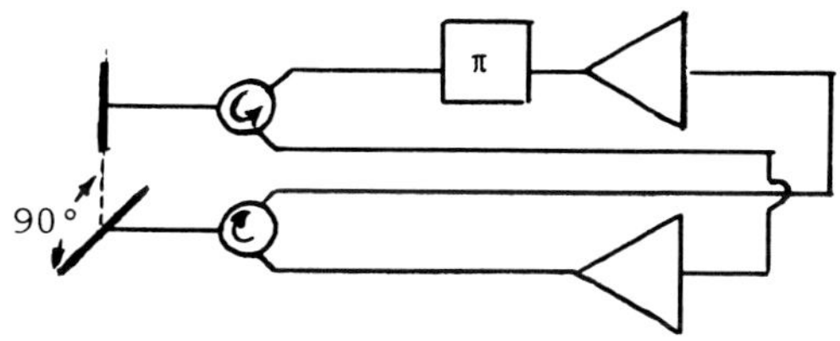

Fig. 3.39 Polarization jammer.

If the jammer is a repeater, it can use the radar illumination to determine the radar polarization and then respond with the orthogonal polarization while the jammer remains on the radar tracking axis. As the tracking axis moves away from the target, the jammer must remember the initial orthogonal polarization to continue developing a tracking error. If the variations in jammer polarization keep up with the variations of the radar the emissions of the jammer will remain (orthogonal) and the radar will never see the jammer.

Figure 3.39 shows a theoretical cross-polarization jammer [28, p. 579]. A signal received on the horizontal antenna is amplified and retransmitted with vertical polarization after a 180 deg shift in phase (or change in sign). The received vertical signal is merely retransmitted as an amplified horizontal signal. Thus, the received signal in Fig. 3.40a becomes the orthogonal signal in Fig. 3.40b on transmission.

Against a scanning radar with a nutating feed, the jammer's orthogonal signal must be stronger than the target return by a ratio greater than the radar antenna's copolarized peak gain to its cross-polarized peak gain (see Figs. 3.38b and 3.38i). For example, copolarized gain might be vertical and cross-polarization horizontal, and the peak gain of the vertical antenna in Fig. 3.38b might be 20 dB above the peak gain of the horizontal antenna in Fig. 3.38i. Then, if the jamming-to-signal ratio has this same ratio of power and if the target lay halfway between the co- and cross-polarization scan axes, the target- and jammer-induced error signals that drive the antenna servo will be of the same magnitude. A larger horizontal jamming signal will tip the scales and cause tracking around the wrong axis.

Range or velocity gate stealing might occur before employing orthogonal polarization since this reduces the level of the competing copolarized signal.

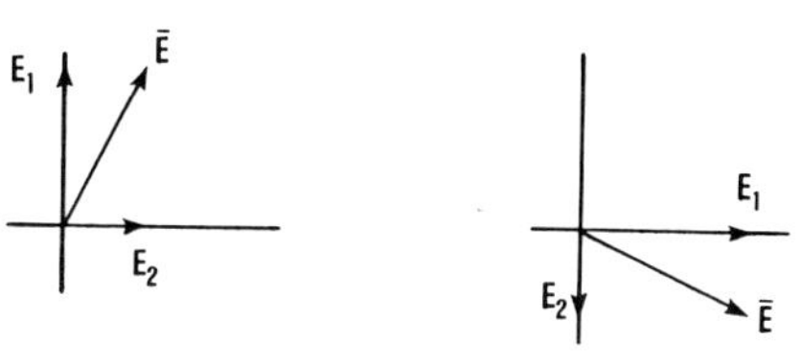

Fig. 3.40 Polarization jammer signals.

(a) Received **(b) Transmitted**

In fact, the purity of the jammer (the ratio of jammer co- to cross-polarization), rather than its power, is easily the limiting factor in generating tracking errors. For such a polarization jammer to be effective against a monopulse radar, the purity of jammer polarization should exceed the purity of any of the radar's four beams on the tracking axis (see Chapter 9).

Multiple Repeater Sources

Multiple unresolved repeaters can induce the same effect in radars as multiple noise sources and force the radar to track the power centroid of a collection of targets and jammers. Because of the target-like nature of the repeater signal, other effects can be generated in the radar as well. False targets can be generated in radars during their acquisition mode and angular tracking errors exceeding the separation of the jammer sources can be generated in monopulse radars. This section briefly reviews these effects.

A radar will track the power centroid of two unresolved incoherent sources just as it will track the power centroid of two noise jammers. An example of two incoherent sources is two transponder jammers radiating at different radio frequencies, but close enough in frequency to be received simultaneously by the same radar. Of course, range is not denied and/or range gate deception can be used to deceive a pulse radar. If a nontarget source is created by scattering the jammer power from the ground, the jamming is sometimes called terrain bounce [28, p. 493]. Figure 3.41 shows the geometry of this type of jamming.

The angular tracking error from the real target can be expressed in terms of the power received from the target, the ground, and the target-ground angular separation. In Fig. 3.41, the radar antenna gain toward the target and toward the ground is G_{R1} and G_{R2}, respectively. If the jammer is a linear repeater, the radiated jamming power depends on the effective area of the jammer receive antenna $\lambda^2 G_{JR}/4\pi$ and the jammer system gain G. The jammer transmit antenna is not pointed directly at the radar and, thus, the jammer antenna peak gain G_{J1} is pointed at the ground. Some value of gain G_{J2} must be assumed toward the radar.

Of the jammer power directed toward the ground, only a fraction ρ^2 is scattered toward the radar. (The quantity ρ is the terrain reflection coefficient.) The power bounced off the ground can be formulated approximately as

$$P_2 = \frac{P_R G_{R1}}{4\pi R^2} \frac{\lambda^2}{4\pi} G_{JR} G G_{J1} \rho^2$$

The power received from the target of cross section σ_T will add to the power radiated by the jammer toward the radar. Thus, the power radiated directly at the radar is

$$P_1 = \frac{P_R G_{R1}}{4\pi R^2} \left(\frac{\lambda^2}{4\pi} G_{JR} G G_{J2} + \sigma_T \right)$$

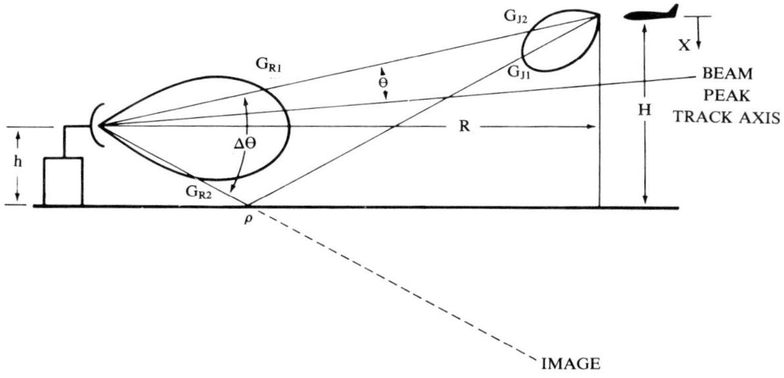

Fig. 3.41 Terrain bounce.

Combining these results into the power centroid formula results in a tracking point away from the real target of

$$\frac{\bar{\theta}}{\Delta\theta} = \frac{P_2}{P_1 + P_2} = \frac{\rho^2 G_{J1}}{\sigma + G_{J2} + \rho^2 G_{J1}}$$

where

$$\sigma = \frac{\sigma_T}{(\lambda_2/4\pi)G_{JR}G} \quad \text{and} \quad \Delta\theta = \frac{2H}{R}$$

Under ideal conditions, the jammer does not radiate directly at the radar, or $G_{J2} = 0$. Then the tracking error is

$$\frac{\bar{\theta}}{\Delta\theta} = \frac{1}{1 + (\sigma/\rho^2 G_{J1})}$$

and errors that are a significant portion of the target-image separation require that

$$\sigma \ll \rho^2 G_{J1}$$

By using previous results, we can rewrite this equation as

$$\frac{1}{(J/S)_o} = \frac{4\pi\sigma_T}{\lambda^2 G_{J_R} G G_{J1}}$$

where $(J/S)_o$ is the previously computed linear repeater jamming-to-signal ratio. Thus,

$$(J/S)_o \gg 1/\rho^2$$

These results suggest that a J/S ratio on the order of 10 dB will result in $\bar{\theta} = 0.9\Delta\theta$ under the assumptions of a Gaussian-shaped antenna pattern of 2 deg, a reflection coefficient of 1, a range of 10 n.mi., and a target height of 500 ft. The achievement of such errors is complicated by jammer transmission directly toward the radar.

If the second source is an actual repeater flying in with the penetrator or being used for standoff jamming, the J/S ratio is simply computed.

The jammer power received by the radar is

$$P_{r_J} = \frac{P_R G_{RJ}}{4\pi R_J^2} \frac{\lambda^2}{4\pi} \frac{G_{JR} G G_{JT}}{4\pi R_J^2} \frac{\lambda^2}{4\pi} G_{RJ}$$

The target power is

$$P_{r_T} = \frac{P_R G_{RT}}{4\pi R_T^2} \frac{\sigma_T}{4\pi R_T^2} \frac{\lambda^2}{4\pi} G_{RT}$$

The resulting J/S ratio for this linear repeater is then

$$\frac{J}{S} = \frac{P_{r_J}}{P_{r_T}} = \left(\frac{G_{RJ}}{G_{RT}}\right)^2 \frac{\lambda^2}{4\pi} \frac{G_{JR} G G_{JT}}{\sigma_T} \left(\frac{R_T}{R_J}\right)^4$$

In many cases, the jammer is designed to keep the high-power amplifier saturated, because the amplifier is then most efficient. Under such conditions, the jammer is a saturated repeater with constant output power and variable gain and, in general, the J/S ratio is

$$\frac{J}{S} = \frac{P_J}{P_R} \frac{G_J G_{RJ}}{G_R^2} \frac{R_T^4}{R_J^2} \frac{4\pi}{\sigma_T}$$

Of course, such computations of the J/S ratio may or may not be significant. With a large J/S ratio, target-like returns will be displayed on a radar screen. Unless these returns have characteristics similar to the characteristics of true targets, such as variation in amplitude or angular extent, the radar operator may not be fooled. Similarly, a large jamming return at a range not being tracked automatically by a range gate normally will not benefit the jammer.

If the phase of the signals from the two-point sources bears a fixed relationship, the tracking point of the radar can be outside the baseline connecting the two sources. Many authors have derived this result under the subject of target glint [16, p. 85] and others have treated it as an ECM technique [4, p. 205] called phase-front distortion or cross-eye [28, p. 555]. By referring to

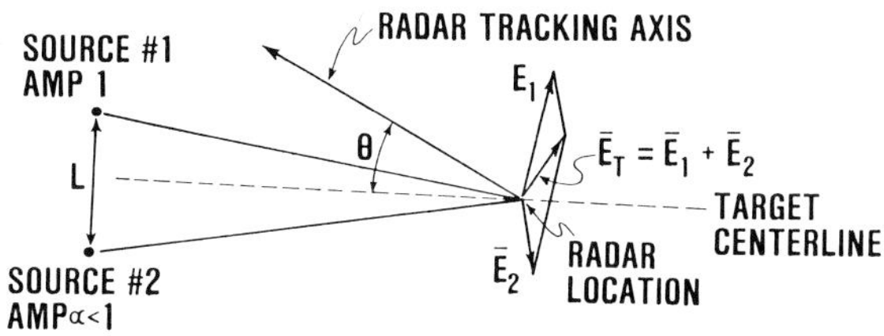

Fig. 3.42 Cross-eye error formation.

Fig. 3.42 and remembering that the sources are too close to be resolved by the jammed radar, one can easily envision the source of this error.

Chapter 2 indicated that the electric field, designated by E, vibrates perpendicularly to the direction of travel. This is depicted by the vectors (arrows) E_1 and E_2 in Fig. 3.42, where each field is radiated by its respective source. If either source were present, a radar would align its tracking axis with the ray to the source. This is equivalent to the parallel alignment of the radar aperture with the electric field from that source. With both sources present, the total electric field E_T is oriented as shown. When the radar aligns its aperture with E_T, its tracking axis points beyond the source with higher power.

The orientation of the electric field vectors in Fig. 3.42 is critical. If the signals from the two sources arrive at the radar as shown in Fig. 3.42, they are said to be "out of phase," that is, when one is maximum "up," the other is maximum "down." If the signals arrive at the radar in phase, the total field is shown in Fig. 3.43. Now the radar track axis lies between the two sources. This latter condition is not desirable since both sources are usually mounted on the same vehicle as an approach to achieving the desired fixed-phase relationship.

Of course, any phase relationship can exist between the two electric fields as they are observed at the radar. For example, even with sources that have a fixed-phase relationship, the phase between the radiated signals sensed by the radar will vary with changes in the location of the radar. In Fig. 3.44, this error has been formulated [4, p. 205] as

$$\theta = \frac{\Delta\theta}{2} \frac{1 - \alpha^2}{1 + 2\alpha \cos\phi + \alpha^2}$$

where

θ = error in tracking measured from middle of sources
$\Delta\theta = L \cos \psi / R$
L = source separation
α = relative source amplitude
φ = relative source RF phase sensed at radar

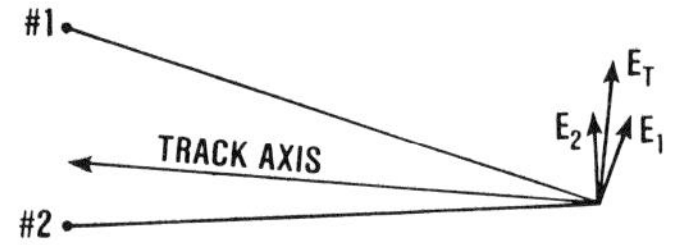

Fig. 3.43 In-phase sources.

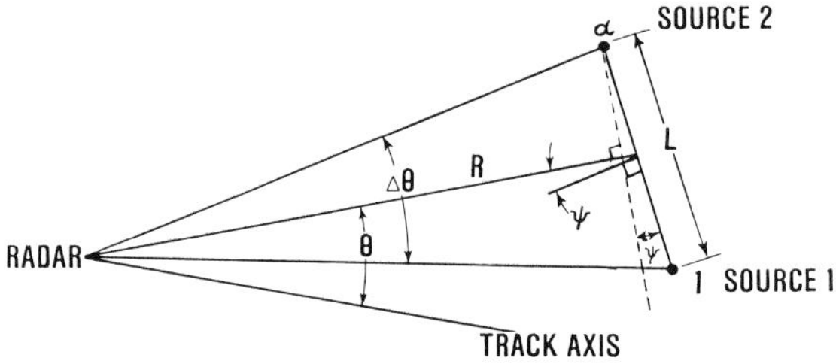

Fig. 3.44 Cross-eye jamming.

The above equation predicts very large errors as the two sources approach equality in amplitude and an out-of-phase relationship. The predicted errors are plotted on Fig. 3.45 for various amplitude ratios and phase relationships. If the source amplitudes are nearly equal, an almost exact out-of-phase relationship is required for any angular error. Very large angular errors result if this relationship is achieved. For more unequal source amplitudes, a greater variation in the observed source phase is possible before the induced angular error decreases to less than the source separation. The largest possible angular error also decreases as the source amplitudes become unequal, but the smallest angular error generated and the allowable source-phase error from 180 deg increases.

Figure 3.46 shows a theoretical cross-eye jammer [28, p. 568]. Signals received by antennas 1 and 3 are reradiated by antennas 2 and 4. Both signals travel the same distance from the radar, through the jammer, and back to the radar. Since the signals travel the same distance, they are normally in phase

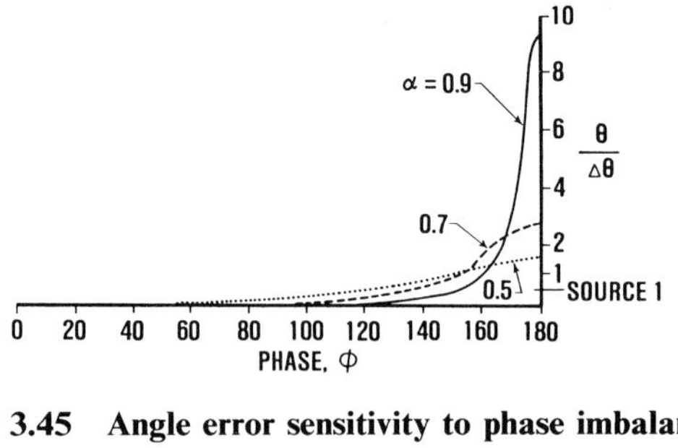

Fig. 3.45 Angle error sensitivity to phase imbalance.

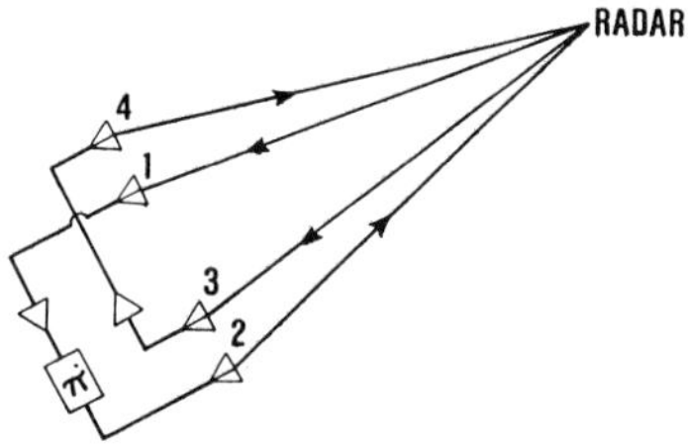

Fig. 3.46 Theoretical cross-eye system.

when they are received by the radar. In path 1-2, however, insertion of a 180 deg phase shifter is equivalent to changing the sign of the signal. Thus, the signals are out of phase and the radar will track with a $0.6\theta_3$ error if there is no skin return. (θ_3 is 3 dB beamwidth of the radar.)

The signals can be expected to differ in power, say P and $\alpha^2 P$. Then, the cross-eye sources will appear to come from a source of power $P(1-\alpha)^2 = P_2$ at an angle from the real jammer center (see Fig. 3.46). This source competes in an incoherent manner with any skin return; thus, the earlier results of this section can be used here to compute the tracking error. Figure 3.47 shows the geometry of this analysis. If the cross-eye sources are out of phase, the position of this apparent source is at an angle $\Delta\theta^1$ given by Fig. 3.45 as

$$\Delta\theta^1 = \frac{\Delta\theta}{2}\frac{1+\alpha}{1-\alpha}$$

If the target return is S, then the tracking error is

$$\frac{\bar{\theta}}{\Delta\theta^1} = \frac{P_2}{P_2+S} = \frac{P(1-\alpha)^2}{S+P(1-\alpha)^2} = \frac{(J/S)(1-\alpha)^2}{1+(J/S)(1-\alpha)^2}$$

or

$$\bar{\theta} = \frac{\Delta\theta}{2}\frac{(J/S)(1-\alpha^2)}{1+(J/S)(1-\alpha)^2}$$

This error is plotted for various conditions in Fig. 3.48.

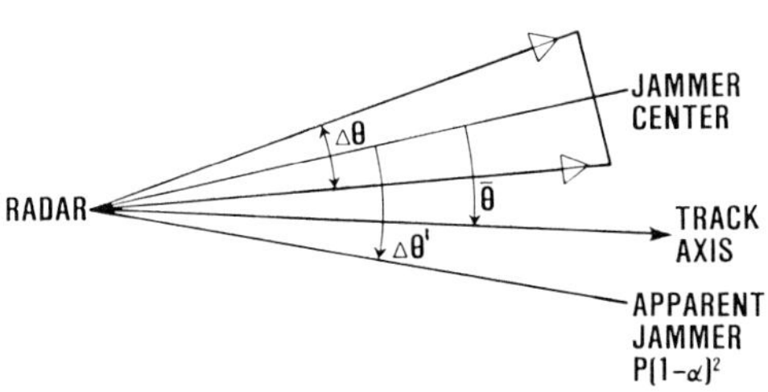

Fig. 3.47 Geometry incorporating target return.

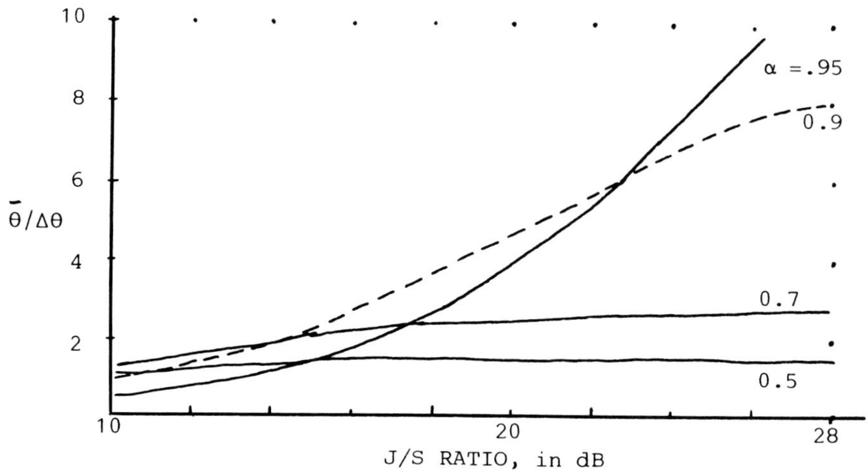

Fig. 3.48 Effect of target signal on cross-eye.

Active ECM Summary

This section has concentrated on winning individual battles but not necessarily the war. It has emphasized one-on-one encounters involving target-tracking radars guided by radio frequency. This was the original intent. However, all of the ECM techniques used at radio frequencies have direct application to the infrared or electro-optical portions of the electromagnetic spectrum, although many would not be considered practical. But applicability means nothing if one is being tracked optically and his jammer operates at RF. A good jammer is useless if tactics require him to fly over threats that can simply set up a barrier of fire if they know he is coming. Again, the many-on-many (mass) scenario and the element of surprise remain important in winning the war.

3.4 PASSIVE ELECTRONIC COUNTERMEASURES

Passive electronic countermeasures consist of those actions taken to reduce the effectiveness of enemy radar-guided weapon systems short of their physical destruction or the radiation or electromagnetic energy. As observed in Chapter 2, radar detection depends on the ratio of the signal power to interference power. The interference could be internal receiver noise or returns from clutter. Radar target tracking is also limited by the signal-to-noise or signal-to-clutter ratio. Thus, passive ECM techniques should clearly modify these ratios in favor of the penetrating aircraft. These ratios are expressed in Fig. 3.49 in a manner that draws attention to the methods of controlling them.

Reducing the penetrating aircraft's radar cross section, sometimes called observable reduction, is desirable whether clutter or noise limits performance. When performance is limited by clutter, increasing the background

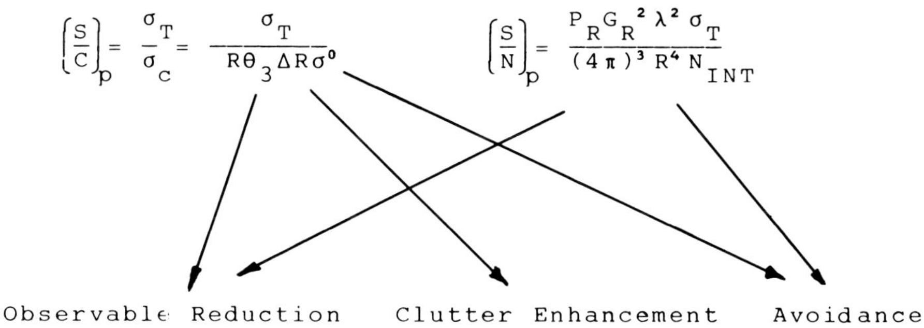

Fig. 3.49 Approaches to passive electronic countermeasures.

clutter even more through the choice of flight path or use of chaff is appropriate. Finally, increasing the distance to the threat is always a positive step.

Reduction of Observables

To assess the desirability of a reduction in radar cross section (RCS), this section more formally defines radar cross section, considers factors affecting the radar cross section of an object, and determines one effect of radar cross section on the performance of a radar system. Reduction of observables generally considers reduction of the target signature across the entire electromagnetic spectrum, including infrared and optical frequencies. Only the radar frequency range is addressed in this discussion.

Reduction in radar cross section can be better understood when one realizes that the radar cross section of an object σ_T generally has little relationship to its physical cross section. The radar cross section [18, p. 33] is that area intercepting an amount of power that, when scattered equally in all directions, produces an echo at the radar equal to the echo from the target.

$$\sigma_T = \lim_{R \to \infty} 4\pi R^2 \frac{\text{reflected power density (at radar)}}{\text{incident power density (at target)}}$$

In this formula the distance between the radar and the target R increases to infinity because the measured cross section becomes constant only at great range. Practically, the distance R should exceed $2D^2/\lambda$, where λ is derived

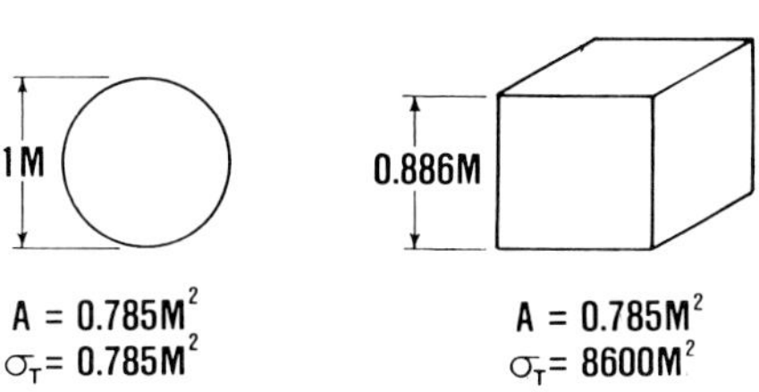

Fig. 3.50 Cross-section example.

from the measurement frequency and D is the target dimension normal to R. That this definition produces a radar cross section unrelated to the physical cross section is apparent from a simple example. A conducting sphere of 1 m in diameter and a cube 0.886 m on a side, shown in Fig. 3.50, have the same physical cross section A when the cube is viewed from the normal to any side. If the sphere is much larger than the wavelength of incident energy, it has a radar cross section equal to its physical cross section, or less than 1 m^2. The cube, however, has a cross section given by $\sigma_T = (4\pi/\lambda^2)A^2$ which, for incident power at 10 GHz, equals 8600 m^2. This example suggests one approach to reducing the cross section: elimination of large flat areas.

The radar cross section of a target at any frequency is a function of the angle at which the target is observed (see Fig. 2.19). If the geometry of observation is fixed, then the cross section is a function of the radar illuminator frequency. These effects are shown in Fig. 3.51 for a flat 1 m^2 plate at 10 GHz. The radar cross section, expressed in decibels relative to 1 m^2, is definitely a function of both frequency and observation angle.

If a reduction in the radar cross section is possible without the elimination of the object in question, what level of reduction should be attempted? Certainly, it is unreasonable to attempt complete elimination of the radar cross section at all frequencies. Over what frequency interval should the reduction of the cross section be attempted? A reasonable goal of any reduction effort is to provide the reduction over a "wide" bandwidth. What is wide? Certainly, the number of anticipated threats should be included in the definition. Since certain reduction schemes are included in the design of aircraft that may be in the inventory for 10–15 years, future threats must be considered. A third goal is to lower the large average values of aircraft cross section around the aircraft rather than to concentrate on narrow specular regions of high cross section (see Fig. 3.52). In accomplishing this last goal,

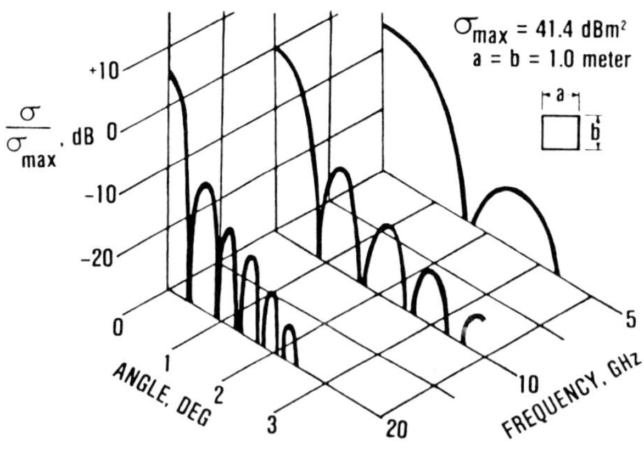

Fig. 3.51 Radar cross section of flat plate.

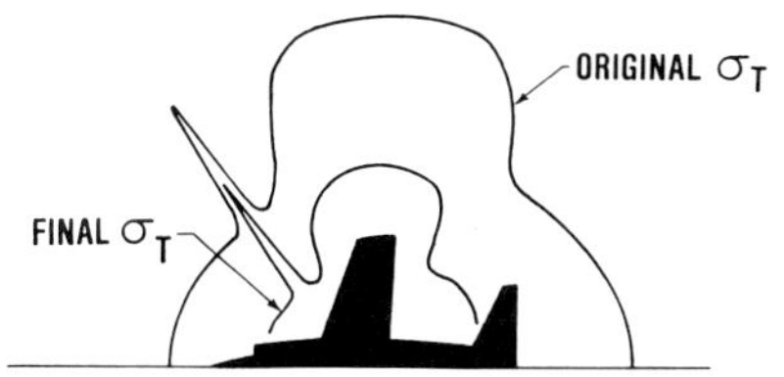

Fig. 3.52 Reduction in cross section.

the reduction in cross section again should be compatible with the mission requirements dictated by the threat. For example, the region immediately below an aircraft will typically have a high cross section, but a threat radar would probably be physically unable to view the aircraft from such a position. Therefore, there is little reason to reduce the cross section of the underside unless the aircraft frequently exposes its underside. (The extent to which the above goals can be accomplished is beyond the scope of this text.)

Just as important as potential accomplishments in the reduction of cross section is the impact of reduction on the performance of threat radars. For the radar limited by noise (jammer or internal), the radar detection range is given by

$$R = \left(\frac{P_R G_R^2 \lambda^2 N}{(4\pi)^3 N_{INT} L} \right)^{1/4} \sigma_T^{1/4}$$

Therefore, if a range R_o is achieved for a cross section of σ_{To}, then, for any other cross section, the radar range is given by

$$R = R_o \left(\frac{\sigma_T}{\sigma_{To}} \right)^{1/4}$$

Figure 3.53 shows this detection range. Obviously, a range reduction of one-half requires a reduction in cross section of 12 dB or approximately 95%. Inherent in this statement, and the above equation, is the assumption that parameters other than radar cross section do not change (e.g., G_R).

Even more dramatic results are achieved in the clutter environment.

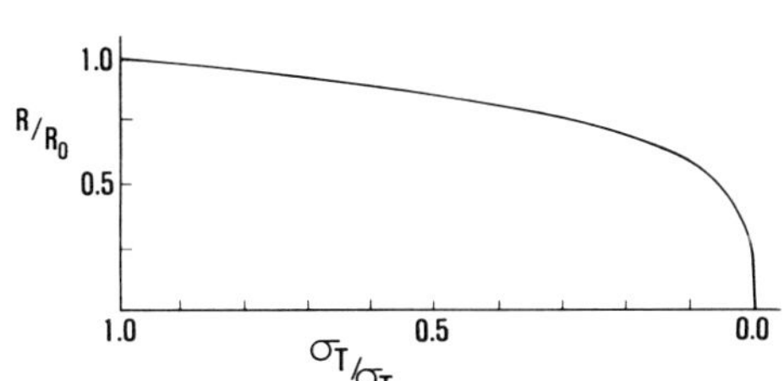

Fig. 3.53 Detection range in noise vs RCS.

For a ground radar with a specified$_{\text{subclutter}}$ visibility (expressed in decibels) in a specific type of ground clutter, the radar range can be written ($V = 3 \times 10^8$ m/s)

$$R_o = \frac{2\sigma_{To}}{\theta_3 V\tau\sigma^o} 10^{SVC/10}$$

and $R = R_o(\sigma_T/\sigma_{To})$ for any other cross section. Because of the assumed variation of the competing clutter cross section with range, a reduction of one-half in radar range requires a reduction of only one-half in the target cross section.

A reduction in cross section also enhances any of the active jamming techniques by reducing the competing target return. There are two approaches to the reduction of cross section: one is to eliminate strong sources of aircraft echo and the other is to apply radar-absorbing material to the control sources of the echo.

We have already observed the effects of using curved rather than flat surfaces on the radar cross section. In discussing these effects, we assumed that high-frequency scattering from a flat surface many wavelengths in size is comparable to light reflected from a mirror. This section continues the analogy to describe multiple plates, antennas, and other aircraft scatterers. The discussion provides only a rudimentary order-of-magnitude description of the various scatterers; far more accurate and complicated analyses can be found elsewhere [52]. Figure 3.54a shows a ray incident on a flat plate at an angle θ relative to the normal to the plate. It is well known that the reflected ray travels in a direction θ away from the normal to the plate and will not be sensed at the source. (A more exact theory accounting for the effects of "diffraction" permits computation of the backscattered field at non-normal incidence. In fact, this precise result is plotted in Fig. 3.51.)

A second plate located at 90 deg relative to the first forms a "corner reflector" that scatters a strong signal back to the source over a wide variety of source positions. If the area of the corner reflector is A as defined in Fig. 3.55, then the radar cross section varies slowly with the angle θ, also redefined in Fig. 3.55 [52, p. 589].

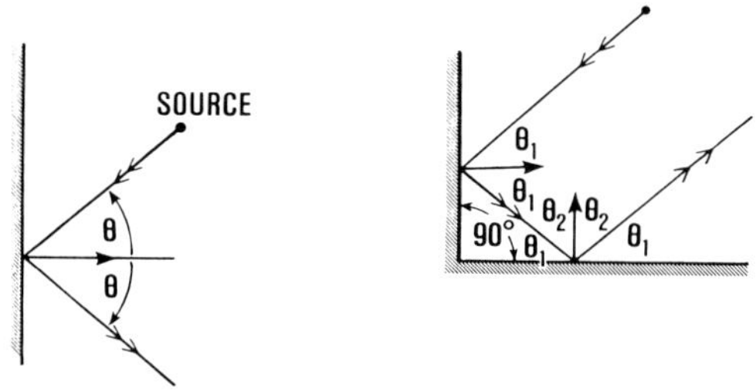

(a) (b)

Fig. 3.54 Ray scattering.

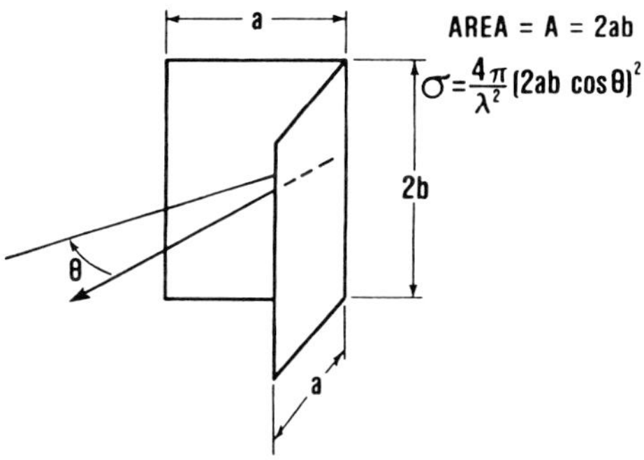

Fig. 3.55 Two-sided (dihedral) corner reflector.

Addition of a third plate forms the three-dimensional corner reflector shown in Fig. 3.56. The radar cross section again is a slowly varying function of the angle from the normal and is proportional to the area that is fully involved in the reflection process [15, p. 27-16].

Figure 3.57 compares the cross sections of a square plate and a two-sided corner reflector. Obviously, the shape of the corner reflector enhances the cross section at large angles off the normal incidence.

Aircraft antennas are a second strong RCS source because they are constructed to focus electromagnetic energy in a particular direction. As shown in Fig. 3.57, for example, a flat-plate array antenna composed of slots in a conducting ground plane will scatter just as a smooth flat plate at other than the operating frequency of the antenna. Even at the antenna frequency, it will scatter in the same manner but at a lower power level. Reflector antennas will act similarly.

Engine inlets are still another scattering source. Each inlet can be modeled as an empty cylinder closed at one end, i.e., a cavity. The actual radar cross section is a strong function of the internal construction [42] as well as the polarization of the illumination and the shape of the cavity opening. A first-order approximation suggested in the literature for scattering from a

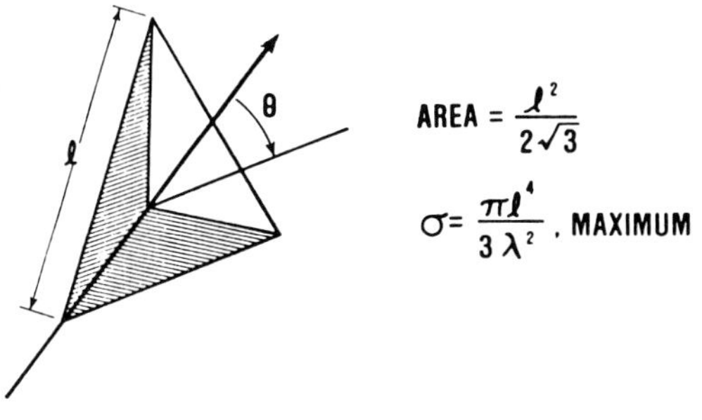

Fig. 3.56 Three-sided (trihedral) corner reflector.

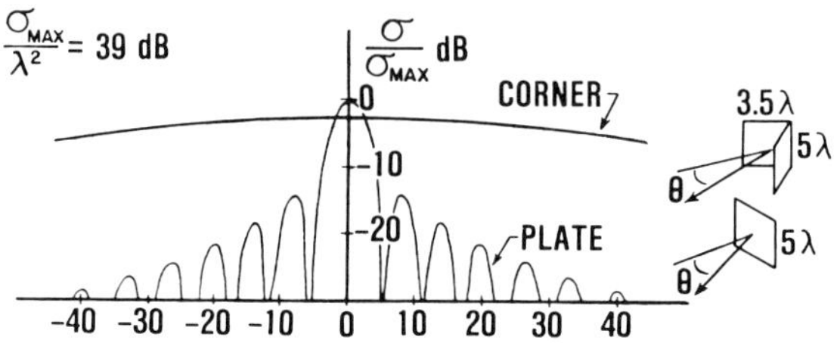

Fig. 3.57 RCS comparison.

cylinder of diameter d is

$$\sigma_T = \frac{0.4(\pi d)^3}{\lambda} \cos\theta \left| \frac{\sin\left(2\dfrac{\pi d}{\lambda}\sin\theta\right)}{\dfrac{2\pi d}{\lambda}\sin\theta} \right|$$

which is plotted in Fig. 3.58 [52]. The angle θ is measured from the normal to the circular opening and is less than 90 deg.

After the major sources of aircraft radar cross sections have been identified, what can be done to eliminate them? This is not a simple task, especially in engine inlets and other necessary cavities. Efforts to reduce the cross section should begin early in the design of airframes to avoid building in corner reflectors and large flat scatterers. Metallic radomes may help hide radar and other antennas at all but the operating frequency [43].

A comparison of the airframes in Figs. 3.59a and 3.59b shows what can be done to reduce radar cross section. The comparison uses a combination of ray theory and diffraction theory and is thus quite rudimentary. Multiple scattering is considered only when ray theory predicts a specular path and scattering from the edges is ignored. The model in Fig. 3.59a has a large rectangular fuselage and the wings join the body at a right angle. The wings are assumed to have a flat leading edge at the 10 GHz illuminating frequency and an antenna at the nose is simulated by the flat end of the fuselage. In Fig. 3.59b, the fuselage has the same volume; it has been changed to a circular

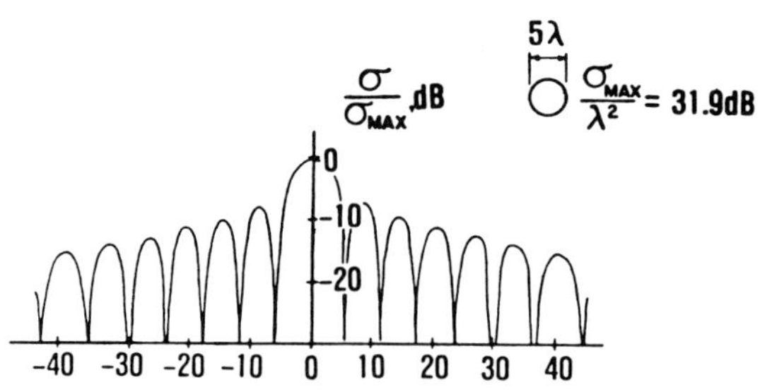

Fig. 3.58 Approximate cross section of an engine inlet.

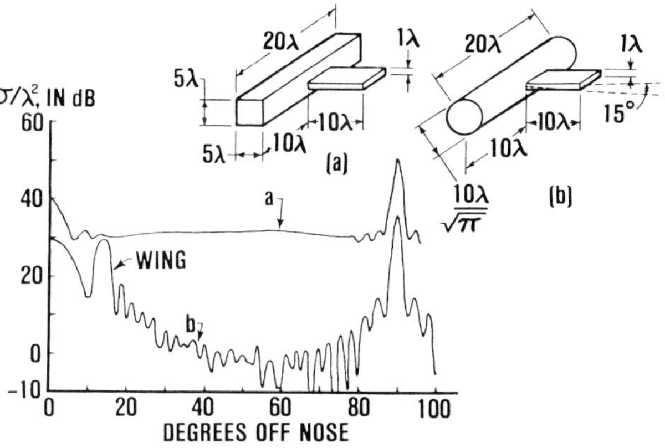

Fig. 3.59 Example of RCS reduction.

cylinder and the wings are swept back at 15 deg (arbitrarily). A metallic radome can be used to cover the antenna.

For an existing aircraft whose structure cannot be altered and for the structural elements of new aircraft that continue to scatter strongly, the application of a radar-absorbing material may help in reducing the radar cross section.

Radar-absorbing material (RAM) for military aircraft differs from the commercial material used in anechoic chambers. The commercial material shown in Fig. 3.60 is typically a carbon-impregnated urethane foam of the thickness of a wavelength at the lowest frequency of interest. Such material employs physical shaping that slowly changes the properties of propagation within a volume of space, such as pyramids, to reduce reflections. Such material often scatters back one-millionth of the incident power when that power is incident at an angle of 90 deg to the surface.

As shown in Fig. 3.61, typical radar-absorbing material used by the military will appear as a multilayered medium, much like plywood, suitable for

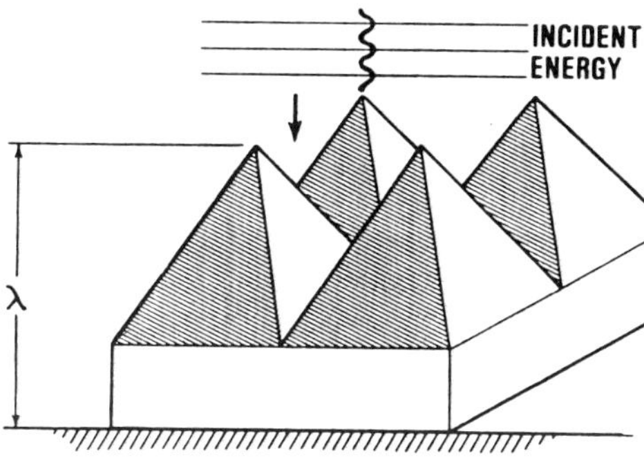

Fig. 3.60 Commercial absorber.

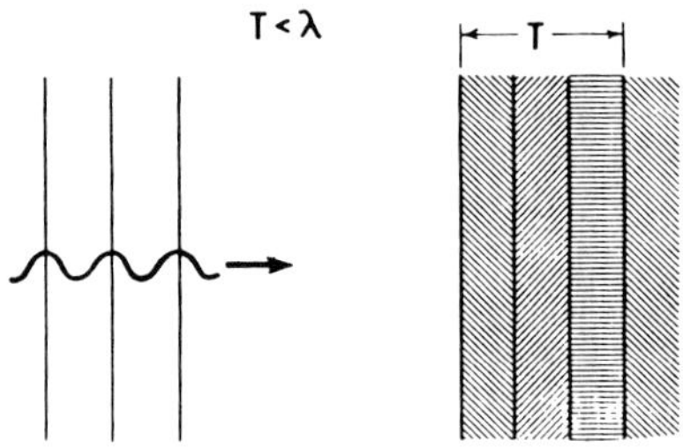

Fig. 3.61 Radar-absorbing material.

application on various aircraft surfaces. Like the commercial absorber, the RAM minimizes the reflection of RF energy at its outer surface and absorbs the incident power in the medium between metal and air. For a more detailed discussion of construction techniques and performance parameters, the reader should refer to Ref. 4 (p. 369).

Background Enhancement

The advantages gained by a penetrator in a high-clutter environment, particularly natural clutter, were discussed in Chapter 2. This section focuses on enhancement of man-made clutter through the use of chaff, as corner reflectors were described in an earlier section.

Chaff consists of thousands of narrow rods of conductive material, such as aluminum, or glass fiber coated with a conductor. As shown in Fig. 3.62, these fiber rods may be 0.5-1 mil in diameter. Each of the rods constitutes a "dipole" with a radar cross section varying between 0 and 0.866 λ^2 square meters at the frequency at which the dipole is one-half wavelength long. The cross-section value observed depends on the viewing angle. The expected radar cross section σ_e, averaged over all viewing angles, is 0.15 λ^2 [44] to 0.20 λ^2 [45], depending on the length-to-diameter ratio of the dipole.

As shown in Fig. 3.63, this expected dipole cross section also depends on frequency [45]. The maximum cross section occurs at chaff lengths that are

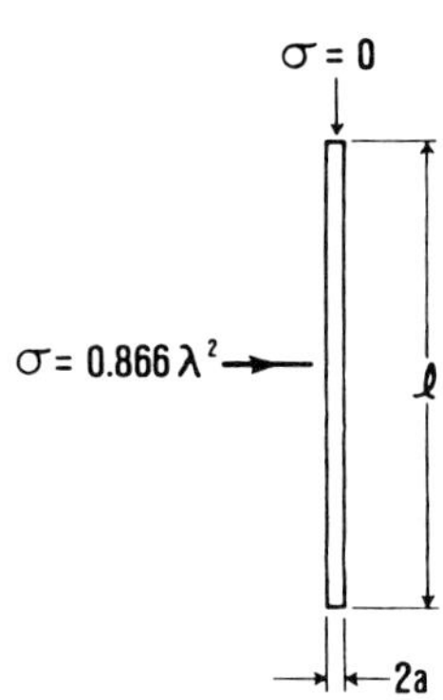

Fig. 3.62 Single-chaff dipole.

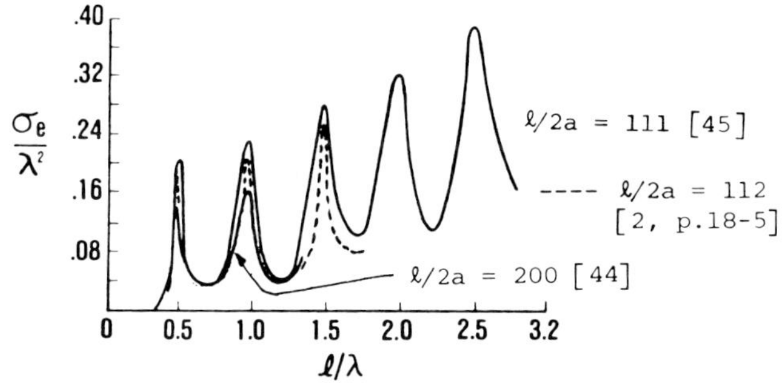

Fig. 3.63 Normalized chaff cross section vs frequency.

approximately integral multiples of one-half the illuminating wavelength. Again, chaff can be characterized by assigning an average cross section to it. The average is formed over a frequency range around the frequency at which the peak cross section occurs and within which the chaff cross section is within 3 dB of the peak value. This characterization then assigns to chaff a given cross section valid within some bandwidth around the frequency associated with the peak value. The bandwidth is usually 10–15% of the center frequency and the chaff is assigned a cross section of zero outside of this bandwidth.

When an aircraft discharges (drops) chaff, the chaff expands to form a chaff cloud (see Fig. 3.64). As viewed by a radar, the radar cross section of this chaff cloud will normally be less than the physical area viewed by the radar. How much less depends on the number N of chaff dipoles visible to the radar. For small dipole densities, one can assume that each dipole displays its average cross section. For higher densities, shielding and bunching reduce the effectiveness of each dipole.

Some sources claim that 1 lb of chaff will yield a 60 m^2 target at all frequencies over 1–10 GHz [19, p. 223; 30]. And, if all chaff is cut to one wavelength, the cross section will be

$$\sigma = 6600 \frac{M}{f_{\mathrm{GHz}}} \mathrm{m}^2 \qquad \frac{M = \text{mass, kg}}{f = \text{frequency, GHz}}$$

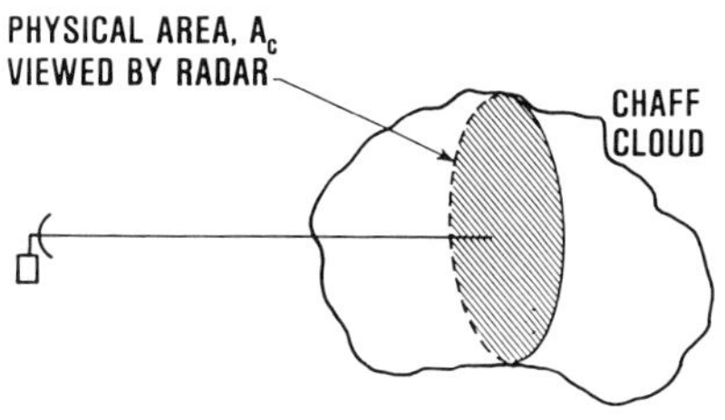

Fig. 3.64 Chaff cloud.

This formula assumes chaff strips 0.01 in. wide and 0.001 in. thick cut to one-half wavelength. Even higher cross sections are possible. For example, aluminum chaff constructed of strips is said to have a minimum width of 0.2 mm, a thickness of 0.01 mm, and a length-to-width ratio of less than 500 [48]. For a frequency greater than, or equal to, 3 GHz, these values yield the following cross section from 1 kg of aluminum:

$$\sigma = \frac{22{,}222}{f_{\mathrm{GHz}}} \ \mathrm{m}^2$$

However, this cross section is the final value after the chaff cloud has dispersed and it presumes that all dipoles are still in the same radar resolution cell. Only the illuminated dipoles can contribute to the backscatter.

Certainly, all dipoles are in the radar resolution cell immediately after disbursement. During these first few moments, the effect of dipole interaction on the radar cross section of the cloud can be explained in the following manner. If N equals the total number of dipoles in the cloud and A_c equals the physical area viewed by the radar, then the number of dipoles per unit of projected area n is

$$n = N/A_c$$

The radar cross section is given by [44]

$$\sigma = A_c[1 - e^{-n\sigma_e}]$$

where $\sigma_e = 0.15\lambda^2$, the expected dipole cross section.

When the number density is large, $n\sigma_e$ is also large and the chaff cross section equals the projected area A_c. When n becomes small, $n\sigma_e$ also becomes small and then, as $e^{-n\sigma_e} \approx 1 - n\sigma_e$, the cloud cross section is

$$\sigma \approx A_c[1 - (1 - n\sigma_e)] = A_c n\sigma_e = N\sigma_e$$

This latter condition occurs only after the dipoles are widely separated. Use of this equation results in the values shown in Table 3.6 depicting the effects of diffuse and dense chaff clouds. This table was developed for chaff at 10 GHz, but similar results are obtained at other frequencies if the following is noted. Each dipole is $\lambda/2$ long and can be assigned an area of this length on a side or an area $\lambda^2/4$. With this area per dipole in a projected area A_c and with all dipoles in the same plane, the number of dipoles is $N = 4A_c/\lambda^2$. In general, diffuse or dense is relative to this value of one dipole per $\lambda^2/4$ of projected area.

The chaff cloud will move with the wind and fall at 0.5–5 ft/s, with the glass chaff falling at the slower rates [46]. Its radar cross section will vary with look angle and time. Chaff in an aircraft windstream must have large variations in cross section during the first minute if turbulence induced by

Table 3.6 Chaff Dipole Efficiency

No. of dipoles/m^2	Cross section as % of physical area	Dipole efficiency, %
200	3	100
20,000	96	30

the aircraft governs dispersion of the cloud [47]. After 1 min, atmospheric turbulence governs cloud spreading. The cloud will reach its final size in approximately 5 min. After the size of the cloud has stabilized, its cross section will still vary in a few tens of milliseconds and will cause a spreading of the Doppler return of the chaff [19, p. 225; 49, p. 69]. There is an additional spread in frequency due to windshear.

Some sources claim that, after several minutes to an hour, the chaff dipoles tend to lie horizontally [2, p. 18-10] because the vertically oriented dipoles fall out of the original cloud and leave only horizontal dipoles. But, measurement data [47] support a different picture of dipole orientation. During the initial expansion of the cloud, the dipoles are primarily oriented vertically. The orientation of each dipole is random, but the average axis orientation is 30 deg from vertical. The initial expansion is presumed to take 5 min and, over an interval of 5–50 min, the dipole orientation will slowly change from vertical to random.

Self-Protective Chaff Employment

Chaff is dispensed by many methods, ranging from pyrotechnic (AN/ALE-40) to mechanical (AN/ALE-38)[50]. Self-protective chaff is typically ejected from cartridges containing an explosive to cause rapid blooming. Each cartridge contains chaff cut for many different frequency bands and may have more than 10^6 dipoles [49]. Construction of a typical chaff cartridge is shown in Fig. 3.65. Upon firing, a squib blows a piston out of the cartridge and takes the many lengths of chaff with it.

For self-protection, the chaff must expand rapidly because it must present a cross section larger than the target while the aircraft is still within the same resolution cell. The time available for this blooming depends on the speed of the aircraft and the size of the radar resolution cell; blooming in 50 ms may be required. The chaff cloud is sometimes modeled with a cross section increasing exponentially in time [51] as

$$\sigma(t) = \sigma_o(1 - e^{-t/\tau_\sigma})$$

where $\sigma_o = N\sigma_e$, as defined previously, and τ_σ is the cross-sectional rise time. Thus, if the rise time is large, a large number of dipoles is required because the chaff must steal the range gate before the cloud has dispersed.

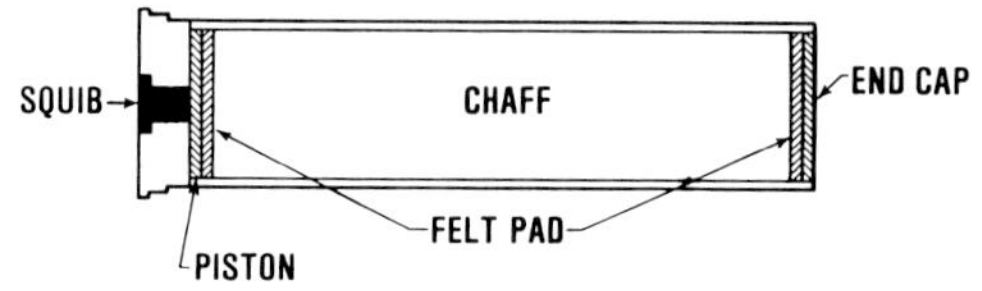

Fig. 3.65 Chaff cartridge.

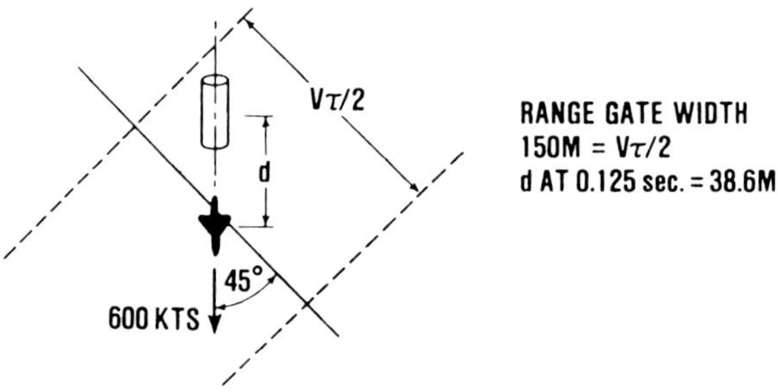

Fig. 3.66 Chaff employment.

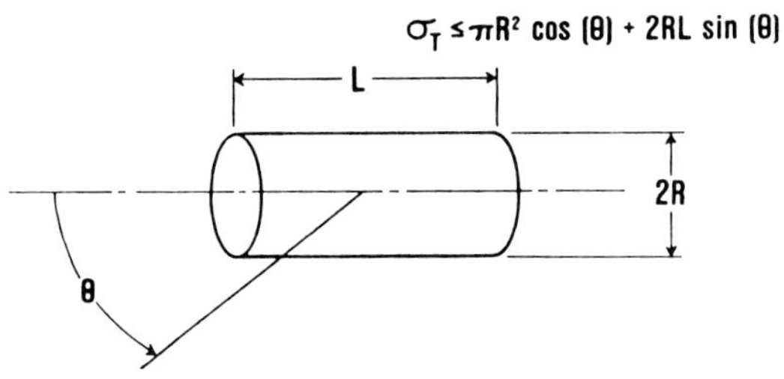

Fig. 3.67 Chaff cloud geometry.

Figure 3.66 shows such a chaff cloud behind an aircraft flying at 600 knots and being tracked by a radar with a 1 μs pulse width. If the aircraft is centered initially in the range gate of the radar, the cloud must be within 106 m of the aircraft to remain in the same gate. If it is within this range cell, its cross section will depend on the viewing angle θ, as shown in Fig. 3.67. If the cloud size were 20 m long by 2 m in diameter, then, for the geometry shown, $\sigma_T = 30.5$ m^2.

Example 3.3: Chaff Self-Protection

Given the radar in Example 2.2, a dense chaff cloud of 20 m^2, and a target cross section of 10 m^2, compute the required MTI improvement factor if an output signal-to-clutter ratio of 20 dB is required for tracking.

Answer: The form of the radar is irrevelant as long as the chaff does not fill the beam. Thus, the input and output signal-to-clutter ratios are

$$I = \frac{(S/C)_o}{(S/C)_p} = \frac{100}{0.5} = 200 \rightarrow 23 \text{ dB}$$

3.5 ACTIVE JAMMING SYSTEMS

Previous sections have discussed the effects of jamming signals on radar systems. This section addresses the devices necessary for generation of such signals. It does not include actual systems or the capabilities of their components, but it does treat the basic jammer configurations, the characteristics of major components, and a typical transmitting system.

A complete jammer consists of much more than the transmitter considered in this discussion. As shown in Fig. 3.68, some mechanism must be available for assessing the electromagnetic environment surrounding the aircraft. This mechanism can range all the way from yesterday's intelligence, through a sophisticated radar warning receiver (RWR), to an even more elaborate surveillance receiver. A decision-making element is also required to select the appropriate jamming response based on the environmental inputs. In Fig. 3.68, this element is an electronic warfare officer (EWO), but it could be a computer or simply an instruction to the pilot to start jamming at a certain geographic location. The final element is the transmitter, which is here addressed first with a description of its major components.

Since a noise jammer radiates noise, either a low-power noise source must be amplified to a high power or a high-power noise signal must be generated directly. The latter can be accomplished with a high-power oscillator. A repeater jammer also requires signal amplification and, in the transponder mode, a low-power signal source. Both noise and repeater jammers require antennas. Thus, the next major considerations are low-power signal and noise sources, high-power devices, and antennas.

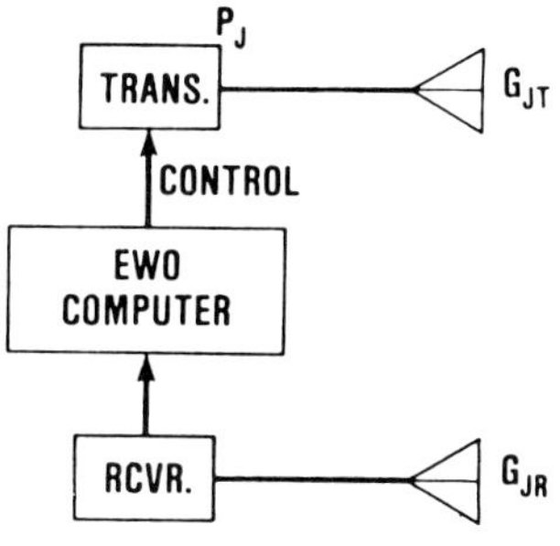

Fig. 3.68 Jammer system components.

Low-Power Sources

A voltage-controlled oscillator (VCO) is an oscillator whose frequency is determined by a control voltage. There are many methods of constructing a VCO, each of which has certain unique advantages. The VCO specifications listed in Table 3.7 can be used to select the most appropriate oscillator for a particular application [53, 54]. The list is not complete, but the author considers them the most important and consistent with the purposes of this discussion. The quantities defined in Table 3.7 are also shown in Fig. 3.69.

The VCO settling time is certainly important if a target-like jammer attempts to counter a multitude of pulsed radar threats by radiating appropriately modulated cover pulses. The jammer frequency probably must change rapidly to keep up with the many radar RFs and PRFs. Of course, if the threat is a CW radar, the VCO settling time may not be the important concern. For both types of radars, post-tuning drift is extremely important since the jammer is probably ineffective if it does not generate a signal within the bandwidth of the radar receiver.

Bushnell [12, p. 331] provides a discussion of typical VCO characteristics and explains how an attempt to improve one characteristic may cause others to degrade. He limits himself to transistor and GUNN diode sources that are varactor tuned and he states that the performance shown in Table 3.8 is typical.

Some noise jammers simply amplify a low-power noise signal to a desired level before coupling the signal to the antenna. The initial noise signal [55]

Table 3.7 VCO Specifications

Frequency range	The tunable bandwidth over which the oscillator meets all specifications.
Minimum power	The minimum RF output power at all output frequencies within the oscillator's tunable bandwidth under all specified conditions.
Output frequency	Of all the frequency components that may be present at the oscillator output port, the output frequency is the frequency of the desired output.
Settling time	The time t_{ST} required for the output frequency to enter and stay within a specified error band Δf_{ST} centered around a reference frequency f_R after application of a step input voltage.
Post-tuning drift	The maximum change in frequency Δf_{PTD} from the frequency measured at the beginning of the time interval, t_1. The time interval $t_2 - t_1$ shall be referenced to the application of a tuning command at t_o. The period of measurement ends at t_2.

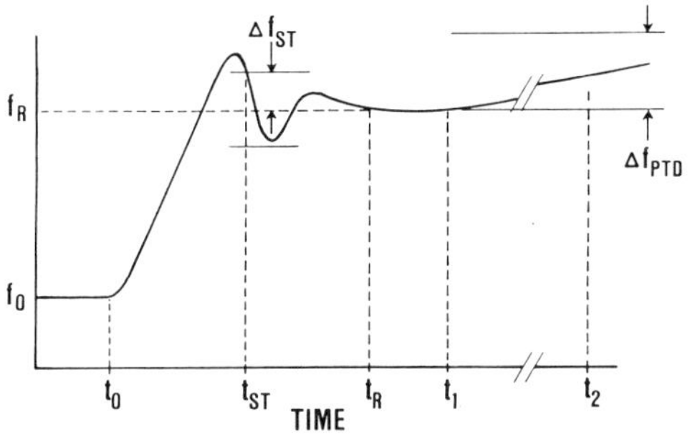

Fig. 3.69 VCO parameters.

can be generated by gas discharges as in a fluorescent light or by passage of a current through a semiconductor diode in a controlled avalanche breakdown. One reason why the noise thermally generated in a simple resistor cannot be used is that it is not truly constant with frequency. A second reason is that a noise source provides power many times larger than the power generated in a resistance at room temperature and less follow-on amplification is required. The noise power of a source is usually expressed in terms of excess noise ratio (ENR) or the ratio by which the noise power of a source exceeds the noise power of a resistor at standard temperature. (Table 3.9 shows typical parameters of noise sources.)

Table 3.8 Typical VCO Parameters

Power, minimum	50 mW	0.25–8 GHz
	20 mW	8–16 GHz
Bandwidth	Octave over 0.25–8 GHz	
	4 GHz over 8–16 GHz	
Step response	10 ns	
Preferred device	Transistor	0.25–12 GHz
	Diode	12–20 GHz
Post-tuning drift	4 MHz in 10 s	

Table 3.9 Low-Level Noise Sources

Frequency range, GHz	Excess noise ratio, dB	Flatness, dB
0.01–0.6	6.0	±1.1
1.0–12.4	25.0	±0.5
12.4–18.0	15.0	±0.5
1–26.5	25.0	±2.0
26.5–40.0	23.0	±3.0

High-Power Devices

The high-power sources of most ECM systems are thermionic in nature rather than solid state. This means that they are more like television picture tubes than transistors. Solid state and other types of sources can be used in specialized applications, but the thermionic sources still retain specific advantages in high-power, broadband, airborne ECM systems.

A thermionic source, or tube, will contain the elements shown in Fig. 3.70. The cathode is heated until it emits electrons that are formed into either a circular- or rectangular-shaped beam by focusing electrodes. The electron beam travels through an interaction region containing "slow wave structures" and electric and magnetic fields. Amplification of the RF signal occurs in this region. Finally, the electron beam impacts a collector that absorbs the remaining beam energy.

Type O Tubes. Electron beam tubes that contain a magnetic field aligned with the electron beam in the drift region are called type O tubes. Traveling wave tubes and most klystrons are examples of these tubes. The magnetic field prevents the spreading of the electron beam and is formed by a selenoid (a long coil of wire containing many turns) or by a periodic structure of permanent magnets (PPM). (Figure 3.71 shows these two types of focusing structures.) Since the selenoid field is more uniform than the PPM field, fewer electrons intercept the slow-wave structure, which is also within the drift space. A selenoid-focused traveling wave tube can thus generate more RF power and operate more efficiently than a PPM-focused tube of an equivalent size. However, a selenoid-focused tube weighs more than a PPM-focused tube and power is consumed in the selenoid field.

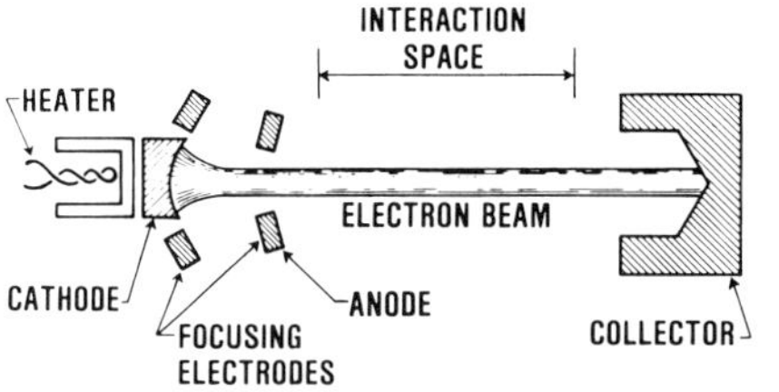

Fig. 3.70 Elements of microwave tube.

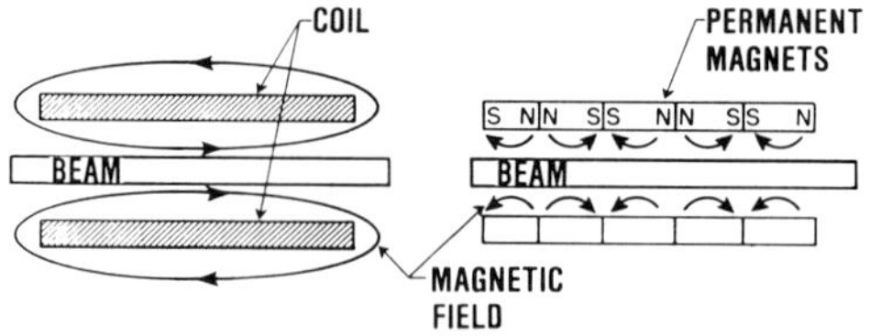

Fig. 3.71 TWT focusing structures.

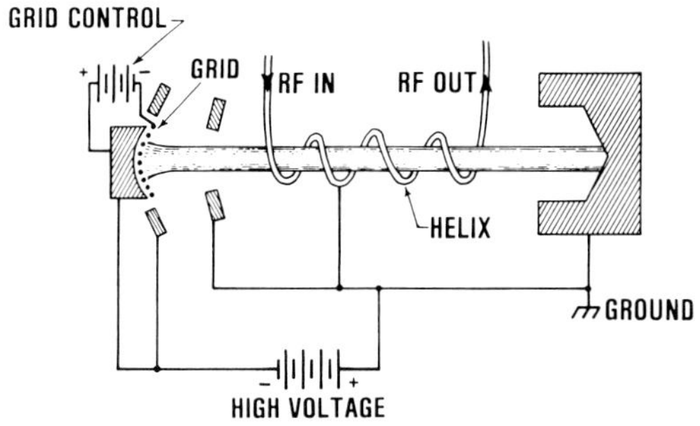

Fig. 3.72 TWT with helix.

Also within the drift region is a slow wave structure on which signal amplification occurs. This structure is normally a helix of wire concentric with the electron beam (see Fig. 3.72). Electromagnetic waves propagate along the wire at the speed of light, but they move in the beam direction at a speed reduced by the pitch of the helix. The electrons in the beam move with a velocity that increases as the accelerating anode voltage is increased. When both the electron beam and electromagnetic wave move with approximately the same velocity, strong interactions occur. In Fig. 3.72, for example, a signal is injected at one end of the slow-wave structure and an amplified signal resulting from interaction with the electron beam is extracted at the other end. This is the normal operating mode for a traveling wave tube. Also shown in Fig. 3.72 are the high-voltage connections to the tube structure and an element called the "grid." The grid is a control element that can turn off or modulate the electron beam with much less voltage than the cathode-to-anode difference. Typical broadband TWT characteristics are given in Table 3.10 for CW tubes (ones capable of continuous operation) and in Table 3.11 for pulsed tubes (ones that operate in a pulse-like manner).

Table 3.10 CW TWT Characteristics [58; 59]

Operating frequency, GHz	Minimum power output, W	Gain at rated power, dB	Focusing		Eff, %	Weight, lb
			PPM	Sel.		
1.5–30	500	30	X		16	12
2.6–5.2	300	30	X		13	9
4.8–9.6	450	37	X		24	9.5
9.0–17.0	100	37	X		11	6
2.0–4.0	1000	30		X	8	22
9.0–18.0	200	30	X		7	8

Table 3.11 Pulsed TWT Characteristics [58; 59]

Operating frequency, GHz	Min peak RF output, kW	Duty cycle, %	Gain at rated power, dB	Focusing		Eff, %	Weight, lb	Max pulse width, μs
				PPM	Sel.			
2.5–8.0	1	10	30	X		7	8	30
2.6–5.2	1	2	50	X		8	6	50
8.0–18.0	1	4	50	X		5	6	20
2.6–5.2	1	4	30	X		5	9	–
8.0–17.0	1	4	35	X		4	5	–
4.0–8.0	1	–	40	X		29	–	–

The tube efficiencies may not be representative of tubes developed specifically for ECM. Carruba [12, p. 390] states that traveling wave tubes have efficiencies of 20–25% when depressed collector techniques are used. As shown in Fig. 3.73, a traveling wave tube with a depressed collector has the collector at a potential less than the accelerating anode or helix. Mendel [60] describes a helix-type traveling wave tube producing 1 kW and operating at 1.6–4.8 GHz with an efficiency of at least 45%.

The efficiency of the tube is the ratio, expressed as a percentage, of the RF output power to the power supplied for operation of the traveling wave tube. Since the power supplied to the traveling wave tube must be at a potential of a few thousand volts, an additional power loss is suffered in the tube power supply as it transforms the aircraft voltage up to the voltage required by the traveling wave tube. Furthermore, this efficiency is achieved only when the maximum RF power is developed. The traveling wave tube consumes most of its power in accelerating the electron beam. Unless this beam is cut off, it consumes power even when no RF signal is being amplified.

Maximum tube efficiency is thus attained by turning the tube off when no RF signal is desired and using the tube at its maximum (or saturated) power when an output signal is desired. Any larger input signal will not increase

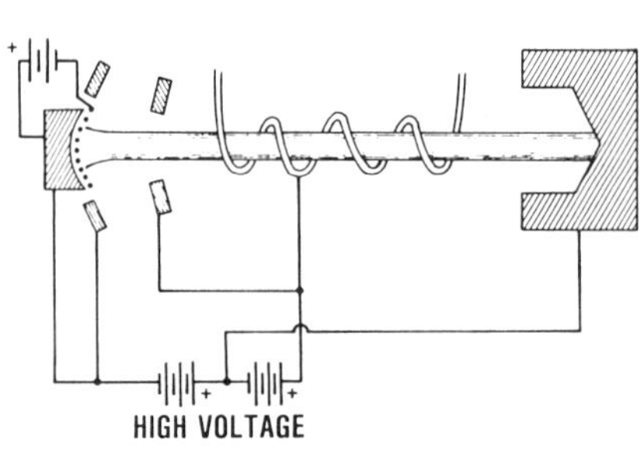

Fig. 3.73 TWT with depressed collector.

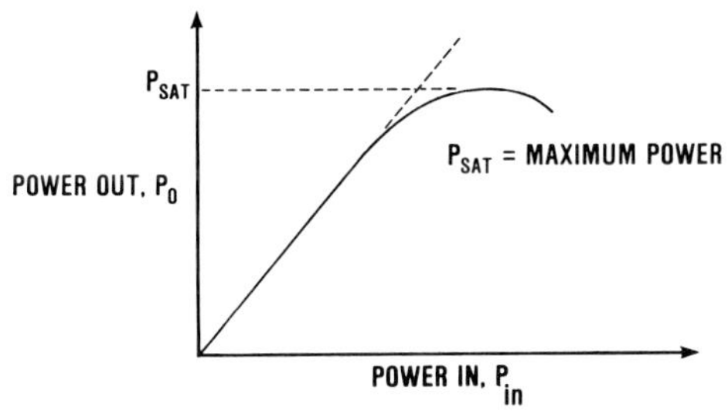

Fig. 3.74 TWT gain variation.

the output power and any smaller input signal causing a smaller output will reduce the efficiency of the tube. A typical power out vs power in curve is shown in Fig. 3.74. This figure shows that the small-signal gain (P_{in} well below that which saturates the traveling wave tube) is larger than the gain at saturation.

Operation in a saturated condition produces some undesired results. For example, if two signals are amplified, they "mix" within the traveling wave tube and more than two signals come out. These additional signals are generated at the expense of the two original signals because the tube power, listed in Table 3.10, is not merely split between the two original signals. As shown in Fig. 3.74, a related effect is that, as the input power increases beyond the value that saturates the tube, the output power of the tube drops. The measured output power is at the frequency of this input. Some power is now generated at twice the input frequency, or at the second harmonic.

In addition to the traveling wave tubes described above, there are various types of "dual-mode" traveling wave tubes. A dual-mode tube can be operated in either a CW or pulsed mode. Of course, a CW tube can always be pulsed, but the power out remains unchanged. In a dual-mode tube, the level of pulsed power is approximately 10 times the CW level. Mendel states that it is difficult to design an electron gun and interaction structure capable of operating efficiently at the two different power levels. Nonetheless, Fig. 3.75 shows the performance of a dual-mode tube that almost achieves the 10:1 pulse-up ratio [60].

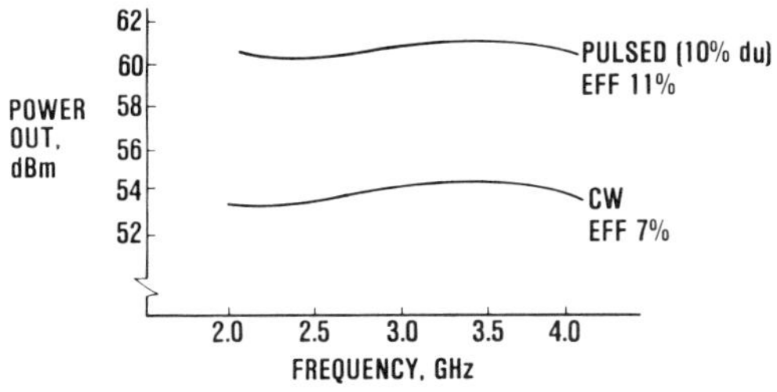

Fig. 3.75 Dual-mode tube performance.

Table 3.12 Dual-Mode Tube Performance

Operating frequency, GHz	Power output, kW	Duty cycle, %	Saturated gain, dB	Eff, %
2.0–4.0	1.0p/0.2 CW	5	30	7/8.5
4.0–8.0	1.0p/0.1 CW	5	40/23	9/3.5
8.0–16.0	1.0p/0.1 CW	5	40/20	7.6/3

The performance of other available dual-mode tubes is shown in Table 3.12. All of these tubes weigh 3.73 kg and use periodic permanent magnets.

Another approach to constructing a dual-mode tube is to build a transparent pulse-up tube [60], as shown in Fig. 3.76. In the CW mode, the beam of the transparent tube is shut off and the tube introduces approximately 1 dB of loss in the RF path. During pulsed operation, the electron beam for the output tube is turned on and the tube now amplifies the signal passing through it.

M-Type Tubes. In an M-type tube, the region of interaction contains perpendicular electric and magnetic fields. These fields are also perpendicular to the direction of motion of the electron beam. Some popular M-type tubes, such as the magnetron and continuous cathode crossed-field amplifier (CFA), contain a continuous rotating cloud of electrons instead of an electron gun or beam. This text does not consider such tubes because they are relatively narrowband, 15–25% [61, 62]. It concentrates instead on the injected-beam crossed-field amplifier that can be used in ECM systems as both an amplifier and an oscillator.

A cross section of an injected beam M-type tube is shown in Fig. 3.77 [64]. Although a linear construction is shown, the sole and delay lines can be wrapped around to form a cylindrically shaped structure shown in Fig. 3.78. The cathode and anode in these structures inject an electron beam of rectangular cross section into the interaction region.

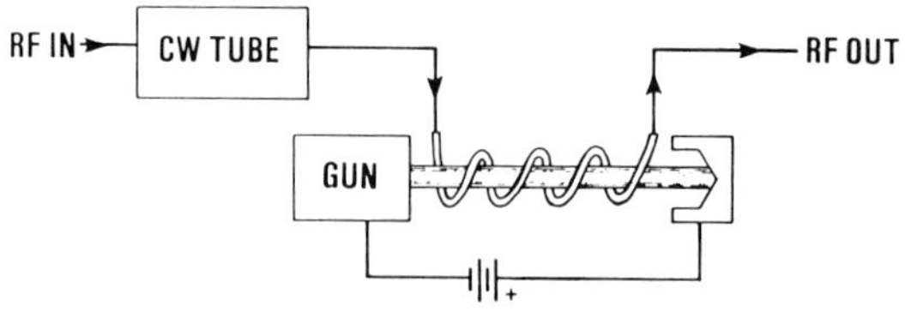

Fig. 3.76 Transparent dual-mode design.

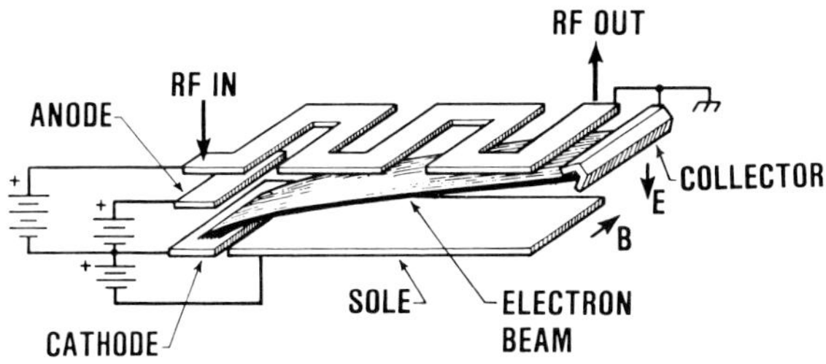

Fig. 3.77 Injected-beam crossed field amplifier.

Both the electric and magnetic fields exist in the interaction region. The magnetic field forces the electrons toward the sole, or bottom, plate and the electric field attracts the electrons toward the slow wave structure. The meander or zigzag line is the typical slow wave structure because it provides large contact areas to a cooling structure. This is important since the slow-wave structure intercepts an important fraction of the current from the cathode. The current not intercepted is finally absorbed by the collector.

Such injected-beam CFA tubes have displayed a continuous power of 1–3 kW over selected bandwidths within 1–7 GHz. The instantaneous bandwidth is only 67% of an octave, but a 2:1 (or octave) bandwidth can be achieved by varying the voltage between the cathode and the sole. Three different sole voltages will provide octave coverage [12, p. 390]. The CFA gain is only 15–20 dB, much less than the gain available in a traveling wave tube, and it is defined differently. CFA gain is the output power divided by the input power at a fixed level of RF input power. If the input signal decreases, the crossed-field amplifier can go into oscillation at the high levels of current necessary for maximum gain [62]. Thus, as the input signal decreases, so must the beam current decrease under the control of a grid element, which results in a region of "constant" small-signal gain. If these are CFA shortcomings, its RF conversion efficiency of 40–60% compensates for them.

If the delay line in the M-type tube is grounded at the RF output and the tube gain is increased, the device becomes an oscillator and generates RF energy. This type of tube is called a backward-wave oscillator (BWO) or carcinotron. The output frequency can be changed rapidly by varying the

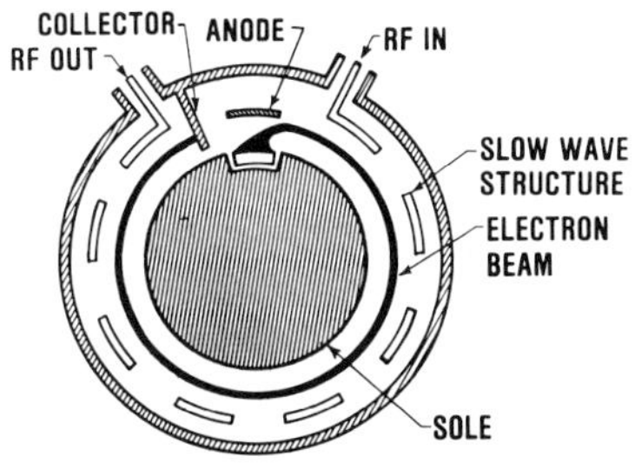

Fig. 3.78 IBCFA circular format.

cathode-to-sole voltage. Typical characteristics from an old source [63, p. 9-47] are CW power generation of several hundred watts at 1.0 GHz and 100 W at 10 GHz, a tuning range (frequency deviation) of 30% and a tuning rate of 30 MHz at 1 GHz varying to 300 MHz at 10 GHz [2, p. 27-26; 63, p. 9-55].

ECM Antennas

Chapter 2 discussed the basic antenna elements most widely used in ECM systems, such as the spiral, dipole, and horn, and provided examples of their gain, radiation pattern, and useful frequency range. This section addresses examples of these elements designed to achieve increased RF bandwidth and their application to airborne ECM systems, especially array antennas.

Basic Antennas

Spiral Antennas. Spiral antennas are very broadband, low-gain, and usually low-power radiators used in radar warning and surveillance receivers. (The specifications for a current spiral antenna are shown in Table 3.13.) The term voltage standing wave ratio (VSWR) describes how well the antenna couples to its feed line. A VSWR of 1 represents perfect coupling. The gain tabulated is relative to an isotropic source radiating circular polarization.

Dipole (Blade) Antennas. A blade antenna is comparable to a very thick dipole (actually one-half of a dipole or a monopole). Because of this thickness, it functions over a greater bandwidth than a thin dipole and it must also be aerodynamically shaped. (Such a structure is shown in Fig. 3.79.) A blade antenna with a voltage standing wave ratio of 2 over a 6:1 bandwidth has been mentioned in the literature [64]. This VSWR means

Table 3.13 Spiral Specifications [104]

Frequency coverage	0.5–18 GHz
VSWR	<2.5
Gain	-15 dB$_{ic}$ at 0.5 GHz
	-3 dB$_{ic}$ at 2.0
	$+1$ dB$_{ic}$ 3–18 GHz
Beamwidth	110 deg, 0.5–2 GHz
	75 deg, 2–18 GHz
Polarization	Circular

*dB$_{ic}$ is the gain (expressed in decibels) relative to an antenna radiating equally in all directions an electromagnetic wave that is circularly polarized and of the same sense.

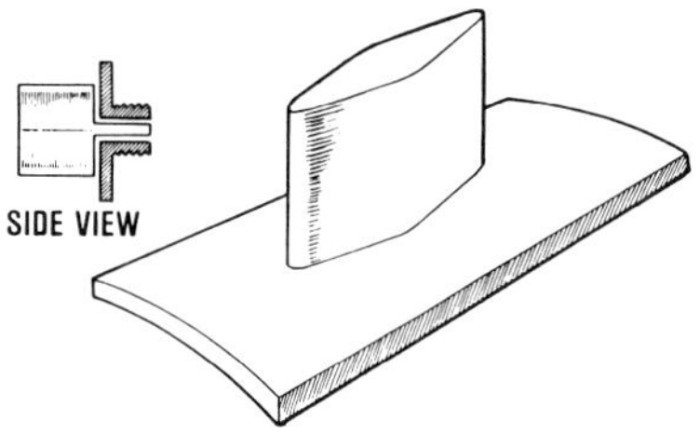

Fig. 3.79 Blade antenna.

that at worst, one-ninth of the power incident on the antenna is reflected back to the ECM tube. This type of antenna is used by jamming transmitters at frequencies less than 2 GHz.

Horn Antennas. Above 2 GHz, the typical ECM transmit antenna is the horn, which is a very broadband device that operates as shown in Fig. 3.80. The incident signal at the throat of the horn simply expands outward, fills the larger opening (aperture), and finally radiates into space. The biggest difficulty is making the desired incident signal propagate down the feeding waveguide over the octave bandwidth of a typical ECM system. (A system whose highest operating frequency is twice its lowest operating frequency is said to operate over an octave.) This problem is solved by the use of a "ridged" waveguide that has a bandwidth of approximately 3:1. Larger bandwidths are possible, but at power levels much reduced from those of a normal waveguide. The shape of the radiation pattern does not remain constant as the frequency changes, but varies from a broad pattern (e.g., 120 deg beamwidth) to a narrower pattern as the frequency increases.

It can be shown that, for a maximum gain from the available horn aperture, the flare angle of the horn should be approximately 28 deg [65]. This gradual flare takes up too much room in an ECM system of limited volume. Increasing the flare angle results in a cylindrical wave incident on the horn aperture. This cylindrical wave results in poor radiation patterns (recall the plane wave surface generated in the paraboloid in Chapter 2) and must be modified to a plane wave. A dielectric lens is used for this purpose. Operating on the same principles as an optical lens, the incident energy from the horn throat is slowed once it enters the lens. Since the energy reaches the center of

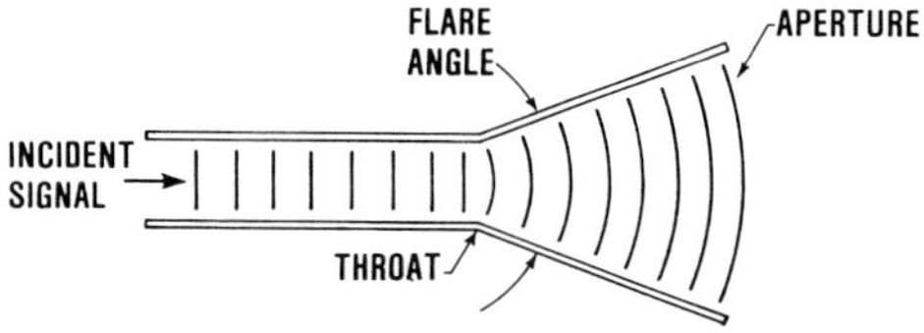

Fig. 3.80 Horn cross section.

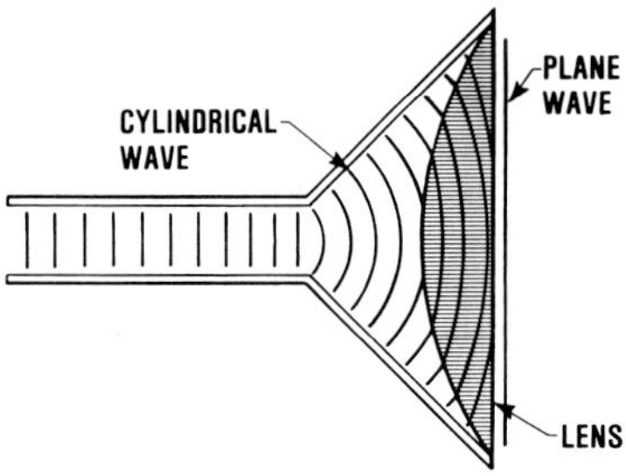

Fig. 3.81 Horn antenna with dielectric lens.

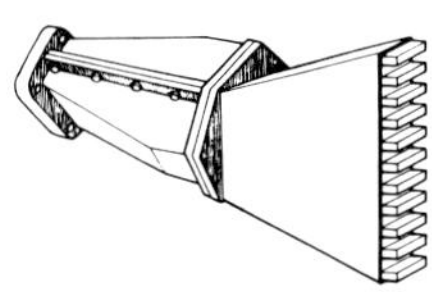

Fig. 3.82 Dielectric loaded horn with corrugations.

the horn first, the lens must be thickest at the center (see Fig. 3.81). The lens, however, will reflect energy back down the waveguide because it represents a physical change in the propagation medium. To minimize these reflections, the surface of the lens is corrugated as shown in Fig. 3.82. A similar horn is described in the literature as radiating efficiently over an octave of bandwidth.

Array Antennas. The constant requirement for greater effective radiated power, coupled with the finite aircraft power available, naturally forces the designer of the ECM system to increase the gain of the transmit antenna. Since an increase in antenna gain is coupled with a reduction in beamwidth (coverage), methods for determining the direction to the threat must be included. Finally, the requirement for antenna coverage around the entire aircraft dictates use of a rapidly steerable antenna to counter multiple threats. How can the requirements of the ECM system be potentially satisfied?

Array antennas may satisfy these requirements. An array antenna consists of many (normally identical) elements, each consisting of a horn, dipole, spiral, etc. The signals emitted by the elements are electromagnetic waves, which are characterized in space by an electric field that changes direction every half wavelength. By overlaying the electric fields in a constructive manner, i.e, by lining up the positive excursions of the electric field, one can increase the power radiated in a particular direction. This alignment can be accomplished either by time delay or element phasing (see Figs. 3.83 and 3.84).

In Fig. 3.84, the beam can be rapidly shifted by switching the power source between the available inputs of the beam-forming network. Formation of all beams at once provides multiple simultaneous beams if multiple power

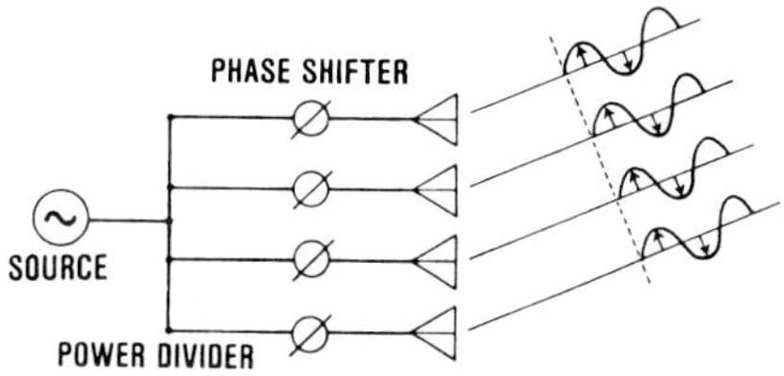

Fig. 3.83 Phase-shifted array.

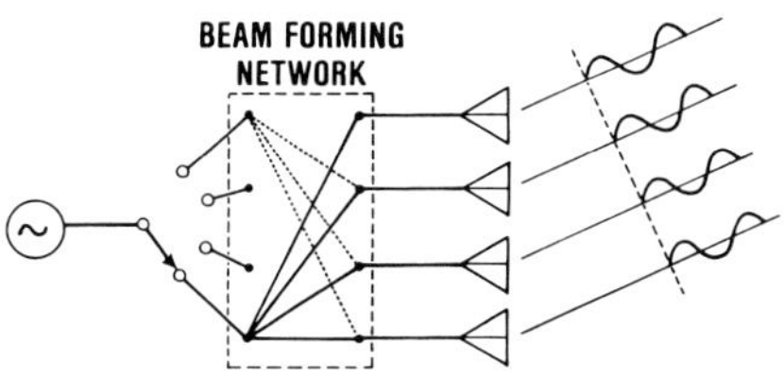

Fig. 3.84 Time delay array.

sources are available. The beam-forming network does introduce some loss, which is circumvented by placing the power sources between the beam former and antenna elements (see Fig. 3.85). Such an array can provide ERP levels from a few kilowatts to tens of megawatts at 100% duty cycle. This level can be achieved over a 3:1 bandwidth within a 120 deg field of view [66]. Of course, multiple simultaneous signals may not be efficiently amplified if a traveling wave tube is the final amplifier of the signal.

The phased array approach is capable of similar performance levels [67]. As described in Ref. 67, a system capable of 800 kW ERP uses a 1660 W tube coupled with a 27 dB gain antenna consisting of a planar array. In contrast to a linear or one-dimensional array, a planar array is two-dimensional. The element phase shifters are made of a ferrite material and introduce less than 1 dB of loss. In the system described, multiple high-power traveling wave tubes provide the input signal. If multiple simultaneous signals are required, the array and the high-power sources are divided, thereby creating essentially two (or more) separate transmit chains.

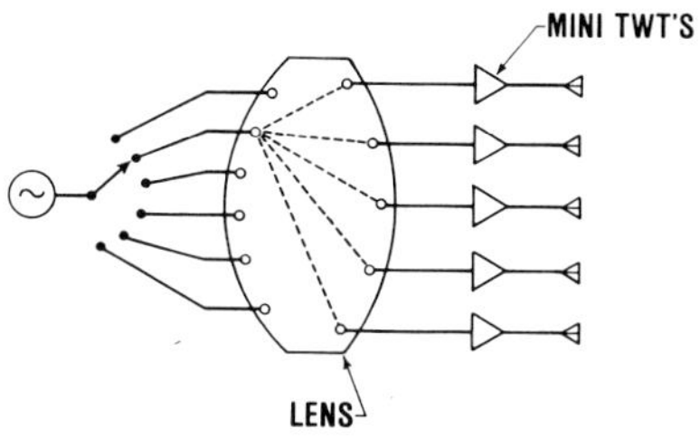

Fig. 3.85 Lens fed multibeam array.

Noise ECM Systems

There are basically two ways to construct a noise jammer: direct noise amplification (DINA) and frequency modulation by sine wave plus noise (FM/ S + N). A DINA jammer can be a combination of one of the previously described noise sources and an amplifier. Since such a source is not very frequency selective, an alternative is shown in Fig. 3.86. Now a noise source with a variable bandwidth can be tuned to the center of a radar band. However, the necessary fluctuation of noise amplitude results in inefficient operation.

Figure 3.87 shows a more efficient noise jammer. In this instance, an M-type BWO is varied in frequency sinusoidally by varying its sole voltage. The center frequency is set by a constant voltage also applied to the sole. Addition of the video noise creates a uniform density of noise power over the frequency range being jammed.

Figure 3.88 should help to explain the operation of this system. A video pulse is generated every time the jammer is scanned through the radar band-pass. If the oscillator can be periodically scanned back and forth rapidly enough, a continuous output will be observed at many radar frequencies. At some frequencies, successive jammer pulses cancel each other. The addition

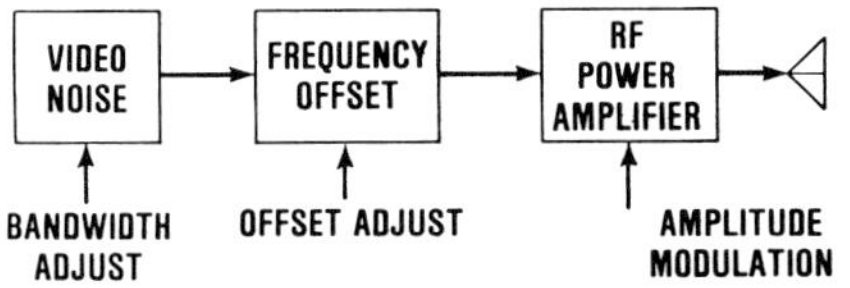

Fig. 3.86 Direct noise amplification (DINA).

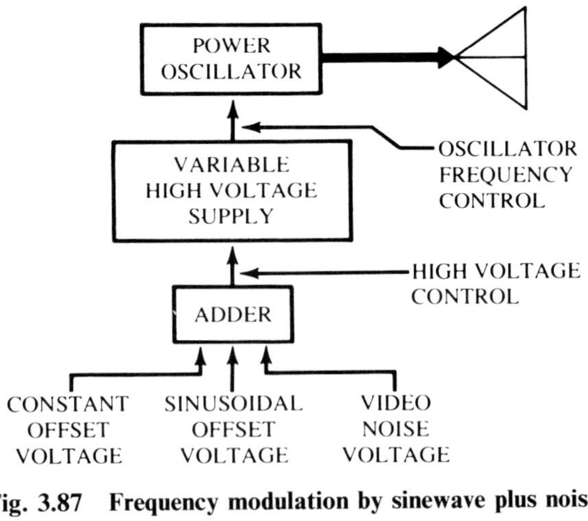

Fig. 3.87 Frequency modulation by sinewave plus noise.

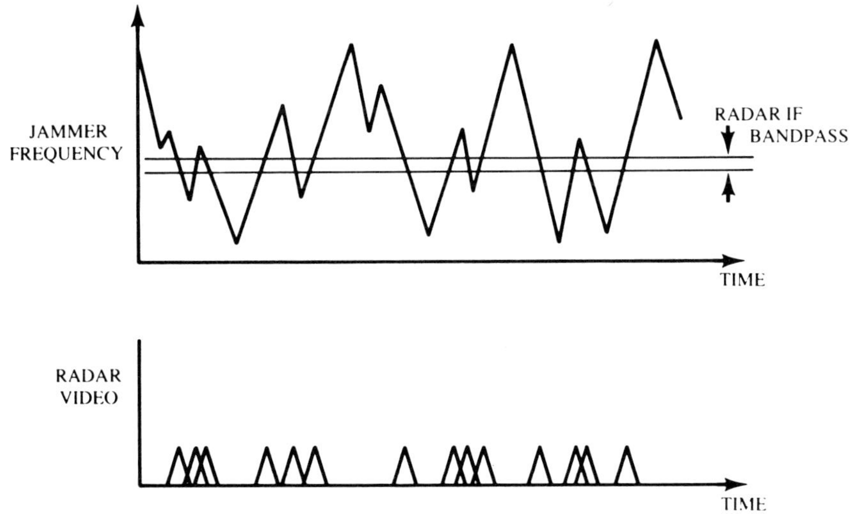

Fig. 3.88 FM/S + N effect on radar output.

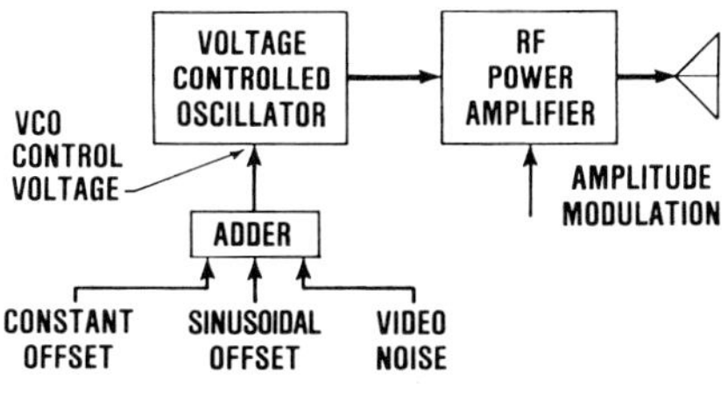

Fig. 3.89 Master oscillator/power amplifier.

of the noise signal prevents this cancellation and a noisy output is present at all frequencies.

Frequency modulation of a high-power oscillator requires variation of a control voltage in the kilovolt range. Since this control requires a specialized modulator designed to match the characteristics of the tube, it is not easily changed. The need for rapid reconfiguration of an ECM system has led to the current dominance of the master oscillator/power amplifier (MOPA) approach shown in Fig. 3.89 [38]. Use of a VCO in this system generates a low-power FM/S + N signal. This constant-amplitude signal is then amplified by the TWT final amplifier at its most efficient level.

Problems

3.1 How are ECM techniques classified herein?

(a) Search, Track (b) Active, Passive (c) Deception, Masking (d) None of these

3.2 Which tracking radars reveal their scan frequencies?

(a) Monopulse (b) CONSCAN (c) LORO (d) COSRO (e) CONSCAN with SWC

3.3 For the jammer and radar in Tables 3.2 and 3.3, respectively, by how much does the jammer power exceed the receiver noise power (assume − 139 dBW) when the jammer operates at 50 n.mi. and is self screening?

(a) 45 dB (b) 60 dB (c) 20 dB

3.4 Continuing Problem 3 above, what would be the J/S ratio if the polarization of the radar remained horizontal but the polarization of the jammer were 30 deg from the vertical? Assume $\sigma_T = 2.5\ \text{m}^2$.

(a) 12.3 dB (b) 15.3 dB (c) 9 dB

3.5 Your target is protected by one airborne interceptor, controlled by a ground-controlled intercept (GCI) radar site. The site contains both an early warning (EW) radar and a height finding (HF) radar with parameters as follows:

EW radar	HF radar
$P_R = 2.75$ MW	$P_R = 1.38$ MW
$G_R = 30$ dB	$G_R = 33$ dB
RF = 1–1.1 GHz	RF = 2.2–2.4 GHz
PRF = 400 Hz	PRF = 525 Hz
$B_R = 240$ kHz	$B_R = 1.2$ MHz
$\theta_{az} = 3$ deg	$\theta_{el} = 1$ deg
rpm = 6	Nodding rate = 10 deg/s
Noise figure = 3 dB	$\tau = 1\ \mu$s
Elevation 200 ft	Elevation 200 ft
$(S/N)_{min} = 13$ dB	

Each radar can be quickly tuned over its entire operating range and both are noncoherent. The interceptor requires 5 min to reach the GCI site.

You have one penetrator, cross section $10\ \text{m}^2$, and one standoff jammer capable of carrying two noise jammers. These two jammers could be any combination of the following two systems:

Jammer 1	Jammer 2
$P_J = 2.0$ kW	$P_J = 1.0$ kW
$G_J = 20$ dB	$G_J = 20$ dB
RF = 1–2 GHz	RF = 2–4 GHz
$B_J = 3$–200 MHz	$B_J = 3$–200 MHz

Your penetrator cannot fly below 1000 ft or faster than 450 knots.

(a) Without a jammer, how close can you fly before detection? What is your time to target? Hint: Assume screening from a smooth Earth (not flat) and detection of target when visible.

(b) If your SOJ is at 100 n.mi., above what altitude must it fly to be effective? Assume noncancellation of the grazing ray.

(c) If a 3 dB J/S ratio is required to screen the penetrator against either radar, to what range will jammer 1 screen the target?

(d) At what range will jammer 2 screen the target? Assume CW barrage jamming in (c) and (d).

(e) What penetration profile would you use?

(f) Would you jam both radars? Support your answer.

Questions 3.6–3.10 concern a conical scan tracking radar with the following parameters:

Transmitter power	50 kW
Antenna gain, maximum	34 dB
Transmitter frequency	5.6 GHz
Pencil beamwidth	2.0 deg
Pulse width	0.4 μs
Pulse repetition frequency	1800 Hz
Polarization	Vertical
Noise figure	6 dB
Scan axis gain	31 dB

3.6 The radar acquires a target in range and elevation by nodding its antenna 20 deg in elevation at 14 scans/s. At what range will it detect a 1 m^2 target if a 13 dB signal-to-noise ratio is required? Assume incoherent integration of all pulses within the 3 dB beamwidth (one way) and zero flyback time on the elevation scan.

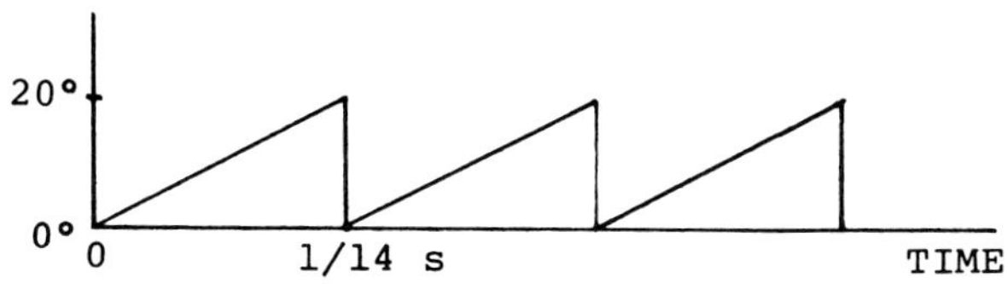

3.7 If transition to the tracking mode requires a 20 dB S/N ratio measured at the IF frequency (i.e., on each pulse), at what range can such a transition occur against the 1 m^2 target?

3.8 Given a 1 kW noise jammer of 20 MHz bandwidth and a 6 dB gain antenna, plot the resultant J/S ratio over the interval 1–20 n.mi. Assume that the radar is in the track mode and that 30 pulses are incoherently integrated on the operator's display. Target is still 1 m^2.

3.9 What would be the J/S ratio at 10 n.mi. if the jammer polarization were 45 deg from the vertical?

3.10 Retain the vertically polarized jammer but increase the cross section to that shown below. Then, realizing that the jammer antenna gain varies at $G(\theta) = G_{max} \cos^2 \theta$, where θ is measured from the nose of the aircraft, plot the range vs angle about the nose when a 10 dB J/S ratio is achieved.

Cross section			
θ, deg	σ_T, m^2	θ, deg	σ_T, m^2
≥ 0	10	60	30
≥ 15	8	75	40
30	6	90	
45	5		

3.11 Two aircraft, flying abreast and separated by a distance L, penetrate an airspace protected by a weapon system using semiactive homing missiles. It is known that a missile fired at the penetrators will approach from directly ahead at a relative closing speed of 2000 m/s. Other missile characteristics are a normal acceleration of 200 m/s^2 and an antenna beamwidth of 10 deg.

(a) At what range will the missile resolve the individual aircraft? Assume the missile initially tracks a point midway between the aircraft.

(b) Assuming the missile accelerates normally at its maximum rate toward one of the aircraft as soon as the aircraft are resolved, what is the approximate miss distance?

(c) Assume a few values for the aircraft separation L and estimate the value that will maximize the miss distance.

4 ELECTRONIC COUNTER-COUNTERMEASURES (ECCM)

One can state that electronic counter-countermeasures are actions taken to insure the use of the EM spectrum by friendly forces *in the presence of man-made interference*. The italicized phrase is significant in the sense that it eliminates from the ECCM list many actions that are necessary for radar performance in natural interference, such as ground clutter. However, many techniques for the reduction of natural interference have a significant impact on man-made interference. For example, monopulse techniques eliminate the angle tracking errors caused by natural fluctuations in target amplitude and experienced by CONSCAN radars, but they also nullify many forms of amplitude-modulated jamming.

Reducing the number of techniques to be classified as electronic counter-countermeasures is the first step in organizing a list of such techniques. It does not seem desirable to list every technique separately by name, because many of these techniques are comparable, if not identical. But separation of techniques into such categories as antenna, transmitter, and receiver electronic counter-countermeasures is desirable [68]. After all, the radar is a system composed of the principal elements discussed in Chapter 2. This classification then identifies the electronic counter-countermeasure with the component modified to implement the electronic counter-countermeasure. However, the objective of the classification procedure described in this chapter is to concentrate on what is being done as opposed to which radar element is tasked to accomplish it. The categories are listed in Table 4.1 and examples in each category in Table 4.2.

The first category consists of all filtering methods by which a known desired signal is extracted from a noisy background. The characteristics of this noise background are also assumed to be known. The most common example is the matched filter receiver discussed in Chapter 2. That receiver was designed to maximize the signal-to-noise ratio of the receiver output in the presence of receiver noise having a uniform power density in frequency. The structure of this optimum receiver changed as the number of pulses processed or target fluctuations changed. A matched filter can also be constructed when the interference is not uniform in frequency. A radar with MTI employs just such a filter.

The antenna can be considered as a spatial filter in the sense that it passes

Table 4.1 ECCM Categories

ECCM Categories
Correlation in known interference
Matched filter in frequency
Spatial filtering
Multiple measurements
Identification of unknown interference
Saturation elimination
Operator displays

signals from one direction more favorably than it does signals from another direction. If a uniform noise background is postulated at all positions in space, then the optimum antenna is the antenna with the maximum gain for the available aperture. Then, like the MTI filter, an antenna with a sidelobe canceller will be optimum if the noise background contains a dominant, localized noise source.

The second category of ECCM techniques consists of the actions necessary for the implementation of the first category. These actions consist of maintaining linear receiver operation (if necessary) and determination of the noise field characteristics necessary to select the proper correlation technique. Admittedly, this division is somewhat arbitrary; however, this breakout is used in the belief that it will enhance comprehension of the individual

Table 4.2 ECCM Techniques

Correlation in known interference	
Matched filtering	Spatial filtering
Simple IF filter	Uniform aperture
Combination filter	Low sidelobe antenna
MTI	Sidelobe cancellation (SLC)
Frequency diversity	Polarization
Phase coding	Home-on-jam (HOJ)
Spread spectrum	Sidelobe blanking (SLB)
Leading-edge tracking	Beam deletion
COSRO/LORO	
Staggered/jittered/sliding PRF	
Diplex operation	Multiple measurements
Pulse width discrimination	Monopulse
Detected pulse interference	Scan with compensation (SWC)
Log fast-time constant (LOG-FTC)	Guard channel
Identification of unknown interference	
Saturation elimination	Operator displays
Gain control	Spectrum analyzer
Logarithmic amplifier	Additional displays
Dicke-fix	

techniques to a somewhat greater extent than a mere listing of electronic counter-countermeasures.

It may appear that this chapter on ECCM techniques commits an injustice to the subject. Certainly, much material can be incorporated into this dicussion (see, for example, Ref. 68). However, it is the author's contention that ECCM has been discussed in Chapter 2 and, furthermore, that ECCM is an inherent part of any radar design considering noncooperative targets in other than a benign background composed of simple receiver noise. The radar design process begins with a description of the target and the interfering signal (usually in the form of mathematical models). The simplest interference is receiver noise. These target and interference models are inherent in many of the receivers described in Chapter 2. A receiver covered in Chapter 3 uses leading edge tracking. This receiver is also designed to estimate signal parameters (i.e., time delay) in the presence of simple receiver noise. The signal, however, was "assumed" in Chapter 3 to have a form not expected from a simple target and to have an initial portion due to backscatter from the target, followed by a large jammer signal. If this signal description is wrong, the receiver may not work at all. The incorporation of ECCM techniques into radar design is accomplished by assuming (measuring) the form taken by both the desired signal and the undesired interference and then constructing (using) a receiver (radar processor) that optimizes the extraction of the desired signal parameters in the presence of the noise. Thus, just as Chapter 2 was divided into sections entitled detection in noise and detection in clutter, another section could have been entitled detection in the presence of jamming.

Any description of specific techniques must be preceded by a few "postulates" of electronic counter-countermeasures. First, it is best to win the power battle. Any radar is fighting a $1/R^4$ power loss and more radiated power "won't hurt." Second, as much variety as possible is desirable in using the EM spectrum and in signal processing. Standardization of defensive forces may save money, but a penetrating force, especially an airborne force, has much difficulty in carrying all it needs to the battle. A variety of defensive forces requires a variety of ECM equipment and makes any airborne attack more difficult. Finally, there is strength in numbers in electronic warfare as in any other form of conflict. Individual radar platforms should have a variety of radar options—and the more *interconnected* the platforms the better. Interconnected is a key word. Without control of emissions through some command structure, too many radars may jam each other and will be only as effective as a mob in accomplishing their objectives.

4.1 SPATIAL FILTERING

As the name of the technique implies, an attempt is made to enhance observation of a desired signal over an undesired signal through the spatially directed radar antenna. This section describes some methods for accomplishing spatial filtering.

Low Sidelobe Antenna

This counter-countermeasure consists of taking the extra steps necessary to minimize antenna sidelobes (see Fig. 4.1) through proper feed design, selection of aperture illumination [13], proper machining, and location of the antenna to minimize scattering from nearby objects. The advantage gained through lower sidelobes is apparent in a review of Chapter 3.

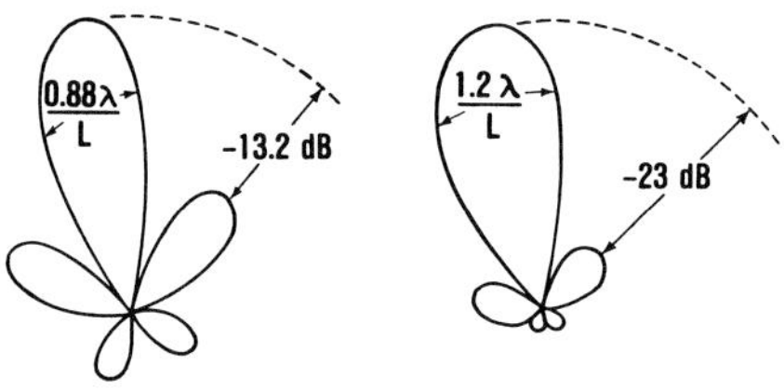

Fig. 4.1 Antenna sidelobe level variation with aperture illumination.

Sidelobe Canceller (SLC)

As opposed to creating uniformly low sidelobes, one can implement localized sidelobe control. This technique amounts to creating antenna pattern nulls, or near nulls, in the direction of undesired signals as shown in Fig. 4.2. Implementation can be accomplished by selecting the weights applied to individual elements of a phased array or by combining the output of the original antenna with the output of an added omnidirectional antenna. The number of jammers nulled is one less than the number of array elements or is equal to the number of additional omni-antennas. Since sidelobe cancellation requires some time for weight adjustment, this technique is most applicable to CW noise or high duty cycle jammers.

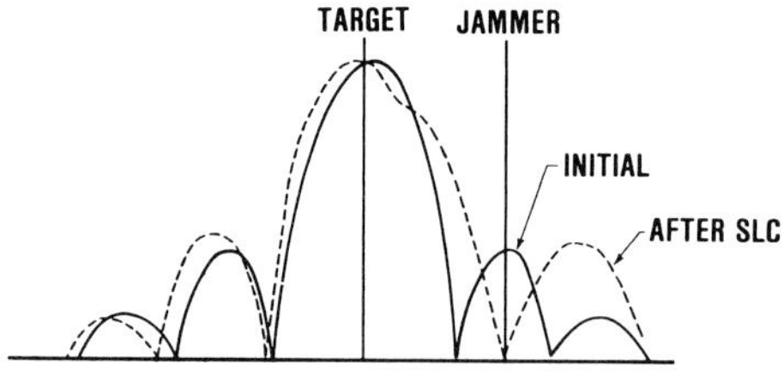

Fig. 4.2 SLC operation.

Sidelobe Blanking (SLB)

This technique applies to pulsed interference and consists of blanking the radar output whenever the output of an auxiliary omnidirectional antenna is

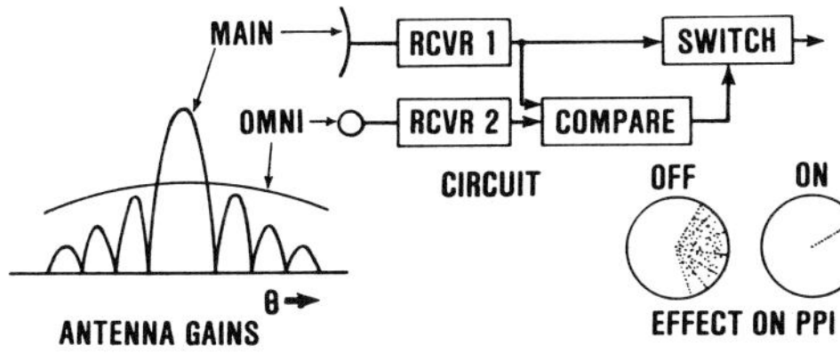

Fig. 4.3 Sidelobe blanking.

greater than the output of the main antenna. As shown in Fig. 4.3, the gain of the omnidirectional channel is adjusted to produce a signal larger than the signal from the sidelobes of the main antenna. This technique not only prevents the display of pulse interference on radar display screens, but also eliminates from the screen any time-coincident target returns. If used to counter a high-power CW jammer, the radar screen will be blank except for a strobe pointing toward the jammer.

Beam Deletion

If the radar has multiple beams (typically elevation beams of an azimuthally scanning search radar) whose outputs are combined for operator viewing on a single display, it is possible to delete any beam that contains jamming. This, of course, eliminates target observation in that direction.

Polarization

As noted in Sec. 3.2, RF energy is typically polarized in some direction when it is transmitted. However, the RF energy initially transmitted from a radar with horizontal polarization will be depolarized to some extent upon reflection from a target; both vertical and horizontal polarizations are now available for detection. If a jammer uses only one polarization, the radar can discriminate against the jammer through the reception of a different polarization. The level of jammer rejection depends on the sophistication of the system, but something less than 30 dB seems achievable.

Home-on-Jam (HOJ)

As opposed to the techniques that attempt to remove jamming signals from the information display or processing channel, home-on-jam attempts to locate the direction from which the interference signal arrives through the angular resolution of the antenna. Home-on-jam is really a concept that is implemented in different ways. For example, the sidelobe blanking described above can be the mechanism used to locate the jammer's angle of arrival. Normally, home-on-jam implies the ability of a missile to guide itself toward a jamming source.

4.2 MATCHED FILTERING

The purpose of the following techniques is to maximize the signal-to-noise or signal-to-clutter ratio by filtering out the frequencies dominated by interference. They require the assignment of characteristics to both the interference and the signal and their effectiveness depends on the quality of the characterization.

Diplex Operation

This technique requires a single radar to simultaneously radiate two widely separated frequencies. If it is a pulsed radar, the emissions are almost time coincident. This capability forces a jammer to use a wider jamming bandwidth if it is a noise system or to handle multiple pulses if it is a repeater [2, p. 14-25].

Frequency Diversity

Frequency diversity is the capability to operate at a large number of frequencies and to change frequencies in a minimal time period. The definition of minimal is qualitative, since it depends on the ability of the jammer to retune. A typical emission history of a radar employing frequency diversity is shown in Fig. 4.4.

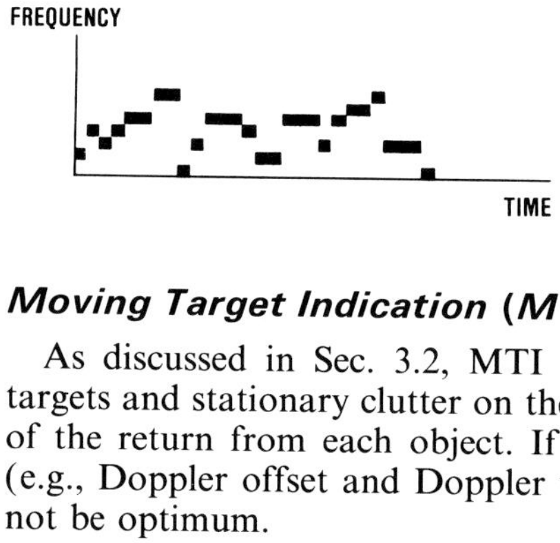

Fig. 4.4 Emission history of radar with frequency diversity.

Moving Target Indication (MTI)

As discussed in Sec. 3.2, MTI techniques distinguish between moving targets and stationary clutter on the basis of the difference in the frequency of the return from each object. If the assumed clutter model is incorrect (e.g., Doppler offset and Doppler width), then the MTI improvement will not be optimum.

Staggered Pulse-Repetition Frequency

A staggered pulse-repetition frequency (PRF) is used in some radars with moving target indication to eliminate blind speeds. As shown in Fig. 4.5, it consists of a PRF that varies from pulse to pulse. However, it has the secondary effect of *inhibiting* repeater jammers in the transponder mode from creating realistic false targets, since the jamming signals do not overlap

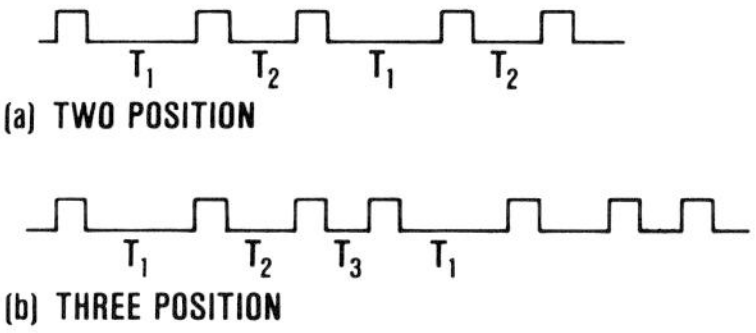

Fig. 4.5 Staggered PRF.

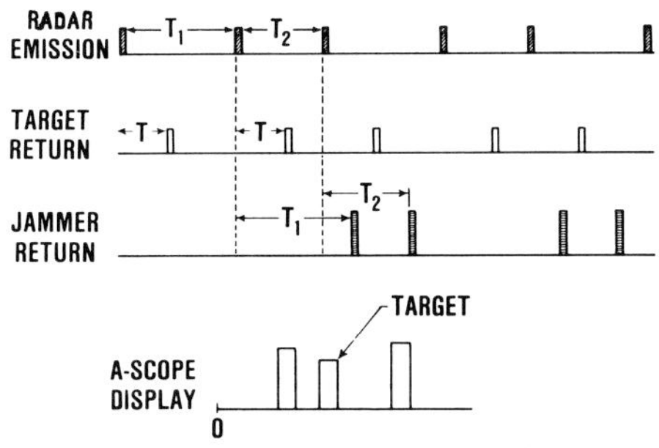

Fig. 4.6 Two-position stagger vs transponder.

on the radar display (see Fig. 4.6). Realistic false targets can be created only with a more complex jammer containing PRF sorters/trackers.

Jittered Pulse Repetition Frequency

This technique is used solely to prevent the creation of false targets at ranges less than the target range. The PRI is changed, much the same as in the staggered PRF; however, the changes are on the order of a few microseconds. These PRI changes from pulse to pulse are too small to enable the elimination of the target blind speeds by MTI techniques.

Sliding Pulse-Repetition Frequency

In this variant, the pulse-repetition interval varies from pulse to pulse in some preprogrammed manner over some continuous range. The variation could be linear, quadratric, or sinusoidal.

Phase Coding

Included in this category are waveforms of the type shown in Fig. 4.7. Basically, a phase modulation is applied to the transmitted waveform, which may be 10–50 μs long, or longer. In the receiver, the phase modulation on the received waveform is compensated for, enabling a range resolution much less than the overall pulse length, typically by a factor of 100. Of importance

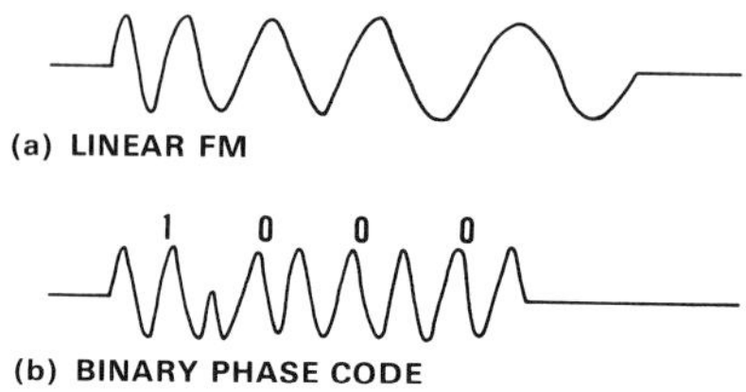

Fig. 4.7 Phase-coded pulses.

here is the fact that wide-band noise is attenuated by this same "compression" ratio, but that a repeated pulse passes unabated.

Example 4.1

Consider a phase-coded pulse composed of 20 segments of 0.5 μs each, for a total pulse length of 10 μs. Each segment (subpulse) occupies a frequency bandwidth of 2.4 MHz, but the actual coherent bandwidth B_R is 120 kHz, as if the radar pulse were a single 10 μs burst. If the jammer bandwidth B_J is 2.4 MHz, the jammer is attenuated by 13 dB.

Pulse Width Discrimination

This technique is applicable in the removal of interfering pulses that differ in duration from the duration of the radar emission. Such interfering pulses can come from pulsed jammers, extended clutter, or hard limited noise jammers. Figure 4.8 demonstrates the implementation of the concept [19, p. 117].

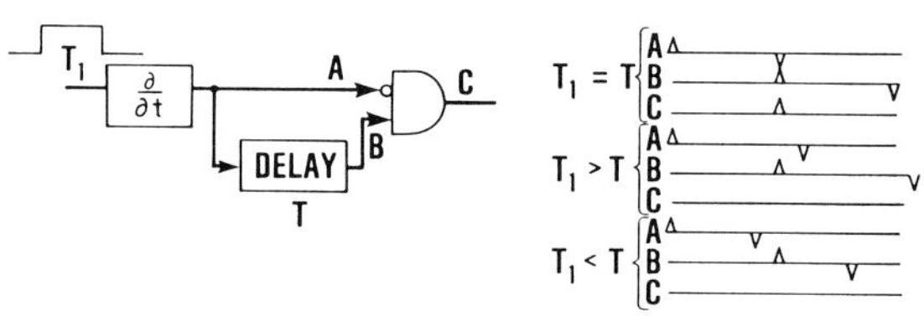

Fig. 4.8 Pulse width discrimination.

Detected Pulse Interference

This technique is comparable to pulse width discrimination with the comparison affected over a PRI interval. The assumption is that a target will not change its position in range too rapidly, and that random pulses will not retain a fixed time separation [15, p. 29-16]. See Fig. 4.9.

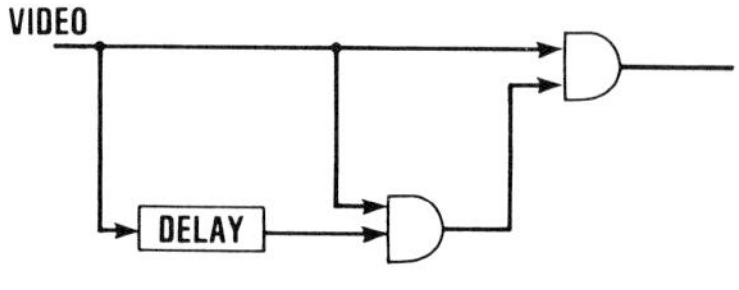

Fig. 4.9 Detected pulse interference.

Leading Edge Tracking

This counter-countermeasure is based on the assumption that any pulsed jamming return will be delayed from the target return by the transit time of the signal through the repeater jammer. Then, if the range gate tracks only the leading edge of the returned pulse, it should not be susceptible to range gate pull-off techniques.

Conical Scan on Receive Only

As discussed in Sec. 3.2, this technique applies to tracking radars. The procedure for the extraction of the angle error can be visualized as matched filtering. By denying the jammer access to the scan frequency, the radar effectively denies knowledge of the filter bandpass frequency.

4.3 MULTIPLE MEASUREMENTS

This is another area that can possibly be incorporated under a generalized matched filter concept. It consists of simultaneous multiple measurements to enhance the accuracy of target location. Monopulse and scan with compensation were addressed in Sec. 3.2 during discussion of tracking techniques. Their ECCM advantages are discussed in Sec. 3.2 as well. Multiple separated netted radar sites can locate jammers through triangulation. Such techniques must overcome the "ghosting" problem created by mass attacks (see Fig. 3.20).

4.4 SATURATION ELIMINATION

For any of the above processing techniques to function properly, the equipment must be operated within some specified dynamic range. Large variations in the signal levels in both the natural and the jamming environment require special circuits and devices to maintain signal levels within prescribed ranges. Three such items are gain control, logarithmic amplifier, and Dicke-fix.

Included under gain control are sensitivity time control (STC), RF attenuation, automatic gain control (AGC), instantaneous automatic gain control (IAGC), and others. All have the property of adjusting the receiver gain at a point in the amplifier chain to prevent saturation in following amplifiers. (See Figs. 2.34 and 3.33 for examples of the need for this gain adjustment.) If the automatic gain control does not reduce the IF gain at short target distances, the IF amplifier will be saturated and the scan modulation lost. With insufficient gain at large distances, the depth of envelope modulation is incorrect and the seeker response will be sluggish.

The logarithmic amplifier is a device whose output voltage is proportional to the logarithm of the input voltage. The result is that a large variation in the input signal will not saturate the output of the amplifier.

The Dicke-fix electronic counter-countermeasure consists of the addition

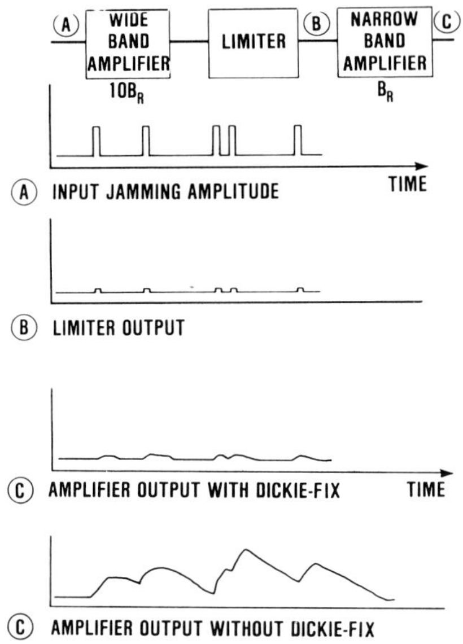

Fig. 4.10 Dicke-fix.

of a wide bandwidth amplifier and limiter ahead of the normal IF amplifier (see Fig. 4.10). One result of incorporating the limiter is that strong jamming signals are clipped before they saturate the receiver. Even though coincident targets are lost, faster receiver recovery allows the viewing of nearby targets [2, p. 14-27].

4.5 OPERATOR DISPLAYS

Included here are such items as A-scopes and panoramic receivers. Operators can use such displays to assess the current ECM environment and take appropriate action. For example, an A-scope can reveal range gate stealing and thus enable operators to track manually in range. An amplitude vs frequency display (spectrum analyzer) reveals clear frequencies to which the radar can be tuned.

Problems

4.1 Can a radar employ pulse-to-pulse RF agility and moving target indication at the same time?

4.2 Can a radar employ burst-to-burst RF agility and moving target indication at the same time?

4.3 Leading edge tracking is an ECCM to range gate deception. True or false?

4.4 A CW noise jammer will blank the screen of a radar employing which technique, SLB or SLC?

4.5 Frequency agility forces the jammer into barrage mode. True or false?

4.6 (**a**) Can Dicke-fix enable the viewing of targets coincident with jamming pulses?

(**b**) Can nearly time-coincident targets?

4.7 What electronic countermeasure may counter leading edge tracking?

(a) Cover Pulse (b) Multiple False Targets (c) Spot Noise?

5 ELECTRONIC SUPPORT MEASURES

"Electronic warfare support measures (ESM) is the division of electronic warfare involving action taken to search for, intercept, identify, and/or locate sources of radiated electromagnetic energy for the purpose of immediate threat recognition.

ESM provides information in a timely fashion and in forms that are readily usable by the activities it supports. The prime consumers of ESM information are the operations personnel engaged in other EW activities: that is, ECM and ECCM.

The importance of ESM as a segment of EW, and as a segment of the total spectrum of conflict, lies in the fact that it acts as the eyes and ears of the commander who engages in, or is engaged by, EW operations. ESM provides information essential to the success of operations in the air combat environment.

The treatment of ESM exclusively as an immediate-response, tactically oriented concept is relatively new. Previously, ESM was more broadly defined to encompass electromagnetic radiation collection efforts for subsequent intelligence exploitation and the support of military operations, as well as an immediate support in the combat environment. These collection efforts, which focus on longer-term intelligence objectives, although they may also support operations in progress, are now generally included in the area of signal intelligence (SIGINT).

The relationship between SIGINT and ESM can perhaps be made less confusing by considering some specific aspects of that relationship. SIGINT and ESM may, and generally do, use the same equipment technology. SIGINT provides intelligence data for a range of consumers and a variety of purposes, including planning, research and development, and operations, among others. While SIGINT may provide indirect or direct support to the tactical commander, ESM directly supports the tactical commander. ESM provides the tactical commander with a capability that can be totally integrated with other force capabilities. ESM assets are totally controlled by the theater commander, while SIGINT assets are responsive to higher-level tasking requirements." [1, p. 4-1].

Another distinction between SIGINT and ESM applications are the types of signals of interest. The SIGINT system operates in an unknown electromagnetic environment and attempts to record sufficient data for subsequent analysis. An ESM system is interested primarily in detecting (and possibly identifying) radar emissions with established signal parameters [36]. In performing this function, an ESM receiver could be employed during a tactical reconnaissance mission or as a radar warning receiver (RWR) on an attacking aircraft. This chapter explores only the RWR role and gives primary emphasis to defining some of the receivers used. (In the current threat environment, a highly desirable system is the so-called multisensor warning receiver (MWR) that also surveys the infrared and optical portions of the electromagnetic spectrum [78].)

5.1 TASKS OF RADAR WARNING RECEIVERS

A radar warning receiver must perform two basic tasks: measure parameters of radar signals incident on the aircraft and analyze the measurements. The parameters measured and the accuracy of measurement vary from system to system, based on aircraft mission requirements, technology, and costs. The parameters available for measurement include time of arrival, pulse width, frequency, amplitude, polarization, and angle of arrival. Once the incident signal parameters are measured, they are analyzed to determine the source of the emissions. The depth of analysis varies, but it usually involves the comparison of the currently measured parameters with threat emitter data collected by previous electronic intelligence (ELINT) or tactical reconnaissance missions. Such comparisons result in the identification of emitters by class (i.e., early warning, antiaircraft artillery) and by type (i.e., SA-5). Finally, the identified threat and its angle of arrival are displayed to the aircrew and possibly used to control ECM systems. Major RWR system characteristics are the range of radio frequencies observed or RF bandwidth, the angular coverage provided about the aircraft, and the number and variety of incident signals which can be measured.

5.2 MAJOR COMPONENTS OF RADAR WARNING RECEIVERS

The radar warning receiver consists of four major components as shown in Fig. 5.1: antenna, receiver, processor, and display. The antenna is often a cavity-backed spiral antenna (see Chapter 2) because the RWR must observe a large RF bandwidth. Usually, multiple spiral antennas, each pointing in a different direction, are used to increase the angular coverage about the air-

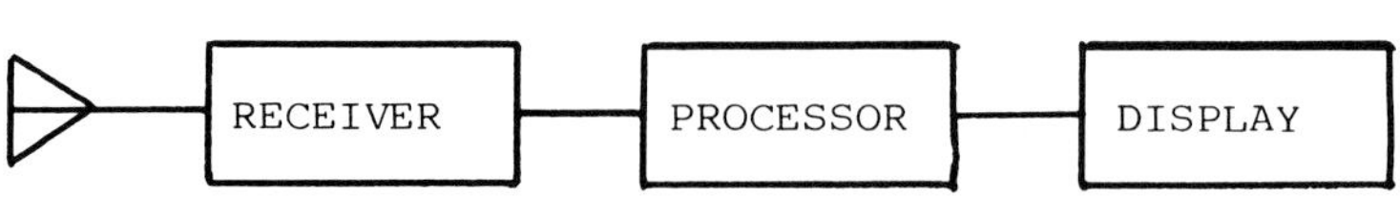

Fig. 5.1 Elements of radar warning receiver.

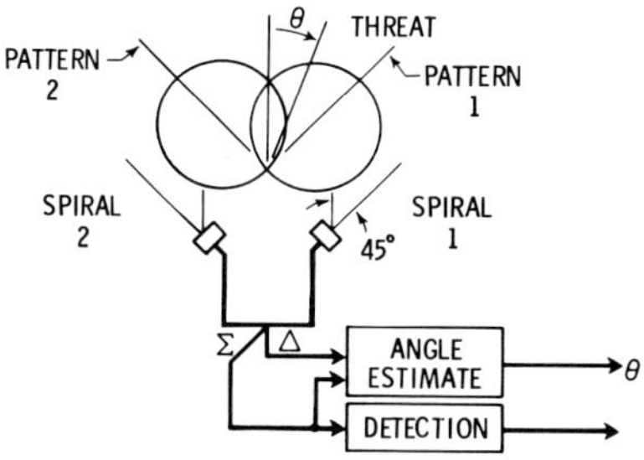

Fig. 5.2 A method of determining angle of arrival.

craft and, possibly, to permit accurate determination of the threat's angular location (see Fig. 5.2).

The signal from a threat will be received by the two antennas with different gains and will result in two signals of different amplitudes at the two antenna outputs. Similar to the amplitude-sensing monopulse radar, this difference in signal amplitude will yield the threat direction after sum-difference processing as shown in Fig. 5.2.

The antenna provides a signal for the receiver portion of the radar warning receiver. The functions of the receiver are to detect and measure some of the available signal parameters. Some important characteristics of the receiver, often called an intercept receiver, are found in Boyd et al. [2, pp. 9-4—9-7] and are summarized in part below.

1. Inability to use integration techniques. Because of the wide variation in the electrical characteristics of signals, the opportunities for aid to weak-signal detection via signal integration (as in a typical radar operation) are absent. There is generally a direct signal amplitude vs noise amplitude competition in the detection process such that practical receiver sensitivity is often much less, for example, than values usable in a radar receiver.
2. Complexity of signal characteristics. The basic intercept tasks become increasingly difficult with time because of trends in signal character. To avoid detection, or to avoid countermeasure action, modern weapons systems signals are frequently subjected to programmed or even random variations in character (in radio frequency, PRF, pulse width, etc.) during a transmission interval; the result is to greatly magnify intercept and identification problems. Thus, the value of precise measurement of radio frequency must be viewed with some qualification in the era of pulse-by-pulse frequency jump transmission, rapid tuning capability in transmitters, etc.
3. Wide frequency ranges to be monitored. In general, the intercept receiver must monitor a total radio frequency band substantially in excess of the frequency ranges of the individual signals of the electronic systems to be detected within this band.

4. Wide dynamic ranges encountered. The wide variations in the received signal level that must be anticipated are enormous. Because of the one-way transmission to the intercept receiver (vs the two-way action that may be involved in the operation of the signal emitting system), signal levels are apt to be high—high sensitivity sometimes is unnecessary (and undesirable because of the possible introduction of lower-level interfering signals). But in contrast, the intercept receiver may, in another circumstance, be faced with the task of intercepting a low-power transmission via radiation from minor lobes of a transmitting antenna and from a great distance—a situation arguing for the maximum sensitivity. An intercept receiver of general utility, then, must be prepared to operate over a very large dynamic range.
5. Presence of false signals. There is always the threat of decoy signals produced by an alert enemy to capture the attention of the intercept system. There is the threat that certain signal characteristics used to "fingerprint" signals (antenna scan rates, PRF, etc.) are being subtly modified by the enemy to lend confusion to the operation. While it is not within the province of the intercept receiver to make fundamental decisions in such matters, it is important that the intercept receiver not introduce further confusion by an inability to handle the received data without further distortion or modification.

The measured parameters of a signal are passed to a signal processor that actually identifies the emitter following development of the relationships between the large quantity of individually sensed pulses. For example, the data passed to the processor for each detected pulse may include angle of arrival, frequency, amplitude, pulse width, and time of arrival. These individual data items may be sorted within the processor, first by angle of arrival and then by frequency, to provide a sequence of signal descriptors displaying the variation in an individual radar's pulse-repetition interval and amplitude. A more likely result would be a sequence of data that must be further sorted by pulse-repetition interval. In any event, the intent is to determine the characteristics of RF signals incident on the aircraft. The measured characteristics are matched against previously determined threat data to obtain a "best" fit and to identify the emitter. (This sorting process is shown in Fig. 5.3.)

The processor provides its findings either to automated controllers of ECM resources or to displays for aircrew observation. Such displays range from a few panel lights indicating the presence and status of a threat to computer-controlled video displays. (A typical display is shown in Fig. 5.4.)

5.3 TYPES OF RECEIVERS

Many approaches are available for implementing the receiver portion of a radar warning receiver, each of which has its own unique advantages and disadvantages. These advantages and disadvantages must be weighed with the

Original Data Stream

AOA	RF	AMP	PW	TOA
10	2	1	1	1
10	1	6	6	1.1
90	3	3	2	1.5
10	2	1.5	1	2
90	3.1	3	2	2.5
10	2	1	1	3
90	3	3	2	3.5
10	2	0.5	1	4
10	1	6	6	4.2
90	3.1	3	3	4.5

After Sorting

by AOA of 10°					and by RF, 2 GHz				
AOA	RF	AMP	PW	TOA	AOA	RF	AMP	PW	TOA
10	2	1	1	1	10	2	1	1	1
10	1	6	6	1.1					
10	2	1.5	1	2	10	2	1.5	1	2
10	2	1	1	3	10	2	1	1	3
10	2	.5	1	4	10	2	.5	1	4
10	1	6	6	4.2					

AOA = Angle of Arrival, in degrees
AMP = Amplitude
TOA = Time of Arrival
PW = Pulse width

Fig. 5.3 A simple example of signal sorting.

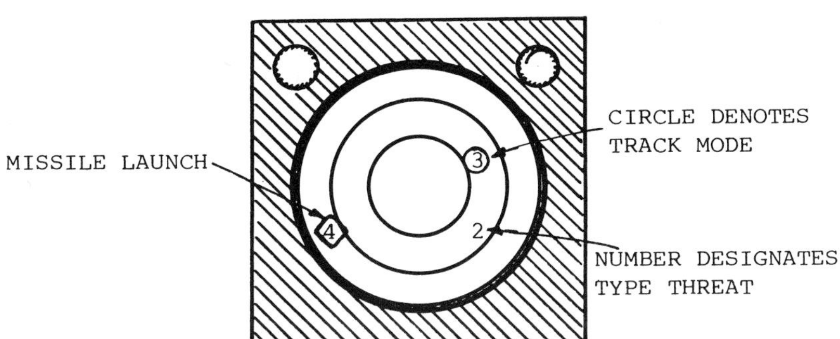

Fig. 5.4 Warning receiver display.

RWR requirements in mind before the "best" approach is selected in any particular RWR design. It is not the purpose of this discussion to provide sufficient information for such a selection, but only to describe each approach and to compare various receivers. The following sources provide more detail and additional references for the interested reader: Refs. 2 (Chapter 9), 12 (p. 321), 36, 48, and 89.

Superheterodyne Receiver

Described in Chapter 2 as the basic radar receiver, the superheterodyne receiver consists of the elements shown in Fig. 5.5. The receiver first amplifies an RF signal and then "mixes" it in the converter with a sinusoidal signal generated in the local oscillator. From the signals available at the converter output, the IF amplifier selects only signals at a frequency of f_{IF}. Thus, only RF signals at a frequency of $f_{LO} \pm f_{IF} = f_{RF}$ plus or minus the half-bandwidth of the IF amplifier B_{IF} are amplified in the IF amplifier, detected, and passed on for further processing. This fixed-frequency IF amplifier permits the selection of signals closely spaced in frequency (i.e., high selectivity) and the detection of weak signals (i.e., high sensitivity). However, the superheterodyne must be modified for RWR purposes, because it senses only RF frequencies over the narrow bandwidth B_{IF}.

Scanning Superheterodyne

If the local oscillator of the superheterodyne is made to sweep over a range of RF frequencies and if the RF amplifier passes all signals over the same RF range B_{RF}, the result is a scanning superheterodyne receiver. The instantaneous bandwidth of this receiver (the acceptance bandwidth) is restricted to the IF amplifier bandwidth B_{IF}, but it will eventually observe the entire RF bandwidth for activity. If the frequency of the local oscillator is abruptly changed by an increment equal to the IF bandwidth, then $B_{RF}/B_{IF} = N$ changes in the local oscillator are required to scan over B_{RF}. Each step must last at least as long as the inverse of the IF bandwidth, or $1/B_{IF}$ seconds. Thus, the minimum time required to search the RF bandwidth is

$$\tau_s = \frac{B_{RF}}{B_{IF}} \frac{1}{B_{IF}} = \frac{B_{RF}}{B_{IF}^2}$$

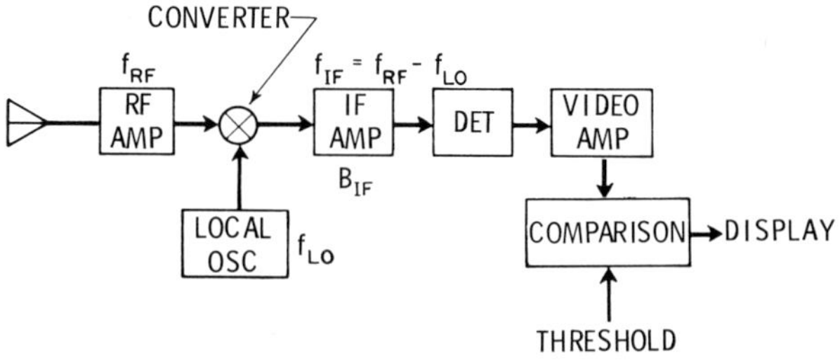

Fig. 5.5 Superheterodyne receiver.

Normally, the local oscillator must dwell at a specific frequency for a time equal to the expected pulse width or pulse-repetition interval; thus, a much greater time is required to search the RF bandwidth. For example, the time at a specific frequency may be 1 ms to insure observation of all radars illuminating the aircraft with a pulse-repetition frequency greater than 1000 Hz. Now a search time of 0.4 s is required if $B_{RF} = 2$ GHz and $B_{IF} = 5$ MHz.

Channelized Receiver

The time required to search a band of frequencies can be reduced if multiple superheterodyne receivers are used in parallel. Current technology in surface acoustic wave devices may permit a sufficient reduction in the size of each superheterodyne receiver to the point that the combined receiver is practical. A channelized receiver covering 2–4 GHz in 200 MHz steps is shown in Fig. 5.6, but there are many other designs [36, 48, 89].

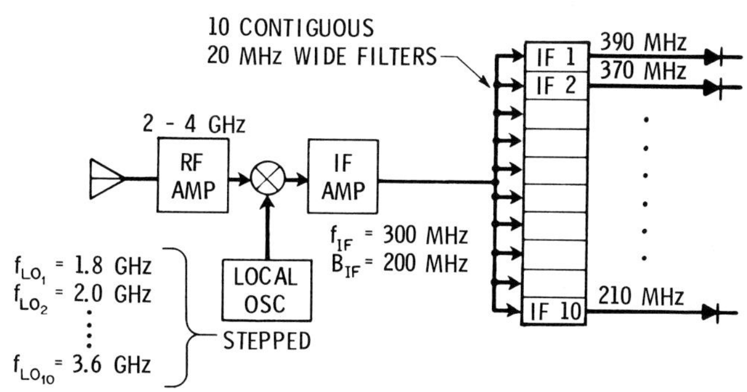

Fig. 5.6 Channelized receiver.

Crystal Video Receiver

Elimination of the converter, IF amplifier, and local oscillator from the superheterodyne results in a crystal video (or direct detection or heterodyne) receiver. If the RF amplifier is also removed, the RF bandwidth is restricted by the bandwidth of the antenna or transmission line to the antenna. Conversely, a narrow-band RF amplifier may sweep over the RF bandwidth of interest. (The latter case would be called a tuned RF receiver if the RF bandwidth were reduced to 5–100 MHz.) Typically, the RF filter (or

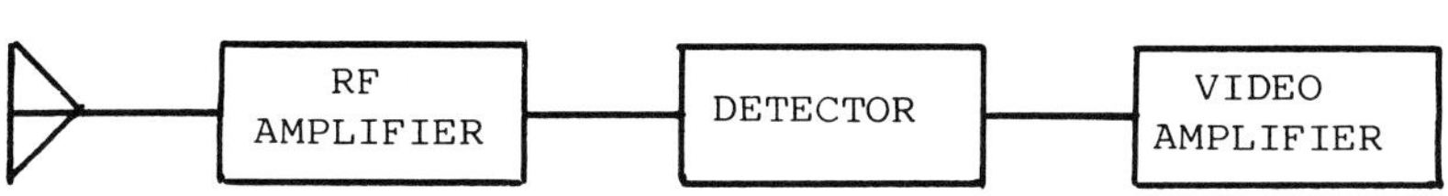

Fig. 5.7 Crystal video receiver.

amplifier) is an octave in bandwidth [89], which provides virtually no RF resolution and less receiver sensitivity than the superheterodyne. (A crystal video receiver is shown in Fig. 5.7.) The crystal video receiver usually does not easily discriminate against CW signals, high-power pulse, or noise jamming environments [89].

Microscan (Compressive) Receiver

A microscan receiver can provide much the same instantaneous observation of gigahertz RF bandwidths available in the crystal video receiver. It can also achieve the superheterodyne's resolution in frequency if pulse-compression techniques are employed.

The basic microscan technique involves a scanning superheterodyne in which the local oscillator scans the equivalent of the RF bandwidth in less time than the smallest pulse width expected. If a signal of fixed frequency is present, it will convert the sweeping signal of the local oscillator to a signal that sweeps through the IF amplifier. Shown in Fig. 5.8 are graphs of the variations in the local oscillator and IF frequency with time. As shown by Hoisington [48, p. 2c-5], if the bandwidth of the IF amplifier is as narrow as that of the normal superheterodyne, the IF will not respond to the swept signal. If the IF bandwidth is increased, the detected noise will also increase. An optimum IF

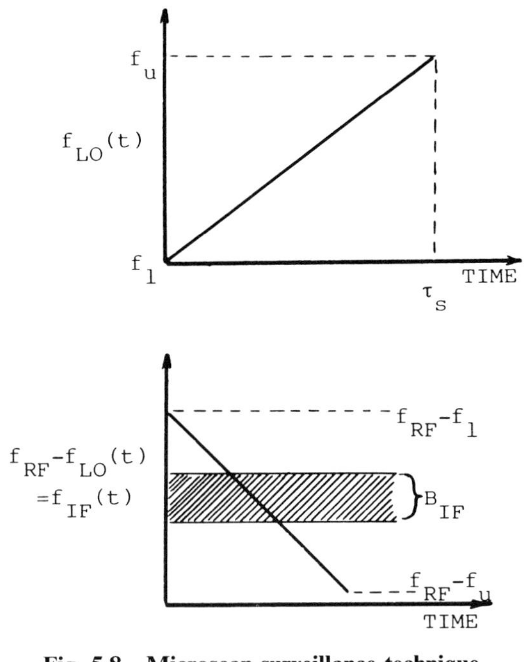

Fig. 5.8 Microscan surveillance technique.

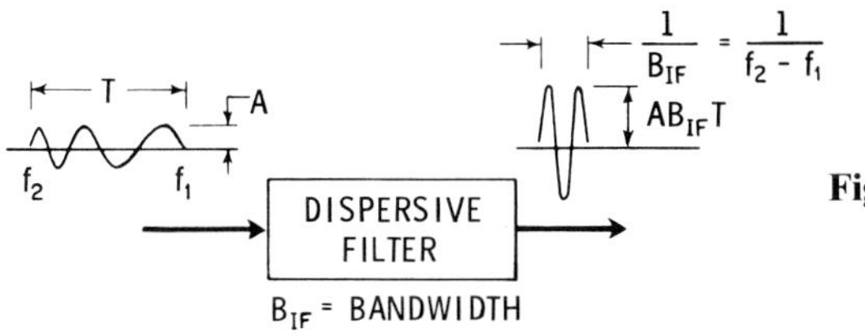

Fig. 5.9 Effect of dispersive filter.

bandwidth B_{IF} for a given RF bandwidth and sweep time τ_s is [48]

$$B_{IF} = \sqrt{\frac{B_{RF}}{\tau_s}}$$

If a 2 GHz RF bandwidth and 0.2 μs pulse width are specified, the result will be a 100 MHz IF bandwidth. Such an IF bandwidth will result in a sensitivity 13 dB less than that of a superheterodyne with a 5 MHz bandwidth and the resolvable difference in frequency increases by a factor of 20.

If the 100 MHz bandwidth IF filter is dispersive, it is possible to regain the sensitivity and selectivity of the superheterodyne. A dispersive filter has the following property. The time it takes for a signal to propagate through the filter varies with the frequency of the signal. For the compressive microscan receiver using a local oscillator that increases in frequency, the IF frequency decreases with time as shown in Fig. 5.8. The dispersive filter in this radar must exhibit a longer propagation time for higher frequencies. (The dispersive filter's input and output signals are shown in Fig. 5.9.) The duration of the signal at the input is determined by the rate of change in the IF signal and the IF bandwidth and is given by $T = B_{IF}(\tau_s/B_{RF})$. The signal out of the dispersive filter will have a duration of $1/B_{IF}$ and an amplitude increased by $B_{IF}T$ [19, p. 501]. The noise at the filter input does not have the proper frequency variation and thus does not increase in power. The signal and noise were initially spread in frequency by mixing with the swept local oscillator. The dispersive filter decreases the duration and increases the amplitude of the spread signal but not that of the noise, for a net peak signal-to-noise ratio increase of

$$B_{IF}T = B_{IF}^2 \frac{\tau_s}{B_{RF}}$$

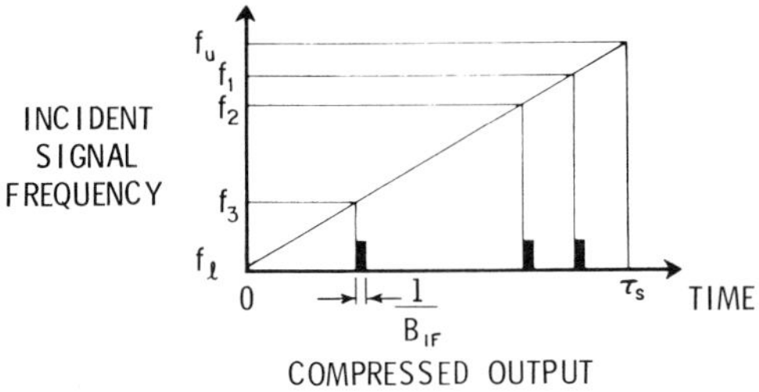

Fig. 5.10 Compressed output.

If the IF bandwidth equals the RF bandwidth, the gain in signal-to-noise ratio is $B_{RF}\tau_s$, usually much greater than one since τ_s is equal to the duration of the shortest pulse expected or to the inverse of the bandwidth of a superheterodyne receiver matched to the same pulse. This S/N increase is just that expected with a superheterodyne receiver. The compressive microscan receiver also has the same selectivity as a superheterodyne. Input signals separated in frequency by $1/\tau_s$ will appear as separate, adjacent output pulses of duration $1/B_{IF}$, as shown in Fig. 5.10. Thus, the time when the output pulse occurs within the short scan time τ_s is directly related to the frequency of the received signal.

Instantaneous Frequency-Measurement Receiver

The instantaneous frequency-measurement receiver is much simpler to construct than the microscan receiver, but it is still capable of measuring the frequency of an incident pulse within a pulse width. It does not, however, accurately process two pulses received simultaneously [48, p. 2a-16]. One approach to construction of an instantaneous frequency-measurement (IFM) receiver is shown in Fig. 5.11. The received signal of amplitude A and frequency ω_o is split into two parts. One part travels a greater distance to the converter and arrives delayed in phase relative to the other part. This difference in phase is directly related to the RF frequency. The converter and filter develop a constant voltage proportional to the phase difference. A measurement of this voltage can be transformed into the associated frequency received.

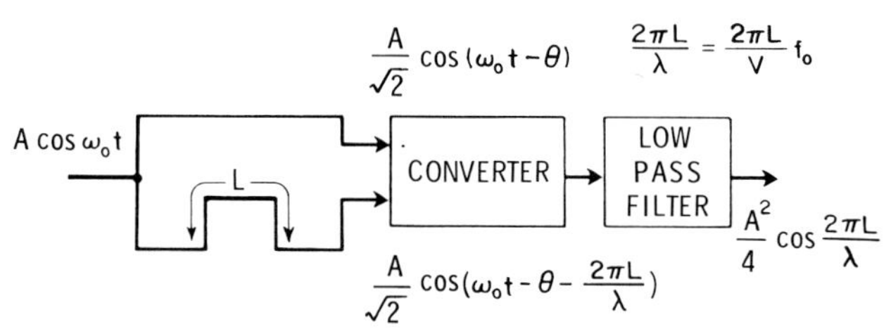

Fig. 5.11 Instantaneous frequency measurement receiver.

Acousto-Optic Receiver

The acousto-optic receiver combines a laser source, a block of crystalline material, and acoustic wave transducers to create the equivalent of a channelized receiver. The principal of operation of this receiver is shown in Fig. 5.12. A received RF signal, after conversion to a lower frequency, is coupled via the acoustic wave transducer to the crystal, called a Bragg cell. The acoustic wave propagates through the crystal as a series of compressions and expansions in the crystal material and coherent light from a laser is scattered by the acoustic pressure waves. The result is a beam of light that diverges

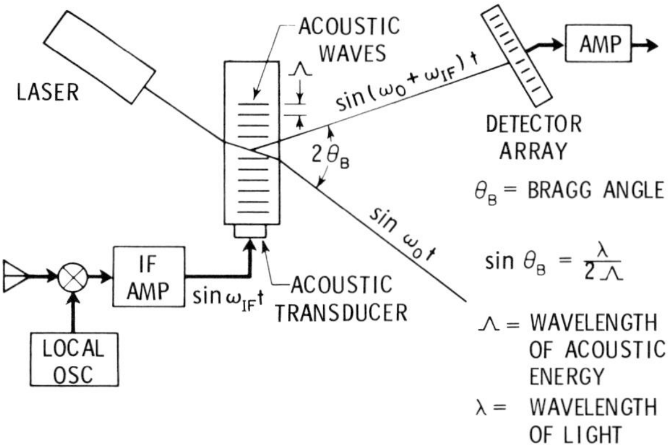

Fig. 5.12 Bulk acoustic-optical receiver.

from the undeflected beam and increases in frequency by an amount equal to the frequency producing the acoustic wave. Since the angle of deflection depends on the frequency producing the acoustic wave, the frequency can be measured by noting which diode in an array is excited. Further information on materials used, other construction approaches, and current device limitations can be found in Refs. 36, 48(pp. 2e-1—2e-5), and 89.

Receiver Comparison

Listed in Table 5.1 are characteristics of various receivers and the sources from which the information was extracted. The relative cost estimates are

Table 5.1 Intercept Receiver Comparison

Receiver	Ref	Sensitivity ($S/N = 1$), dBm	Dynamic range, dB	Frequency resolution	Acquisition time	Instantaneous RF bandwidth	Relative cost [36]
Scanning superheterodyne	36	−62	55	5 MHz	s	5 MHz	Med
Channelized	36	−79	40	25 MHz	Instant	2–4 GHz	High
Crystal video	12	−53*	75†	2 GHz	Instant	2–4 GHz	Low
Instantaneous frequency measurement	69	−75	30	20 MHz	Instant	2–4 GHz	Low–Med
Microscan (compressive)	36	−69	30	1 MHz	μs	1 GHz	Med
Acousto-optical	36 69	−90	30–40	1 MHz	Instant	1 GHz	Low

*2 MHz video bandwidth with 3 dB NF.
†Difference between sensitivity and burnout [12, p.322].

Hoffman's [36]. For reasonable false alarm rates, the sensitivity must be increased by 13 dB if automatic detection by electronic means is used [48, p. 2c-1].

5.4 RECEIVER SYSTEMS CONSIDERATIONS

It is instructive to consider a typical receiver problem to determine what factors most impact the performance of the receiver. For example, is receiver sensitivity or terrain screening the dominant factor preventing the detection of a radar? Finally, the concept of "probability of intercept" must be addressed as it is one criterion of receiver performance.

Example 5.1

A radar warning receiver has a circularly polarized receiving antenna with a gain of 3 dB. The antenna feeds a scanned superheterodyne receiver with a 20 MHz bandwidth and a sensitivity ($S/N = 13$ dB) of -63 dBm. The superheterodyne searches 1–20 GHz by stepping its 20 MHz bandwidth every 4 ms.

a) How much time does it take for the receiver to scan through the 1–20 GHz frequency band?

b) For the radar in Chapter 3, Problem 3.6, determine the range at which detection of the radar by the receiver will occur.

c) If an onboard jammer radiates 200 W of noise over a 100 MHz bandwidth centered on the transmit frequency of the radar in part (b), and if the isolation between the jammer and receiver is 60 dB, at what range will the radar be detected?

Solution

a) The number of individual 20 MHz intervals searched is

$$\frac{\text{RF bandwidth}}{\text{IF bandwidth}} = \frac{19 \times 10^9}{20 \times 10^6} = 950 = N$$

The search time is then $N(0.004) = 3.8$ s.

b) The power at the intercept receiver antenna is given by

$$P_r = \frac{P_R G_R}{4\pi R^2} \frac{\lambda^2}{4\pi} G_{REC} p \geqslant -63 \text{ dBm} = -93 \text{ dBW}$$

where G_{REC} is the gain of the receiver antenna, G_R is the gain on the track axis of the radar, and p represents the polarization mismatch. Solving for the range yields

$$R = 1.51 \times 10^6 \text{ m} = 815 \text{ n.mi.}$$

Apparently, detection will occur at any range at which the radar itself is a threat and terrain screening will normally limit the receiver detection range.

c) The noise background generated by the jammer and sensed by the receiver is

$$\text{Noise [dBW]} = P_J + B_R - B_J - I = 23\ \text{dBW} + 13 - 20 - 60 = -44\ \text{dBW}$$

The received signal for a 13 dB S/N ratio is now -31 dBW. The new detection range is $R = 1202$ m $= 0.65$ n.mi. Obviously, with the parameters used, the jammer has disabled the intercept receiver.

In part (b) of the previous example, the radar emissions would have been detected at a range of 8.2 n.mi. if the sidelobes of the radar antenna are 40 dB below the antenna gain on the track axis and if detection of the radar via its sidelobe emissions is attempted. If the aircraft is located at a distance of 50 n.mi. and desires to detect the radar before the radar detects it, two events must occur simultaneously. The radar antenna must point at the aircraft and the intercept receiver must be tuned to the radar frequency. The probability of such a detection within a specified period is called the "probability of intercept" and achieving a high intercept probability is "... the most important design objective influencing the development of intercept receivers" [2, p. 9-9].

What constitutes an intercept (i.e., detection vs identification) is not universally agreed upon. Assuming that simple detection suffices, the time required to achieve a specified probability of intercept can be computed for the previous example. The scan pattern used by the radar during the acquisition mode is given in Problem 3.6. This pattern is a 2 deg beam scanning vertically 20 deg at a rate of 14 scans/s. Figure 5.13 plots the location in frequency of the intercept receivers acceptance bandwidth as a function of time.

Also shown in Fig. 5.13 are dots representing the brief periods of time over which the main beam of the radar illuminates the intercept receiver. The location in time of the first period of illumination is arbitrary, but, once chosen, it determines the location of subsequent periods of illumination. The size of the radar illumination dots are not to scale for ease of graphic construction (i.e., 53 are required in one receiver search period τ_s).

The first time the intercept receiver dwells at the frequency of the radar, there is only a 10% chance that the radar is also illuminating the receiver. If the receiver is not being illuminated, a definite period of time will elapse until the two events are simultaneous. (If the scan period of the radar in the acquisition mode T_s is related through an integer to the scan period τ_s of the intercept receiver, the two events will never coincide.) The length of time

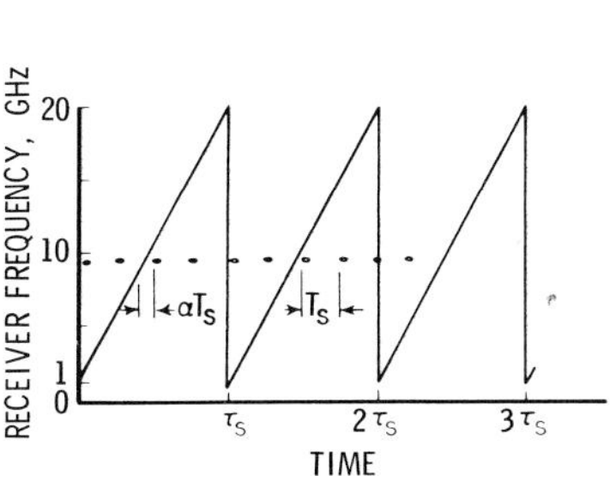

Fig. 5.13 Plot of time variation of acceptance bandwidth of intercept receiver relative to periods of illumination.

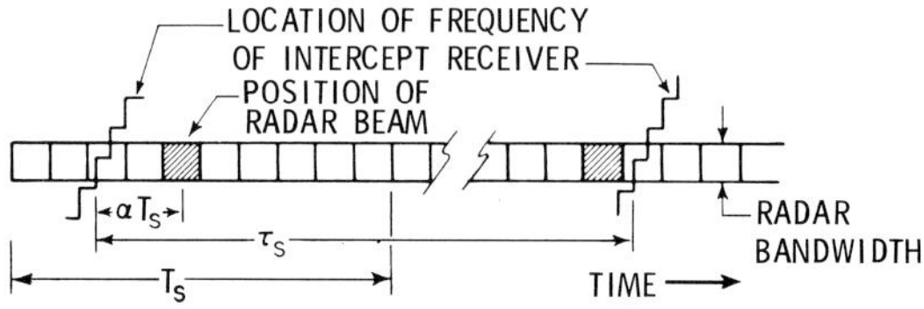

Fig. 5.14 Expanded view of radar illumination and receiver search positions.

depends on the initial displacement of the two events αT_s as shown in Fig. 5.14. Also shown is the receiver stepping in frequency through the frequency of the radar in 20 MHz increments of 0.004 s duration. On the initial sweep, the receiver was tuned to the frequency of the radar at a time αT_s before the receiver was illuminated. These events will occur simultaneously when

$$N\tau_S = MT_s + \alpha T_s \pm 0.05T_s$$

The last term assumes that detection occurs if the receiver is only momentarily illuminated by the radar (i.e., less than 4 ms). For the possible values of α, the number of receiver sweeps before detection is given in Table 5.2. Since any value of α is equally likely, the average time before detection is the sum of the expected value of N times the sweep time plus the initial sweep time to achieve the radar frequency or

$$\text{Expected search time} = 1.7 + 5.5(3.8) = 22.6 \text{ s}$$

The result of the previous computation is extremely sensitive to the accuracy of the values used for the search period of the receiver and the time between illumination periods of the radar. For example, if T_s is changed to 0.07143, the largest value of N, now 21, becomes 3. The use of such accuracy for radar parameters is usually unrealistic; a 1% tolerance on the measured time between radar illuminations is more reasonable. A 1% tolerance in turn implies the inability to predict the occurrence of the 100th illumination period. Under these conditions, it is reasonable to assume the probability of intercept on each receiver sweep $P_{i,1}$ to be 0.1 (the illumination period divided by the time between illuminations) and independent of the intercept probability on other sweeps. The probability of intercepting the radar on one or more of N sweeps is then $P_{i,N} = 1 - (1 - P_{i,1})^N$.

Table 5.2 Number of Sweeps of Receiver before Intercept ($T_s = 0.0714$ s, $\tau_s = 3.8$ s)

Initial miss α	No. of sweeps N	Initial miss α	No. of sweeps N
0	0	0.5	5
0.1	5	0.6	21*
0.2	1	0.7	3
0.3	6	0.8	8
0.4	2	0.9	4

*N becomes 3 if T_s is changed to 0.09143 s.

PART II

6 ANALYTICAL AND ENGINEERING TOOLS

Some of the mathematical and engineering tools used in subsequent chapters are described here, including Laplace transforms and their application to linear control theory, elements of probability and their relation to parameter estimation, and formulation of electromagnetic radiation from a known distribution of sources.

6.1 LAPLACE TRANSFORMS*

The Laplace transform of a function $f(t)$ is defined as

$$F(s) \equiv L[f(t)] \equiv \int_0^{\infty} f(t)e^{-st}\,dt = \lim_{\substack{\varepsilon \to 0 \\ T \to \infty}} \int_{\varepsilon}^{T} f(t)e^{-st}\,dt \qquad (6.1)$$

if the integral exists. The transform is a function of the complex variable $s = \sigma + j\omega$. If $|f(t)|e^{-\sigma t}$ is bounded for $t > t_o$, the integral exists.

Examples

Let $f(t)$ be the unit step $u(t)$ where

$$\begin{aligned} u(t) &= 1, \qquad t > 0 \\ &= 0, \qquad t = 0 \end{aligned}$$

Then

$$F(s) = \int_0^{\infty} e^{-st}\,dt = \frac{1}{s}$$

Let $f(t)$ be the unit ramp

$$\begin{aligned} f(t) &= t, \qquad t > 0 \\ &= 0, \qquad t < 0 \end{aligned}$$

*Most of this section can be found in any book on control theory, for example, D'Azzo and Houpis [90, Chap. 14].

Then

$$F(s) = \int_0^\infty te^{-st}\,\mathrm{d}t = \frac{1}{s^2}$$

A unique relationship exists between a function and its transform. Thus, if a function's transform is known, the function itself can be derived through an inverse-transform relationship. Normally, the inversion is avoided by the use of "transform pair" tables, an example of which is shown in Table 6.1.

A transform or its inverse can often be derived through the application of the theorems listed below. In all cases, $F(s) = L[f(t)]$ and is assumed to exist.

Linearity:

$$L[\alpha f(t)] = \alpha F(s)$$

Superposition:

$$L[f_1(t) + f_2(t)] = F_1(s) + F_2(s)$$

Table 6.1 Laplace transform pairs[a]

$f(t)$	$F(s)$
$\delta(t)$	1
e^{-at}	$1/(s+a)$
$\frac{1}{a}(1-e^{-at})$	$\frac{1}{s(s+a)}$
$\frac{1}{b-a}[(\alpha-a)e^{-at}-(\alpha-b)e^{-bt}]$	$\frac{s+\alpha}{(s+a)(s+b)}$
$\frac{1}{\omega_n\sqrt{(1-\xi^2)}}e^{-\xi\omega_n t}\sin[\omega_n\sqrt{(1-\xi^2)}t]$	$\frac{1}{s^2+2\xi\omega_n s+\omega_n}$
$\frac{1}{\omega_n^2}-\frac{1}{\omega_n^2\sqrt{(1-\xi^2)}}e^{-\xi\omega_n t}\sin[\omega_n\sqrt{(1-\xi^2)}t+\varphi]$ (where $\varphi = \cos^{-1}\xi$)	$\frac{1}{s(s^2+2\xi\omega_n+\omega_n^2)}$

[a]From Ref. 14 (pp. 497–498).

Translation of time:

$$L[f(t-\tau)u(t-\tau)] = e^{-s\tau}F(s)$$

Complex differentiation:

$$L[tf(t)] = -\frac{d}{ds}F(s)$$

Transforms of derivatives:

$$L\left[\frac{\mathrm{d}f(t)}{\mathrm{d}t}\right] = sF(s) - f(o+)$$

where

$$f(o+) \equiv \lim_{\substack{\varepsilon \to 0 \\ \varepsilon > 0}} f(\varepsilon)$$

Transform of integrals:

$$L\left[\int_0^t f(\tau)\,\mathrm{d}\tau\right] = \frac{F(s)}{s}$$

A powerful method for expanding a Laplace transform into a sum of simpler terms for which the inverse transform is available is called the Heaviside partial fraction expansion. The transform $F(s)$ must be expressible as the ratio of two polynomials. For example,

$$F(s) = \frac{P(s)}{Q(s)} = \frac{P(s)}{s(s-s_1)(s-s_2)}$$

$$= \frac{A_o}{s} + \frac{A_1}{s-s_1} + \frac{A_2}{s-s_2}$$

At the zeros of $Q(s)$, $s = 0$, $s = s_1$, and $s = s_2$, $F(s)$ is unbounded and is said to possess a pole. The constants A_i, called residues, can be evaluated by multiplying through by $(s - s_i)$ and letting $s \to s_i$,

$$A_i = \lim_{s \to s_i} (s - s_i)F(s)$$

For repeated roots, the polynominal expansion would become

$$F(s) = \frac{P(s)}{(s-s_1)^2 Q_1(s)} = \frac{A}{(s-s_1)^2} + \frac{B}{(s-s_1)} + \cdots$$

The coefficient B is found by multiplying through by $(s - s_1)^2$ and differentiating with respect to s before allowing $s \to s_1$,

$$B = \lim_{s \to s_1} \frac{d}{ds} [(s - s_1)^2 F(s)]$$

Of great use are the following two theorems:

1) Final Value Theorem

If the Laplace transform of $df(t)/dt$ exists, then

$$\lim_{t \to \infty} f(t) = \lim_{s \to 0} sF(s)$$

2) Initial Value Theorem

$$\lim_{t \to 0} f(t) = \lim_{s \to \infty} sF(s)$$

6.2 AUTOMATIC CONTROL THEORY

In the open-loop servo control system shown in Fig. 6.1, an input signal $r(t)$ causes the dynamic element to respond with an output $c(t)$. If a sample of the output is used as part of the input, as shown in Fig. 6.2, a closed-loop servo system results.

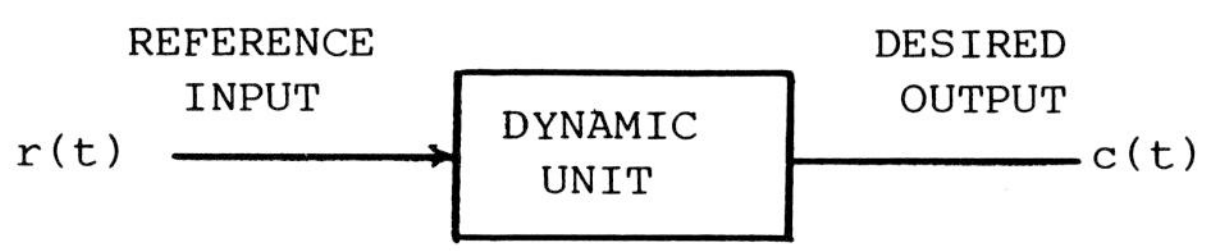

Fig. 6.1 Open-loop servo system.

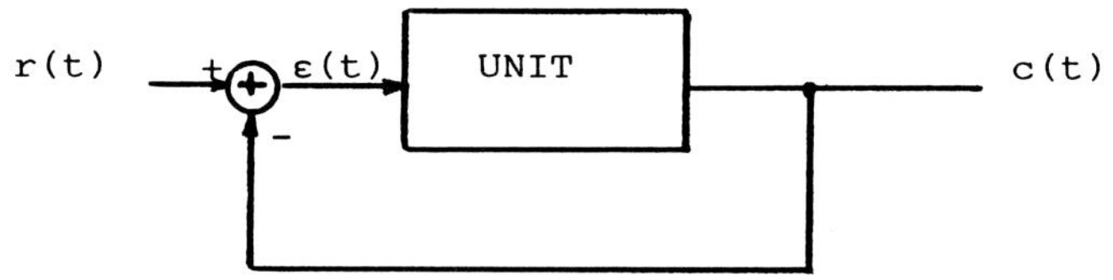

Fig. 6.2 Closed-loop servo system.

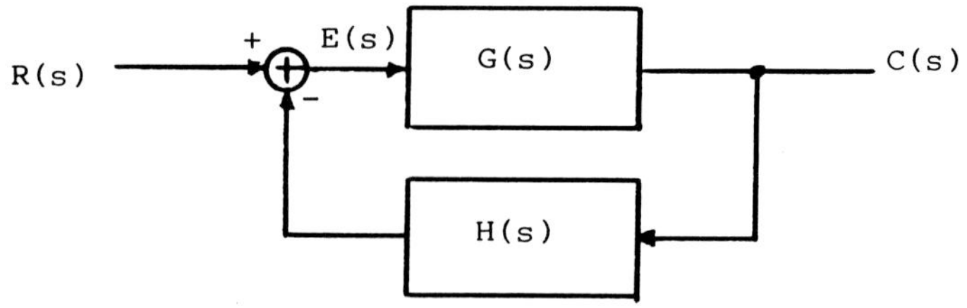

Fig. 6.3 Transform representation of closed-loop servo.

If the dynamic unit can be modeled by a linear differential equation with constant coefficients, then the input and output functions have Laplace transforms related through a closed-loop transfer function. As shown in Fig. 6.3, $G(s)$ and $H(s)$ are the Laplace transforms of the impulse responses of the dynamic unit and an assumed feedback element (filter). The relationship between input and output is

$$\frac{C(s)}{R(s)} = \frac{G(s)}{1 + G(s)H(s)} \tag{6.2}$$

When there is unity feedback [i.e., $H(s) = 1$], the actuating error is defined as $\varepsilon(t)$ and is given by

$$\frac{E(s)}{R(s)} = \frac{1}{1 + G(s)} \tag{6.3}$$

Stability of Closed-Loop Servo

The closed-loop servo will be unstable if any root of the denominator, $1 + GH$, has a positive real part. The existence of such a root can be determined by plotting the open-loop transfer function, defined by $G(s)H(s)$, that results upon the opening of the feedback path. From Eq. (6.2) or (6.3), it is apparent that difficulties arise whenever $G(s)H(s) = -1$. If neither $G(s)$ nor $H(s)$ has a pole in the right-half s plane, the open-loop transfer function is stable and the closed-loop transfer function has no zeros in the right-half s plane.

It can then also be shown [90] that the closed-loop servo is unstable (i.e., unbounded output as time increases for bounded input) if a plot of $G(j\omega)H(j\omega)$ encircles the point $(-1,0)$ in the complex plane as s varies from $-j\infty$ to $+j\infty$. Such a plot of $G(j\omega)H(j\omega)$ is called a Nyquist stability diagram, two examples of which are shown in Fig. 6.4. Figure 6.4a shows the plot for a stable system and Fig. 6.4b indicates that for which the closed-loop servo will be unstable. The $(-1,0)$ point is encircled if it is on the right of the plot of $G(j\omega)H(j\omega)$ as the curve is transversed while ω varies from 0 to ∞.

In addition to indicating the absolute stability of a servo system, the Nyquist diagram provides information on the transient response of stable systems. The closer the plot of $G(s)H(s)$ gets to point $(-1, j0)$, the more the transient response of the system to a step input will possess a large overshoot and will require a long time to achieve a steady-state value. Normally, gain and phase margins are defined to characterize the transient response of a system. The gain margin is given by the magnitude of $1 - G(s)H(s)$ when $G(s)H(s)$ crosses the negative real axis. The phase margin is given by $\arg[G(s)H(s)] - \pi$ when the magnitude of $G(s)H(s)$ equals one.

Another method of determining the stability of a servo system is to generate a Bode plot of the open-loop transfer function. This plot consists of

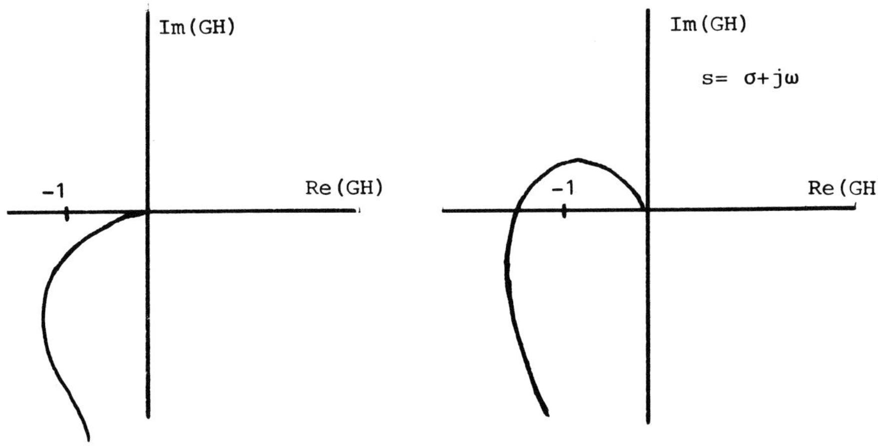

a) Stable. b) Unstable.

Fig. 6.4 Nyquist stability diagrams.

separate graphs of the magnitude and phase of $G(j\omega)$ as a function of frequency. The plot is useful because it is easy to construct. Since the frequency scale is logarithmic and the amplitude scale is in decibels, polynomial transfer functions can be expressed in terms of their roots according to simple rules. For example, if

$$G(s) = \frac{K}{s(s+a)}$$

then, for small ω, $G(j\omega)$ varies as $K/j\omega a$. Thus, $|G(j\omega)| = K/\omega a$ and

$$|G(j\omega|_{dB} = 20 \log_{10} |G(j\omega)| = \left(\frac{K}{a}\right)_{dB} - \omega_{dB}$$

This relationship results in a straight line on the log–log scale of the Bode plot as shown in Fig. 6.5.

For large $j\omega$, $|G(j\omega)| \approx K/\omega^2 a$, which plots as a straight line of twice the slope as for small ω. In particular,

$$|G(j\omega)|_{dB} = K_{dB} - 20 \log \omega - 20 \log |a + j\omega|$$

and the phase of $G(j\omega)$ is

$$\arg G(j\omega) = -\frac{\pi}{2} - \tan^{-1}\left(\frac{\omega}{a}\right)$$

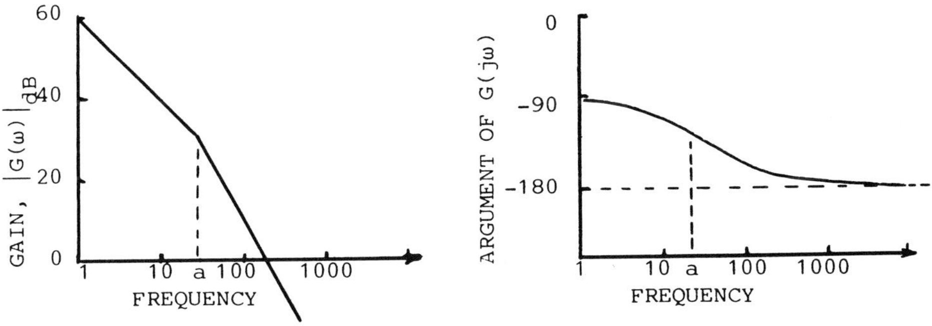

Fig. 6.5 Example of a Bode plot.

The encirclement of $(-1, j0)$ on the Nyquist diagram transforms to the Bode plot as a negative phase shift greater than 180 deg as the gain decreases through unity (or 0 dB).

Performance Criteria

It is necessary to characterize the performance of a system in order to compare different systems and select the optimum system. Various indices of performance show significant differences in their level of complexity, realism, and insight. Since the purpose of this text is to outline basic operating concepts underlying selected ECM techniques, it seems appropriate to consider basic performance criteria applicable to the servo loops of radars. (We assume unity feedback.) The performance criteria considered are stability (as just discussed); response to step, ramp, and parabolic inputs; steady-state error; and the generalized steady-state error.

Once the response of a servo to a particular input is found, it is "judged" in the light of other servo responses and the response desired. For example, a classical transfer function (appropriate for a loaded resonant circuit) is given by

$$G(s) = \frac{1}{s^2 + 2\xi\omega_n s + \omega_n^2}$$

and its response to a unit step is found in Table 6.1. A plot of this response is shown in Fig. 6.6 for a particular value of damping factor ξ.

Two terms, overshoot and one form of response time are defined in Fig. 6.6. Is this servo response the "best" available? Which criterion is most important, small response time or small overshoot? How do other transfer functions or the same transfer function with different constants compare? Other criteria for goodness, such as minimizing the root-mean-square error may be more appropriate than overshoot and response time. Thus, system responses to classical inputs are used to develop a basic awareness of system

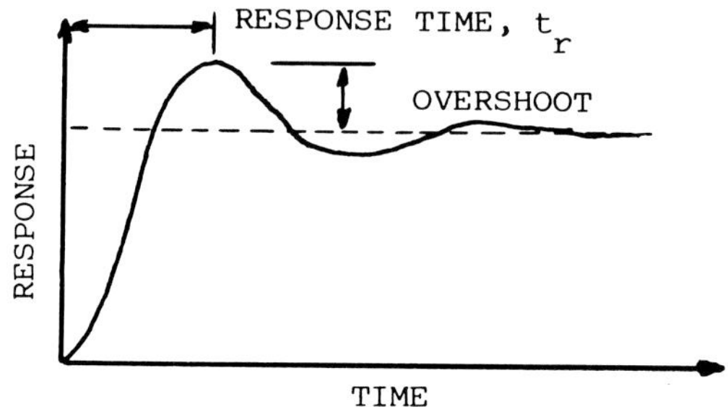

Fig. 6.6 Response to step input.

performance because such responses are easily computed, but they do not unequivocally determine the best system.

Another characterization of system performance is the steady-state error, the value of the difference between input and output $\varepsilon(t)$ as t increases. From Eq. (6.3) and the final value theorem, this error is

$$\lim_{t \to \infty} \varepsilon(t) = \lim_{s \to 0} \frac{sR(s)}{1 + G(s)} \tag{6.4}$$

This steady-state error is usually specified for open-loop transfer functions of particular systems subjected to the classical inputs of step, ramp, etc. The open-loop functions are themselves classified by "type" or the power to which the variable s is raised in the denominator of $G(s)$. Examples of functions of types 0, 1, and 2 are shown in Table 6.2.

The steady-state error of the type 1 system resulting from a ramp input is

$$\lim_{t \to \infty} \varepsilon(t) = \lim_{s \to 0} \frac{s \dfrac{1}{s^2}}{1 + \dfrac{1}{s(s+a)}} = \lim_{s \to 0} \frac{s + a}{s(s+a) + 1} = a = \frac{1}{K_v} \tag{6.5}$$

and to a step input

$$\lim_{t \to \infty} \varepsilon(t) = \lim_{s \to 0} \frac{s(s+a)}{s(s+a) + 1} = 0 = \frac{1}{K_p} \tag{6.6}$$

Table 6.2 Open-loop functions

Type	$G(s)$
0	$1/(s+a)$
1	$1/s(s+b)$
2	$(s+b)/s^2(s+a)$

Table 6.3 Steady-state coefficients

		Input	
	Step	Ramp	Paraboloid
System type	$u(t)$	t	$\frac{1}{2}t^2$
0	K_p	0	0
1	∞	K_v	0
2	∞	∞	K_a
3	∞	∞	∞

The constants K_p and K_v are called steady-state error coefficients and are, in this instance, $1/a$ and ∞. The steady-state error in response to a parabolic input is infinite, or $K_a = 0$. The steady-state error coefficients of various systems to the classical inputs are listed in Table 6.3.

The pattern observed in Table 6.3 for a given type of system is a zero steady-state error up to a particular power of time as an input function. Then, a finite error is achieved and an infinite error is created by all of the higher powers of time as input functions.

As opposed to transient or steady-state responses to impulse, step, or ramp inputs, a generalized steady-state response to an arbitrary input can be defined [16, Chap. 9; 77] as the system response after the transients have died out but before the steady-state value is achieved. The final value theorem states that

$$\lim_{t \to \infty} f(t) = \lim_{s \to 0} sF(s) \tag{6.7}$$

if $\lim_{t \to \infty} f(t)$ exists. A system's actuating error is given by

$$\frac{E(s)}{R(s)} = \frac{1}{1 + G(s)} = F(s) \tag{6.8}$$

F(s) also represents the Laplace transform of the actuating error in response to an impulsive input. For a stable system, $\lim_{t \to \infty} f(t)$ exists and $F(s)$ can be expanded in a Maclaurin series about $s = 0$ as

$$F(s) = F(0) + \sum_{n=1}^{\infty} \frac{1}{n!} \frac{d^n F(s)}{ds^n} \bigg|_{s=0} s^n \tag{6.9}$$

$$= \frac{1}{K_0} + \frac{s}{K_1} + \frac{s^2}{K_2} + \cdots$$

From Eq. (6.9), generalized error coefficients are defined by

$$\frac{1}{K_n} = \frac{1}{n!}\frac{\mathrm{d}^n F(s)}{\mathrm{d}s^n}\bigg|_{s=0} \tag{6.10}$$

and the following definitions are normally used:

$$K_1 = K_v \approx \text{velocity constant}$$

$$K_2 = K_a \approx \text{acceleration constant}$$

For a given type of system, these generalized error coefficients are identical to the steady-state error coefficients defined earlier up to and including the last nonzero error coefficient. It is usually easier to derive the generalized error coefficients by this equivalence to the steady-state coefficients. Error coefficients of higher order can then be found by using Eq. (6.10).

After initial transients have decayed, the generalized steady-state error for an arbitrary input is then

$$E(s) = \frac{R(s)}{K_o} + \frac{sR(s)}{K_v} + \frac{s^2R(s)}{K_a} + \cdots \tag{6.11}$$

or

$$\varepsilon(t) = \frac{r(t)}{K_o} + \frac{1}{K_v}\frac{\mathrm{d}r}{\mathrm{d}t} + \frac{1}{K_a}\frac{\mathrm{d}^2r}{\mathrm{d}t^2} + \cdots \tag{6.12}$$

Equations (6.11) and (6.12) are identical for all time only if $r(t)$ and all its derivatives are zero as $t \to 0+$ (see the Laplace transform of $\mathrm{d}f/\mathrm{d}t$). However, equivalence is not desired for all time, but only for time greater than the time it takes for the system transients to decay. Nonzero initial conditions contribute only to the transient response.

Example

The system shown in Fig. 6.3 has $G(s) = K/s$. Table 6.3 shows that the position error coefficient is infinite. The velocity coefficient is finite and nonzero and thus equals the generalized velocity coefficient. The velocity coefficient is the steady-state error resulting from a ramp input, $R(s) = 1/s^2$. As

$$\frac{E(s)}{R(s)} = F(s) = \frac{s}{s+K}$$

the error resulting from a ramp input is

$$\lim_{t\to\infty} \varepsilon(t) = \lim_{s\to 0} sE(s) = \lim_{s\to 0}\frac{1}{s+K} = \frac{1}{K} = \frac{1}{K_v} \tag{6.13}$$

Higher coefficients are found by use of Eq. (6.10). Thus,

$$\frac{1}{K_a} = \frac{1}{2}\frac{d^2}{ds^2}\left(\frac{s}{s+K}\right)\Big|_{s=0} = \frac{-1}{K^2} \tag{6.14}$$

The error resulting from a general input,

$$r(t) = a_0 + a_1 t + a_2 t^2$$

after decay of transients is given by Eqs. (6.12–6.14) as

$$\varepsilon(t) = \frac{a_1}{K} + \frac{2a_2}{K}\left(t - \frac{1}{K}\right) \tag{6.15}$$

The exact solution of the same problem is

$$\varepsilon(t) = a_0 e^{-Kt} + \frac{a_1}{K}(1 - e^{-Kt}) + \frac{2a_2}{K}\left[t - \frac{1}{K}(1 - e^{-Kt})\right] \tag{6.16}$$

The transient response can be associated with terms containing e^{-Kt}. Thus, for $Kt > 3$, Eqs. (6.15) and (6.16) are essentially identical.

6.3 PROBABILITY AND ESTIMATION*

Probability Distributions and Moments

Often, the outcome of an endeavor is unpredictable and only the probability of a particular outcome is known. For example, in throwing a fair die, one cannot state before the throw which number from one to six will appear on the top face. But, one can state that the probability of a two appearing is one-sixth or

$$P\{2\} = 1/6$$

Also, since a number from one to six must result and no other outcome is possible, there is a 100% probability that a number from one to six shall appear, or

$$P\{1 \text{ to } 6\} = \sum_{i=1}^{6} P\{i\} = 1$$

Throwing a die is an example of an experiment resulting in a discrete set of outcomes, but many experiments have an infinite number of possible outcomes. For example, measurement of the voltage of a battery may yield

*More detail can be found in Whalen [82].

different results on subsequent experiments because of measurement noise. The absolute bounds on possible outcomes may be known, but the probability of measuring any specific value is zero. Thus, if x_0 and x_1 are the lower and upper bounds on the outcome,

$$P\{x_i\} = 0 \qquad x_0 \le x_i \le x_1$$

$$P\{x_0 \le x \le x_1\} = 1 \tag{6.17}$$

The last formula is interpreted as the probability of measuring *any* x between x_0 and x_1.

A new quantity can be defined and related to the probability that the measurement results in *any* x between x_a and $x_a + \Delta x$ where Δx is small. This quantity is the probability density function $p(x)$,

$$P\{x_a \le \tilde{x} \le x_a + \Delta x\} \equiv p(x_a)\Delta x \tag{6.18}$$

where the tilde "$\sim$" denotes a random variable. In general

$$P\{x_a \le \tilde{x} \le x_b\} = \int_{x_a}^{x_b} p(x)\, \mathrm{d}x \tag{6.19}$$

In terms of a probability density, the outcome of the die experiment can be expressed as a sum of delta functions

$$p(x) = \sum_{i=1}^{6} \frac{1}{6}\, \delta(x - i)$$

A probability density observed in many experiments is the Gaussian distribution

$$p(x) = \frac{1}{\sqrt{2\pi}\sigma}\, e^{-(x-\bar{x})^2/2\sigma^2}, \qquad |x| < \infty \tag{6.20}$$

Other distributions often observed are the Rayleigh

$$\begin{aligned} p(x) &= \frac{x}{\sigma^2}\, e^{-x^2/2\sigma^2} && 0 \le x \le \infty \\ &= 0 && \text{else} \end{aligned} \tag{6.21}$$

and the exponential

$$\begin{aligned} p(x) &= \frac{1}{2\sigma^2}\, e^{-x/2\sigma^2} && 0 \le x \le \infty \\ &= 0 && \text{else} \end{aligned} \tag{6.22}$$

Even though the actual outcome of an experiment cannot be predicted, an expected or mean value (or first moment) can be defined as

$$E\{\tilde{x}\} = \int_{x_1}^{x_2} xp(x)\, \mathrm{d}x \tag{6.23}$$

This expected value may never occur in N attempts of the experiment, but, presumably, the sample mean will, in most cases, eventually equal the expected value,

$$\lim_{N\to\infty} \frac{1}{N} \sum_{i=1}^{N} x_i = E\{\tilde{x}_i\} \tag{6.24}$$

If an experiment has an outcome y, which is a function of a random variable $\tilde{x}$, or $\tilde{y} = f(\tilde{x})$, then the expected value of $\tilde{y}$ is

$$E\{\tilde{y}\} = \int_{x_1}^{x_2} f(x)p(x)\, \mathrm{d}x \tag{6.25}$$

The probability density function associated with $\tilde{y}$ can similarly be found to be

$$p(y) = p(x)\left|\frac{\mathrm{d}x}{\mathrm{d}y}\right| \tag{6.26}$$

when $f(x)$ is single valued as

$$p(y)\ |\mathrm{d}y| = p(x)\ |\mathrm{d}x| \tag{6.27}$$

In addition to the expected outcome of an experiment, it is desirable to have a measure of the expected variation in the output from the mean value. One measure of this variation is called the variance, second central moment, or covariance and is defined as

$$\mathrm{Var}\ \{\tilde{x}\} = E\{(\tilde{x} - E\{\tilde{x}\})^2\} = E\{\tilde{x}^2\} - E^2\{\tilde{x}\} \tag{6.28}$$

In Eq. (6.20), the Gaussian distribution, it can be shown that σ^2 is the variance on x and that $\bar{x} = E\{\tilde{x}\}$.

Finally, when an experiment is repeated N times, a joint probability can be defined that gives the probability of a specific set of the results of the N experiments. For example, the probability on the ith experiment that the outcome will result in $x_i < \tilde{x}_i < x_i + \Delta x$ is $p(x_i)\ \mathrm{d}x$. The probability that the set $\bar{x}$ occurs for N experiments, where $\bar{x} = \{x_1, x_2, \ldots, x_N\}$ and each experiment is independent, is given by $p(\bar{x})\ \mathrm{d}\bar{x} = \Pi_{i=1}^{N}\ p(x_i)\ \mathrm{d}x_i$.

Maximum Likelihood Estimation

If one assumes that a particular probability distribution is appropriate in a given experiment, then it is possible to estimate the parameters of the distribution with the data measured during the experiment. For example, if the Gaussian distribution is assumed to describe the distribution of the measurements, the distribution depends on two parameters—the mean $\bar{x}$ and the variance σ^2. If the variance is assumed to be known, the resulting density depends on the mean or

$$p(x|\bar{x}) = \frac{1}{\sqrt{2\pi\sigma}} e^{-(x-\bar{x})^2/2\sigma^2} \tag{6.29}$$

An estimate of $\bar{x}$, depends on how the estimate is defined in terms of the data and a variety of approaches can be followed. The one considered here is the maximum likelihood estimate, denoted $\hat{x}_{ML}$, and is defined as the value that maximizes the probability density given the measured data. Formally,

$$\frac{\partial p(x|\bar{x})}{\partial \bar{x}} \Bigg|_{\bar{x}=\hat{x}_{ML}} = 0 \tag{6.30}$$

In Eq. (6.30), x represents measured data. The density function is differentiated with respect to the parameter to be estimated $\bar{x}$ and the derivative evaluated for various values of $\bar{x}$. The value of $\bar{x}$ that causes the derivative to be zero, given the measured data, marks the maximum or most likely value of the density function. Hence, that value of $\bar{x}$ is the maximum likelihood estimate $\bar{x} = \hat{x}_{ML}$.

It is often easier to differentiate the natural logarithm of the density function. The maximum likelihood estimate is then

$$\frac{\partial}{\partial a} [\ell n p(x|a)] \Bigg|_{a=\hat{a}_{ML}} = 0 \tag{6.31}$$

Thus, if only one measurement is made and Eq. (6.31) is applied to the density of Eq. (6.29), then

$$\ell n\, p(x|\bar{x}) = \ell n \sqrt{2\pi\sigma} - \frac{(x-\bar{x})^2}{2\sigma^2} \tag{6.32}$$

$$\frac{\partial}{\partial \bar{x}} \{\ell n\, [p(x|\bar{x})]\} = \frac{x-\bar{x}}{\sigma^2} = 0 \tag{6.33}$$

Thus, the maximum likelihood estimate is just the value measured

$$\hat{x}_{ML} = x \tag{6.34}$$

Example

A signal is measured N times, yielding N independent power readings. The probability density is assumed to be exponential.

What is the maximum likelihood (ML) estimate of the average power of the signal?

Solution: For the exponential density given by Eq. (6.22) with $\alpha = 2\sigma^2$,

$$p(x|\alpha) = \frac{1}{\alpha} e^{-x/\alpha}$$

and the average value of x is α.

The joint density is

$$p(\bar{x}|\alpha) = \prod_{i=1}^{N} p(x_i|\alpha)$$

Converting to natural logarithms and differentiating with respect to the unknown average value α yields

$$\frac{\partial}{\partial\alpha} \ell n\, p(\bar{x}|\alpha) = \sum_{i=1}^{N} \frac{\partial}{\partial\alpha} \ell n\, p(x_i|\alpha)$$

$$= \sum_{i=1}^{N} \left[-\frac{1}{\alpha} + \frac{x_i}{\alpha^2} \right]$$

Setting this sum equal to zero and solving for α yields the ML estimate of α as

$$\hat{\alpha}_{ML} = \frac{1}{N} \sum_{i=1}^{N} x_i$$

6.4 ELECTROMAGNETIC RADIATION AND ANTENNA PATTERNS

Numerous texts on electromagnetic theory [34, 65, 84] show that the electric field radiated from a volume-current distribution $\bar{J}(\bar{r}')$, at a great distance from the source, in the Fraunhofer or the far zone, is given by

$$\bar{E}(\bar{r}) = \frac{-j\omega\mu}{4\pi} \int_{v'} [\bar{J} - (\bar{J} \cdot \hat{r})\hat{r}] e^{+jk\bar{r}' \cdot \hat{r}}\, dv' \, \frac{e^{-jkr}}{r} \tag{6.35}$$

In Eq. (6.35), $k = 2\pi/\lambda$ and the distance r at which the field is evaluated is such that $r \gg 2D^2/\lambda$, where D is the maximum dimension of the volume V containing the source. The vector $\bar{r}$ to the field point is given by $\bar{r} = r\hat{r}$.

Similarly, if the tangential electric field is known over the surface of a half-space (e.g., the x-y plane), the electric field radiated in the $+z$ direction by the sources in the $-\hat{z}$ half-space producing the tangential field $\bar{E}$ is given by

$$\bar{E}(\bar{r}) = \frac{1}{2\pi} \int_{S'} \nabla \times [\hat{z} \times \bar{E}(\bar{r}')] \frac{e^{-jk|\bar{r} - \bar{r}'|}}{|\bar{r} - \bar{r}'|} \mathrm{d}S' \tag{6.36}$$

If, as usual, one assumes that $\hat{z} \times \bar{E}$ is nonzero over only a finite region of the x-y plane, then the far-zone radiated field is approximately

$$\bar{E}(\bar{r}) \approx \frac{j}{\lambda} \int_{S'} \hat{r} \times (\hat{z} \times \bar{E}) e^{jk\bar{r}' \cdot \hat{r}} \, \mathrm{d}S' \, \frac{e^{-jkr}}{r} \tag{6.37}$$

Example

Find the electric field radiated by the current element $\bar{J} = \hat{x} I \Delta\ell \, \delta(\bar{r}') \, \mathrm{d}V'$.

Hint: It is often essential to express the unit vectors in one coordinate system in terms of the spherical unit vectors. From Ref. 34 (p. 40), we have

$$\hat{x} = \hat{r} \sin\theta \cos\varphi + \hat{\theta} \cos\theta \cos\varphi - \hat{\phi} \sin\varphi$$

$$\hat{y} = \hat{r} \sin\theta \sin\varphi + \hat{\theta} \cos\theta \sin\varphi + \hat{\phi} \cos\varphi$$

$$\hat{z} = \hat{r} \cos\theta - \hat{\theta} \sin\theta$$

Solution: Using the above unit-vector relationships, the intergrand in Eq. (6.35) can be expressed as

$$\bar{J} - (\bar{J} \cdot \hat{r})\hat{r} = I\Delta\ell \, \delta(\bar{r}') \, \mathrm{d}V' [\hat{\theta} \cos\theta \cos\varphi - \hat{\varphi} \sin\varphi]$$

Equation (6.35) can then be integrated to yield the electric field

$$\bar{E}(\bar{r}) = -\frac{j\omega\mu}{4\pi} I\Delta\ell \frac{e^{-jkr}}{r} [\hat{\theta} \cos\theta \cos\varphi - \hat{\varphi} \sin\phi]$$

In the far zone, the electric field has a negligible radial component and varies as e^{-jkr}/r. The radiated field can always be divided into two orthogonal components as

$$\begin{aligned} \bar{E} &= (\bar{E} \cdot \hat{e}_n^*)\hat{e}_n + (\bar{E} \cdot \hat{e}_p^*)\hat{e}_p \\ &= E_n \hat{e}_n + E_p \hat{e}_p \end{aligned} \tag{6.38}$$

The component E_n is the nominal or working polarization, and E_p the cross polarization as

$$\hat{e}_n \cdot \hat{e}_p^* \equiv 0 \tag{6.39}$$

The power P_R radiated by the antenna must equal [34, Chap. 5]

$$P_R = \int_{4\pi} \frac{|\bar{E}|^2}{2\eta} r^2 \, d\Omega \tag{6.40}$$

where $\eta = \sqrt{\mu/\varepsilon}$ is the wave impedance of the space into which the antenna radiates.

From Eq. (6.38), the radiated field can be expressed in terms of a single polarization $\hat{h}$ as

$$\bar{E} = E_n \left| \hat{e}_n + \frac{E_p}{E_n} \hat{e}_p \right| \frac{\hat{e}_n + \dfrac{E_p}{E_n} \hat{e}_p}{\left| \hat{e}_n + \dfrac{E_p}{E_n} \hat{e}_p \right|} = E(\bar{r})\hat{h} \tag{6.41}$$

in which both the scalar $E(\bar{r})$ and the unit vector $\hat{h}$ are generally complex. Proceeding further, the scalar E can be written

$$E(\bar{r}) = \sqrt{\frac{P_R \eta}{2\pi}} \, g(\theta, \varphi) \frac{e^{-jkr}}{r} \tag{6.42}$$

Inserting Eq. (6.42) into Eq. (6.40) results in the requirement

$$\int_{4\pi} |g(\theta, \varphi)|^2 \, d\Omega = 4\pi \tag{6.43}$$

Thus, subject to Eq. (6.43), in general the field radiated by an antenna can be expressed as

$$\bar{E}(\bar{r}) = \sqrt{\frac{P_R \eta}{2\pi}} \frac{e^{-jkr}}{r} g(\theta, \varphi)\hat{h} \tag{6.44}$$

Since the antenna gain G_R is defined as actual power density over the power density from an isotropic source,

$$G_R = |g(\theta, \varphi)|^2 \tag{6.45}$$

From Eqs. (6.42) and (6.41)

$$|E_n|^2 \left| \hat{e}_n + \frac{E_p}{E_n} \hat{e}_p \right|^2 = \frac{P_R \eta}{2\pi r^2} |g(\theta, \varphi)|^2$$

Using Eq. (6.39) yields

$$|g(\theta, \varphi)|^2 = \frac{2\pi r^2}{P_R \eta} \{|E_n|^2 + |E_p|^2\}$$

which in light of Eq. (6.42) can be expressed as

$$|g(\theta, \varphi)|^2 = |g_n(\theta, \varphi)|^2 + |g_p(\theta, \varphi)|^2 \tag{6.46}$$

The total gain is the sum of the nominal and cross-polarized gains.

A Gaussian-shaped radiation pattern has been used extensively here because of its analytic simplicity; but, to determine its range of usefulness, the Gaussian pattern must be compared to more realistic patterns. The patterns considered are the $\sin x/x$ pattern resulting from a uniformly illuminated rectangular aperture and the $J_1(x)/x$ pattern created by the uniformly illuminated circular aperture [34].

The rectangular aperture, shown in Fig. 6.7, of length $D \gg \lambda$ creates an electric field variation in the $\hat{x}$-$\hat{z}$ plane given by

$$g(\theta) = \frac{\sin\left(\dfrac{\pi D}{\lambda}\theta\right)}{\dfrac{\pi D}{\lambda}\theta} \tag{6.47}$$

The 3 dB beamwidth of the rectangular aperture is given by

$$\theta_3 = 2.78\frac{\lambda}{\pi D} \tag{6.48}$$

permitting expression of Eq. (6.47) in terms of the 3 dB beamwidth as

$$g_s(\theta) = \frac{\sin\left(2.78\dfrac{\theta}{\theta_3}\right)}{2.78\dfrac{\theta}{\theta_3}} \tag{6.49}$$

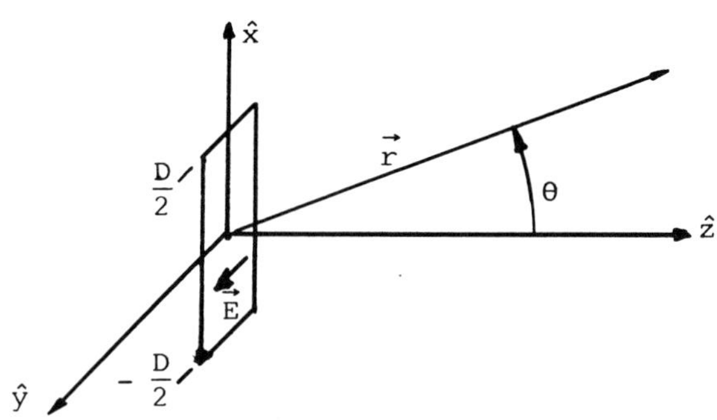

Fig. 6.7 Geometry of rectangular aperture.

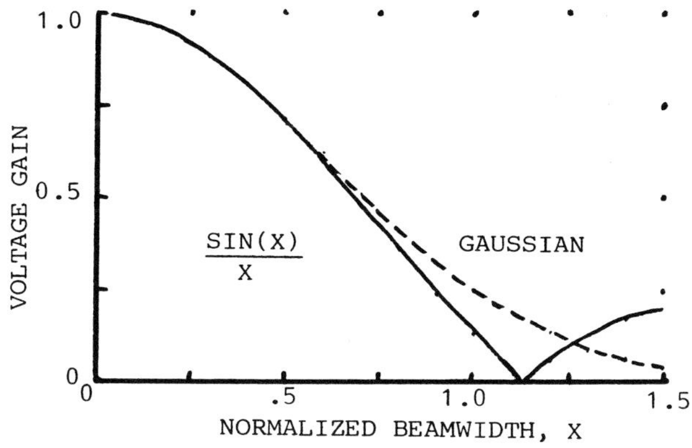

Fig. 6.8 Comparison of Gaussian pattern and pattern from rectangular aperture.

The variation of the electric field for the Gaussian pattern can be written as

$$g_G(\theta) = e^{-2\ell n 2\left(\frac{\theta}{\theta_3}\right)^2} \tag{6.50}$$

These two patterns, plotted in Fig. 6.8, are almost identical out to one-half beamwidth.

The field radiated from a uniformly illuminated circular aperture of diameter D, which is many wavelengths in size and in the x-y plane, is given by

$$g(\theta) = 2\frac{J_1\left(\frac{\pi D}{\lambda}\theta\right)}{\frac{\pi D}{\lambda}\theta} \tag{6.51}$$

In this case, the 3 dB beamwidth is

$$\theta_3 = 1.02\frac{\lambda}{D} \tag{6.52}$$

resulting in a gain expression

$$g(\theta) = 2\frac{J_1\left(3.2\frac{\theta}{\theta_3}\right)}{3.2\frac{\theta}{\theta_3}} \tag{6.53}$$

The Gaussian pattern is compared with the radiation from the circular aperture in Fig. 6.9. Again, very good matching is apparent out to one-half beamwidth.

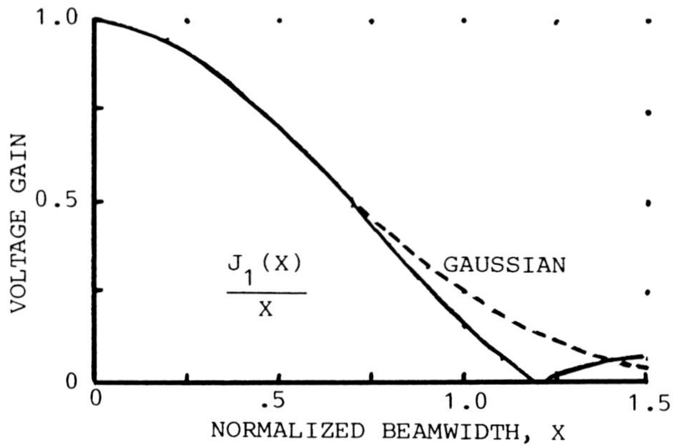

Fig. 6.9 Comparison of Gaussian pattern and pattern from circular aperture.

As a final comparison, the slope of the circular aperture power pattern ($G = g^2$) and its gain relative to the peak gain are tabulated in Table 6.4. Also tabulated are the same quantities for a Gaussian pattern. The slope function is defined as

$$k_s = -\frac{\theta_3}{G(\theta_3)}\frac{dG}{d\theta}\bigg|_{\theta=\theta_3} \tag{6.54}$$

and the loss function

$$L_k = -\left[\frac{G(\theta)}{G(0)}\right]^2_{dB} \tag{6.55}$$

Again, excellent agreement is apparent.

Table 6.4 Comparison of error slope and loss coefficients of a circular aperture pattern and Gaussian pattern

$\frac{\theta}{\theta_3}$	Circular k_s	Circular L_k	Gaussian k_s	Gaussian L_k, dB
0.1	0.523	0.23	0.55	0.24
0.2	1.05	0.91	1.10	0.96
0.3	1.607	2.06	1.66	2.17
0.4	2.209	3.7	2.216	3.85

Problems

6.1 Determine the stability of the control loop shown below by both a Bode plot and a Nyquist diagram.

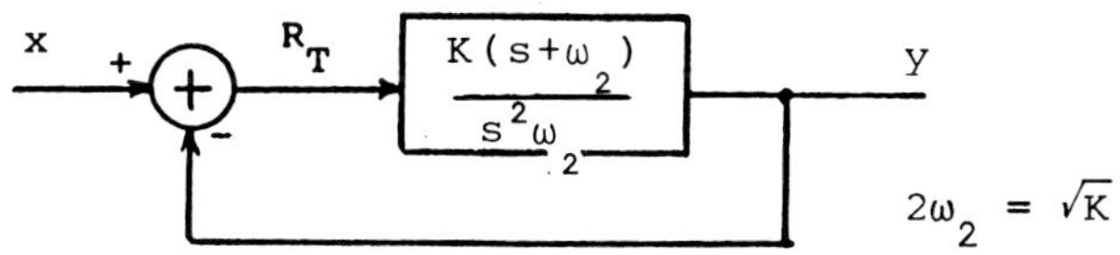

Assume the double pole of the open-loop transfer function, which occurs at $s = 0$, is not in the right-half s plane.

6.2 For the servo of problem 6.1, show that the first four generalized error coefficients are

$$K_0 = \infty, \quad K_v = \infty, \quad K_a = K, \quad K_3 = -K^{3/2}/2$$

6.3 Show that the impulse response of the servo of problem 6.1 is

$$y(t) = 2\sqrt{K}\, e^{-\sqrt{K}t} - Kte^{-\sqrt{K}t}$$

6.4 Assume a random variable $\tilde{x}$ has a Raleigh distribution. Show that the expected value of and variance of $\tilde{x}$ are

$$E\{\tilde{x}\} = \sqrt{\frac{\pi}{2}}\,\sigma, \quad \mathrm{Var}\{\tilde{x}\} = \left(2 - \frac{\pi}{2}\right)\sigma^2$$

6.5 If the input to the servo in problem 6.1 is

$$x(t) = at^3, \qquad t \geq 0$$
$$= 0, \qquad t < 0$$

show that the generalized error for $t \geq 0$ is

$$\varepsilon(t) \equiv x(t) - y(t) = \frac{6a}{K}\left(t - \frac{2}{\sqrt{K}}\right)$$

6.6 An array antenna is formed from two short dipoles with equal current as shown in the sketch below.

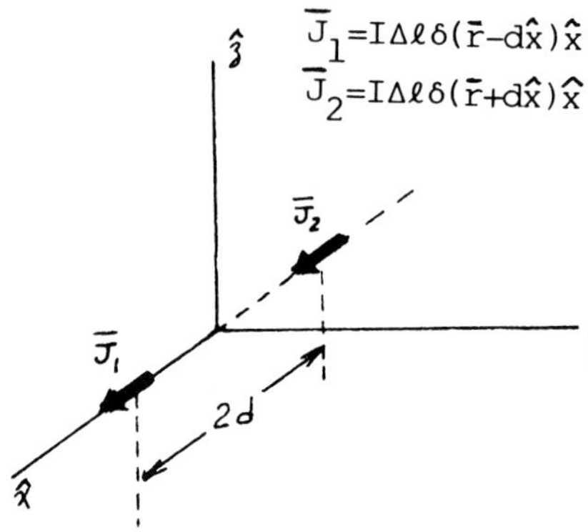

The nominal, or working, polarization is taken to be the polarization of the wave radiating along the $\hat{y}$ axis. Compute the voltage received if the incident field is $\bar{E}_{in} = E_{in}\hat{\theta}$.

Assume the received voltage to be given by

$$v_r = \bar{E}_{in} \cdot \bar{E}_a$$

where $\bar{E}_a$ is the radiation pattern of the array.

6.7 Show that the voltage received in problem 6.6 is maximum in the direction $\theta' = \pm\phi'$, when $\theta = \pi/2 + \theta'$ and $\phi = \pi/2 + \phi'$ are substituted into the solution of problem 6.6 to describe the antenna response near the $\hat{y}$ axis. Assume $k\,d\phi' \ll 1$.

6.8 Repeat problem 6.6 with the current elements oriented in the $\hat{z}$ direction and $\bar{E}_{in} = E_{in}\hat{\phi}$. Note how your results depend on the definition of the coordinate system and the definition of crossed polarization.

7
ANALYTICAL MODELS OF TRACKING RADARS

Before the effectiveness of an ECM technique against a tracking radar can be assessed, the performance of the radar in the absence of ECM must be determined. One approach to this determination of radar performance is to develop an analytical model of the radar and then to attribute the capability of the model to that of the actual radar. Such models often provide an insight into a radar's weaknesses and are much cheaper to use in analytical experiments than the actual radar. However, any cost-effective model is always limited in the scope of the conditions that can be represented and, since these conditions are not always obvious, hardware tests are desirable for verification of the model analysis. Some simple radar models are described in this chapter; the response of the same models when subjected to ECM signals will be analyzed in later chapters.

A tracking radar can be divided into two functional parts. The first part estimates the position of the target within the resolution cell of the radar. For example, it estimates the target angle relative to the present track axis of the antenna by scanning the radiation pattern or by using monopulse techniques. The second part of the radar centers the radar resolution cell on the target, typically with the aid of servomechanisms. Although this discussion emphasizes models of radars capable of tracking only single targets, many elements of these models are relevant in analyzing "track-while-scan" radar systems capable of tracking multiple targets. In either situation, radar tracking is the repetition of the functions performed by these two radar elements: estimating and repositioning.

The result of radar tracking is an estimate of target position in a fixed frame of reference, as shown in Fig. 7.1. The angle and range estimates of the target are $\hat{\theta}_{T_o}$ and $\hat{R}_T$, respectively. The estimated angle of the target is the tracking axis of the radar antenna. The estimated range of the target is the position of the center of the range gate. A Doppler radar estimates the target velocity rather than the range.

The radar model shown in Fig. 7.1 is expanded into a block diagram in Fig. 7.2 to show the relationship between estimates of the angle and range. The RF/IF section merely amplifies the target return to a useful level. The range-error detector estimates the target position in a range relative to the current center of the radar range gate. The range-gate servo centers the range gate on the target, and thus maintains the target return within the range gate.

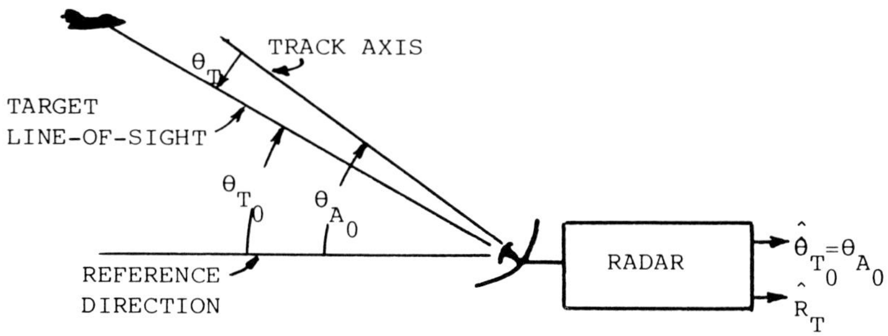

Fig. 7.1 Estimates from model of radar.

If the target return is not within the range gate, the angle detector will not have a signal to process. Likewise, the angle detector must provide an error voltage to the angle servos, repositioning the antenna boresight on the target, or the receiver and range gate will not have a signal to process. The approach to analysis of radar tracking used here is to model estimates of the angle (or range) under the assumption of perfect range (or angle) measurement. Specific mechanizations of estimators will be analyzed and then generalizations made by recognizing that each mechanization is an approximation to the maximum likelihood estimator described in Chapter 6.

The problem of analysis can be simplified by assuming that estimation of the angle or range of the target relative to the center of the resolution cell is developed instantaneously in comparison to the response time of the servo controlling the antenna boresight. Under this assumption, a tracking radar can be modeled as shown in Fig. 7.3, if the radar receiver (RF/IF plus AGC)

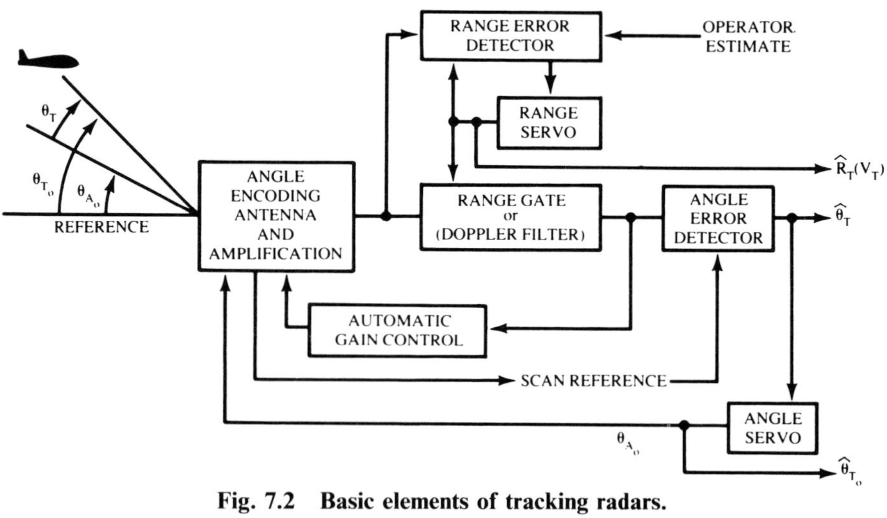

Fig. 7.2 Basic elements of tracking radars.

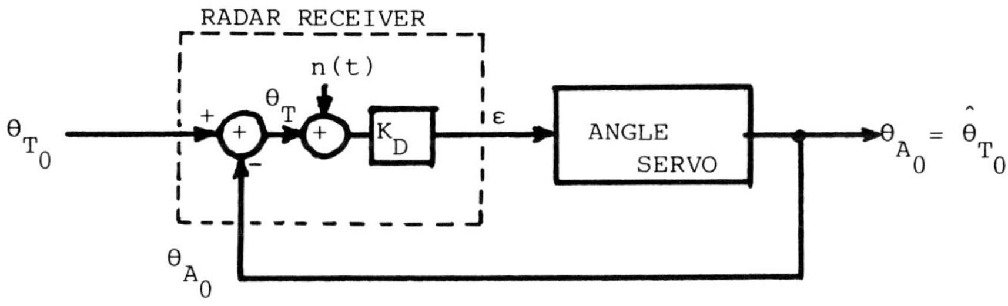

Fig. 7.3 Model of radar angle tracking loop.

and ranging circuits are working properly. In Fig. 7.3, the actual tracking error θ_T is related through the constant K_D to the error voltage ε whenever the noise n developed in the front end of the receiver is zero. In such a model, even though estimation of the range or angle of the target with respect to the current center of the range gate or tracking axis is not required due to the absence of noise, the process of developing these errors is referred to as estimation. If the range and angle of the target are random processes, the radar (receiver plus servo plus antenna) still must estimate these parameters. Such a model has only a limited range of validity, but it provides a useful first-order analysis.

7.1 ANGLE ESTIMATION BY CONICAL SCAN

Since the principles of operation of a conical scan radar such as that shown in Fig. 7.4 were discussed in Part I, this section will focus on a

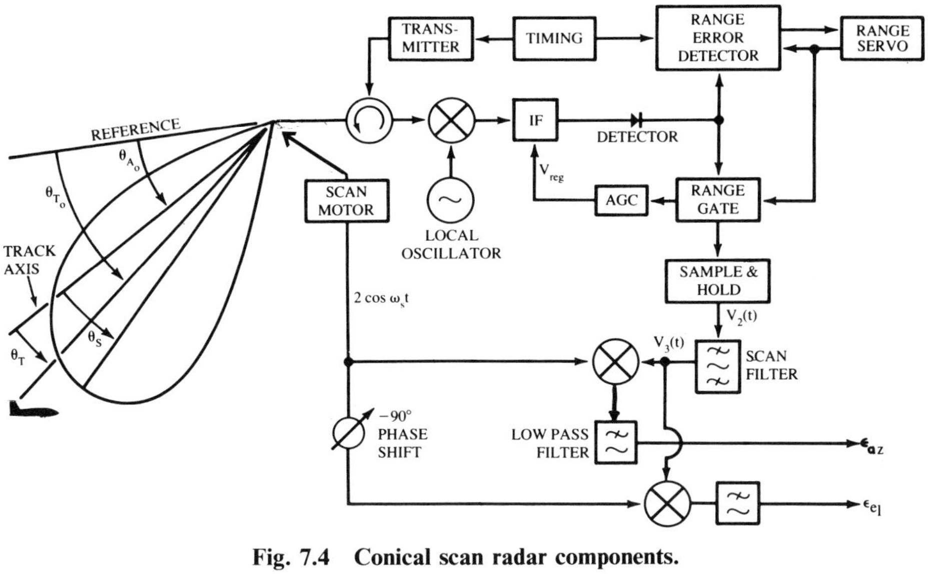

Fig. 7.4 Conical scan radar components.

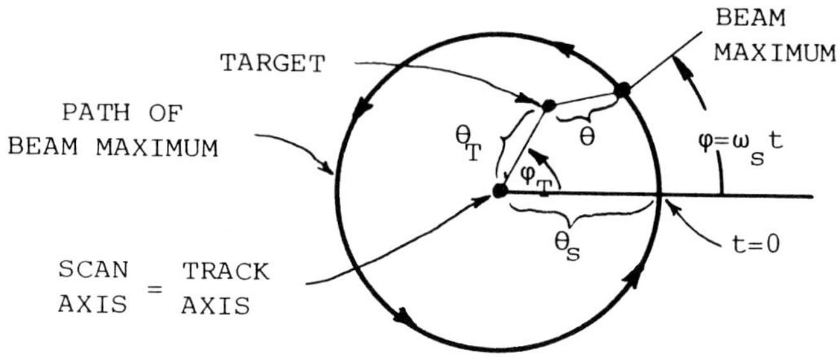

Fig. 7.5 Motion of beam maximum.

mathematical analysis relating the tracking error θ_T to the error voltage ε. The analysis is on the following assumptions: 1) a continuous (i.e., CW) signal is transmitted and 2) with perfect isolation in the circulator, the signal received is modulated in amplitude by the scanning of the antenna.

The maximum gain point of the antenna scans the circular pattern shown in Fig. 7.5 with the noted zero time reference. Figure 7.6 shows the antenna radiation pattern at two instances of time separated by one-half the scan period T_s. Also shown in Fig. 7.6 is the variation of the received signal in time, which is analyzed in this chapter for small tracking errors ($\theta_T/\theta_3 \ll 1$) and in later chapters for large angular errors.

Let $G_R(\theta)$ define the power gain of the radar antenna as a function of angle from the direction of maximum gain. The power received from a target of constant cross section at a fixed range is then

$$P_r = \frac{\sigma_T \lambda^2}{(4\pi)^3 R^4} P_R G_R^2 = K_1^2 P_R G_R^2$$

The amplitude of the received voltage $|V_r|$ will be proportional to the square

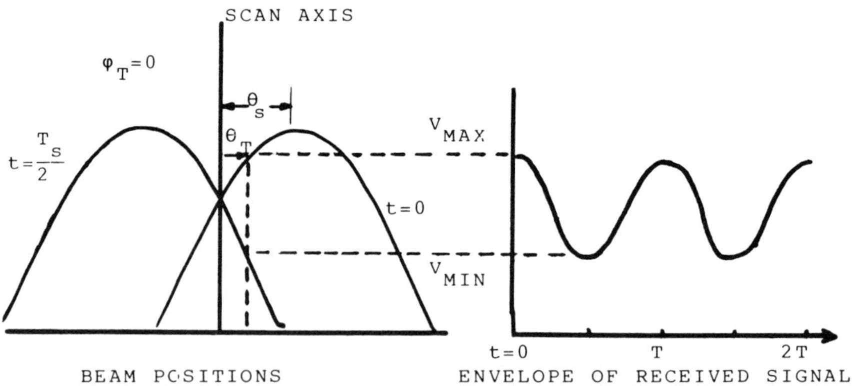

Fig. 7.6 Generation of error signal.

root of the power; thus,

$$|V_r| = \sqrt{2P_r} = \sqrt{2P_R K_1 G_R(\theta)}$$

Prior to its normalization by AGC mechanisms, the RF or IF signal can be written as

$$V_r(t) = \sqrt{2P_R K_1 G_R[\theta(t){:}\theta_T,\varphi_T]}\cos\omega_{IF}t \tag{7.1}$$

explicitly showing the time variation induced by scanning of the antenna. From Fig. 7.5, the law of cosines, and the assumption that $(\theta_T/\theta_S)^2 \ll 1$, it can be shown that $\theta(t) \approx \theta_S - \theta_T \cos(\omega_S t - \varphi_T)$. Use of this result in a two-term Taylor series expansion of $G_R(\theta)$ about $\theta = \theta_S$ yields

$$\begin{aligned} G_R(\theta) &= G_R(\theta_S) + \frac{dG_R}{d\theta}\bigg|_{\theta=\theta_S}(\theta - \theta_S) \\ &= G_R(\theta_S) - \frac{dG_R}{d\theta}\bigg|_{\theta=\theta_S}\theta_T\cos(\omega_S t - \varphi_T) \end{aligned}$$

Defining an error slope coefficient [16, p. 268] by

$$k_s = -\frac{\theta_3}{G_R(\theta_S)}\frac{d}{d\theta}G_R(\theta)\bigg|_{\theta=\theta_s} \tag{7.2}$$

the IF signal can finally be expressed as the simple amplitude modulated signal

$$V_r(t) = \sqrt{2P_R K_1 G_R(\theta_S)}\left[1 + \frac{k_s\theta_T}{\theta_3}\cos(\omega_s t - \varphi_T)\right]\cos\omega_{IF}t \tag{7.3}$$

If it is assumed that the AGC system adjusts the average amplitude of the IF output signal to V_o, the IF signal spectrum takes the form shown in Fig. 7.7.

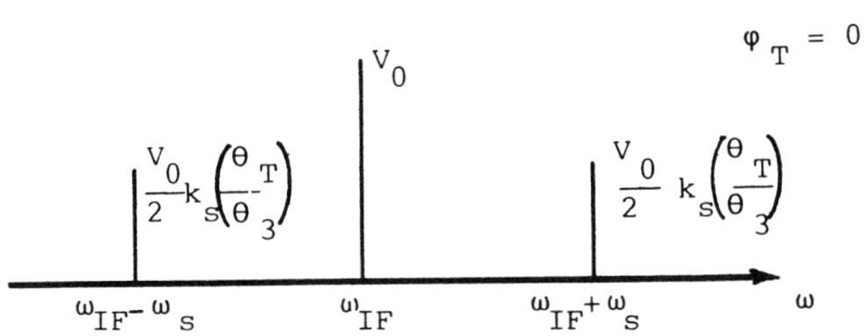

Fig. 7.7 Spectrum of CONSCAN radar IF amplifier output.

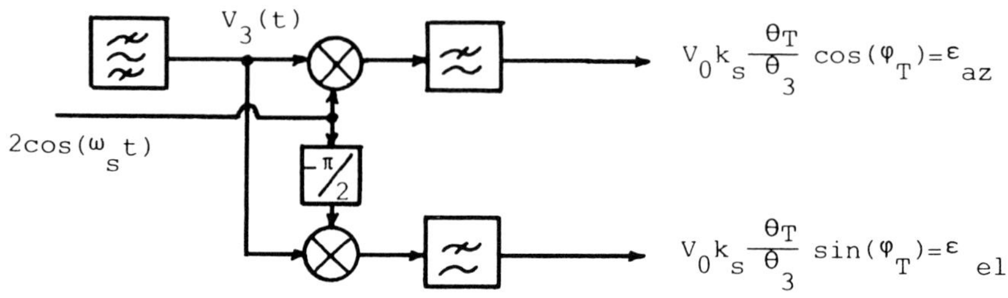

Fig. 7.8 Error voltage generation.

Having assumed a CW transmitted signal, the range gate and sample-and-hold circuit are unnecessary and are thus ignored. As small angular errors and $(k_s\theta_T/\theta_3)^2 \ll 1$ are also assumed, an envelope detector output $V_1(t)$ can be approximated as

$$V_1(t) = K|V_r(t)| \approx V_o\left[1 + \frac{k_s\theta_T}{\theta_3}\cos(\omega_S t - \varphi_T)\right] \tag{7.4}$$

where K is the gain of the IF amplifier.

The scan filter passes the component of the detected signal at the scan frequency. This component is mixed in the angle error detectors with a reference signal derived from the antenna scanning mechanism. The low-frequency mixer products are the error signals that drive the antenna servos and are expressed in Fig. 7.8.

As shown in Fig. 7.5, the angular tracking error in azimuth θ_{az} can be expressed as $\theta_{az} = \theta_T \cos\varphi_T$. Similarly, $\theta_{el} = \theta_T \sin\varphi_T$. The error voltages driving the antenna servos are related linearly to the angular error, provided V_o is held to a constant value. If V_o is constant, the linear model outlined in Fig. 7.3 is valid with the radar receiver replaced by the constant $K_D = V_o(k_s/\theta_3)$ volts/radian. How the radar AGC circuit maintains a constant value of the average detected IF output voltage is quite important and is discussed in the next section.

A Steady-State AGC Model

The AGC circuit considered here is shown in Fig. 7.9 and analysis of this circuit is based on a study by Shustov and Vakin [4, p. 141]. The purpose of an AGC system is to maintain the amplitude of the signal at the IF frequency out of the IF amplifier at a specified level, no matter how large or small the received signal level is. Sometimes called signal normalization, this task is accomplished by varying the gain of the IF amplifier through the use of a control voltage V_{reg}. Typically, the control voltage is not applied until the output of the IF amplifier exceeds some minimum value in order that maximum IF amplifier gain is available for weak targets. Withholding application of the AGC control voltage in this way is called AGC delay and is

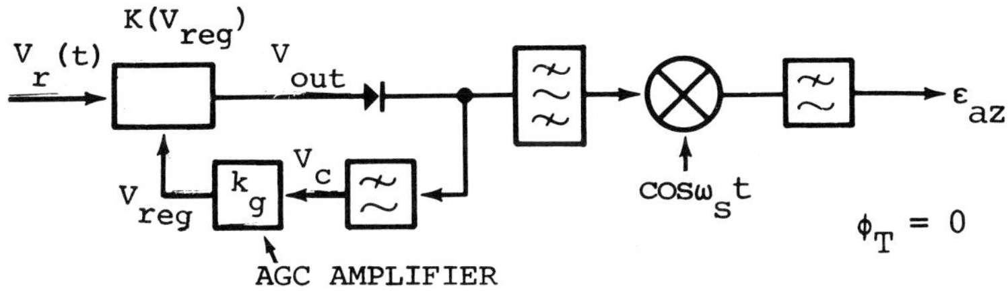

Fig. 7.9 Circuit for automatic gain control.

implemented via an AGC delay voltage. Such a delay in AGC implementation is not included in the present model, but it is considered in Chapter 10. Similarly, the relationship between the control voltage V_{reg} and IF gain $K(V_{reg})$ is normally exponential, but the linear relationship shown in Fig. 7.10 is used here.

The received signal in Eq. (7.3) can be expressed as

$$V_r(t) = V_{ino}\left[1 + \frac{k_s\theta_T}{\theta_3}\cos\omega_s t\right]\cos\omega_{IF}t \tag{7.5}$$

where $(k\theta_T/\theta_3)^2 \ll 1$. If the corner frequency ω_c of the low-pass filter in the AGC feedback loop is much less than the scan frequency, the voltage into the AGC amplifier V_c is approximately

$$V_c = K(V_{reg})V_{ino}$$

where V_{ino} is the square root of the power received from the target at the squint angle antenna gain. As $V_{reg} = k_g V_c = k_g K(V_{reg})V_{ino}$ and $K(V_{reg}) = K_o - \alpha V_{reg}$, the voltage used for the regulation and gain of the IF amplifier are the following functions of the amplitude of the received signal:

$$V_{reg} = \frac{k_g K_o V_{ino}}{1 + \alpha k_g V_{ino}} \tag{7.6}$$

$$K = \frac{K_o}{1 + \alpha k_g V_{ino}} \tag{7.7}$$

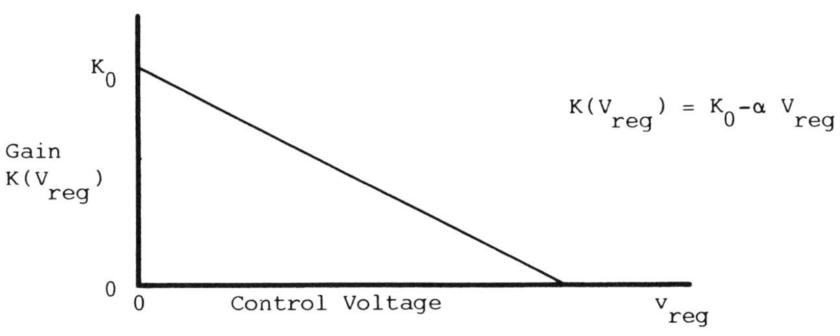

Fig. 7.10 IF amplifier gain vs control voltage.

Use of Eqs. (7.3) and (7.6)indicate that the voltage driving the azimuth antenna servo can be related to the tracking error in angle and input signal level as

$$\varepsilon_{az} = \frac{K_o V_{ino}}{1 + \alpha k_g V_{ino}} k_s \frac{\theta_T}{\theta_3} \cos(\phi_T) \tag{7.8}$$

$$V_{ino} = \sqrt{2P_R K_1 G_R(\theta_S)} \tag{7.9}$$

The original reason for this analysis was to determine the conditions under which the relationship $\varepsilon_{az} = K_D \theta_T$ can be considered valid. Equation (7.8) shows that, as the power received from the target becomes infinite (assume $\phi_T = 0$),

$$\varepsilon_{az} = \frac{K_o k_s}{\alpha k_g \theta_3} \theta_T = \frac{V_o k_s}{\theta_3} \theta_T$$

yielding the following relationship between the actual gain of the radar model K'_D and the input signal level:

$$\varepsilon_{az} = K_D \frac{V_{ino}}{\dfrac{1}{\alpha k_g} + V_{ino}} \theta_T = K'_D \theta_T \tag{7.10}$$

The relationship expressed in Eq. (7.10) is plotted in Fig. 7.11, showing that the desired linear relationship is valid for $\alpha k_g V_{ino} > 10$.

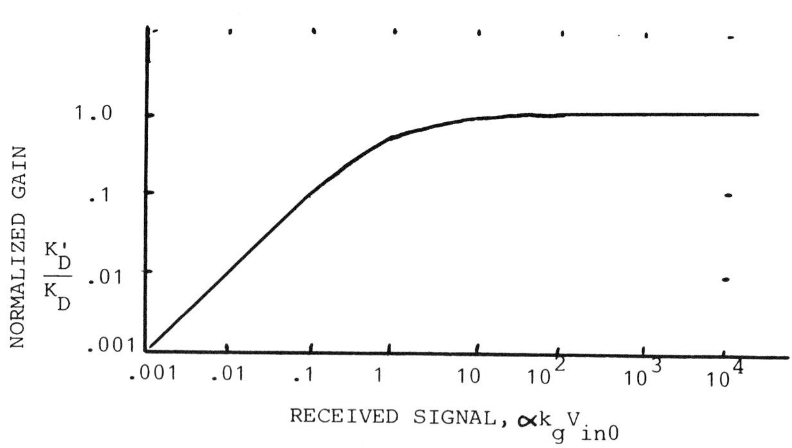

Fig. 7.11 Normalized IF gain variation with input signal.

Analysis of Pulsed Signal

A CW transmit signal has been assumed up to this point in this chapter and will be used extensively in later chapters. If a pulsed transmit signal had been assumed, the linear relationship of Eq. (7.10) between the error voltage and the angular tracking error would again result, provided that the range gate [or sample-and-hold (S&H)] was centered on and sampled the received pulses. The received CW signal $V_r(t)$, given by Eq. (7.3), had the spectral components shown in Fig. 7.7 after IF amplification. The pulsations of a simple pulsed radar can be represented by the infinite sampling waveform,

$$S(t) = \sum_{n=-\infty}^{\infty} P_\tau(t - nT) \tag{7.11}$$

where

$$P_\tau(t) = 1 \qquad 0 < t < \tau$$
$$0 \qquad \text{else}$$

and τ is the pulse duration and T the pulse repetition interval.

The received pulsed waveform is then

$$V_{rs}(t) = V_r(t)\, S(t)$$

which has the spectrum shown in Fig. 7.12 after an IF gain of K. As shown in Fig. 7.12, the spectral components are given by a $\sin x/x$ weighting of an infinite repetition of the nonpulsed spectrum. If $\tau \ll T$ and the envelope of $V_r(t)$ changes slowly enough over the interval τ to be considered constant, the

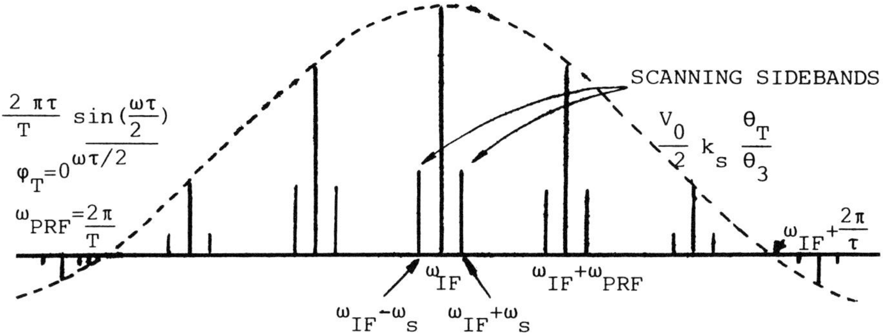

Fig. 7.12 Spectrum of pulsed, amplitude-modulated carrier.

sampled waveform from the detector is, from Eq. (7.4),

$$V_{1s}(t) = V_1(t)V_s(t) \approx \sum_{n=-\infty}^{\infty} V_1(nT)P_\tau(t - nT) \tag{7.12}$$

The range-gate and sample-and-hold circuits simply extend the duration of each sample V_1 (nT) and present the following signal to the scan filter:

$$V_{2s}(t) = \sum_{n=-\infty}^{\infty} V_1(nT)P_T(t - nT) \tag{7.13}$$

where

$$\begin{aligned} P_T(t) &= 1 \qquad 0 \le t \le T \\ &= 0 \qquad \text{else} \end{aligned}$$

The Fourier transform of this signal is

$$V_{2s}(\omega) = T\frac{\sin\dfrac{\omega T}{2}}{\omega\dfrac{T}{2}} \sum_{n=-\infty}^{\infty} V_1(nT)e^{-jn\omega T}e^{-j(\omega T/2)} \tag{7.14}$$

This Fourier transform can be expressed differently, as a minor digression will show. If $V_1(\omega)$ is the Fourier transform of $V_1(t)$ and the Fourier transform of a periodic sequence of impulses is [74]

$$\sum \delta(t - nT) \leftrightarrow \frac{2\pi}{T} \sum \delta\left(\omega - n\frac{2\pi}{T}\right) \tag{7.15}$$

then the Fourier transform of the product of the two signals that are functions of time is

$$V_1(t) \sum \delta(t - nT)$$

$$= \sum V_1(nT)\delta(t - nT) \leftrightarrow \frac{2\pi}{T} \sum V_1\left(\omega - n\frac{2\pi}{T}\right) \tag{7.16}$$

If the left side of Eq. (7.16) is convolved with the pulse $P_T(t)$, Eq. (7.13) results. The Fourier transform of Eq. (7.13) must then be the product of the right side of Eq. (7.16) and the Fourier transform of a pulse

$$P_T(t) \leftrightarrow T\frac{\sin(\omega T/2)}{\omega T/2} e^{-j(\omega T/2)} \tag{7.17}$$

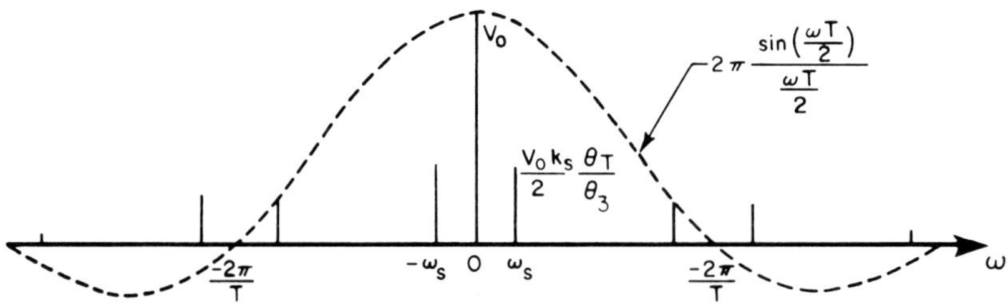

Fig. 7.13 Spectrum at input of scan filter.

Another representation of the signal into the scan filter is

$$\sum V_1(nT)P_T(t-nT) \leftrightarrow 2\pi \frac{\sin(\omega T/2)}{\omega T/2} e^{-j(\omega T/2)} \sum V_1\left(\omega - n\frac{2\pi}{T}\right)$$

$$= V_{2s}(\omega) \tag{7.18}$$

Each term of this sum is the spectrum of $V_1(t)$ weighted by $2\pi \sin(\omega T/2)/(\omega T/2)$ and offset by an integer multiple of $2\pi/T$, the radian pulse-repetition frequency. The spectrum presented to the scan filter is shown in Fig. 7.13.

The scan filter input is very close to the input of the CW case as $V_{2s}(\omega)$ is dominated by the terms near zero frequency. If the scan filter selects only the terms at the scan frequency, the output of the scan filter and the remainder of the analysis will be identical to that of the CW case.

Conical Scan Squint Angle

Two important characteristics of a conical scan radar are related to the squint angle of the antenna: the error slope coefficient and the crossover loss. A Gaussian-shaped antenna pattern is used to show the relationship. As shown in Chapter 6, such a gain function closely approximates either the more realistic $\sin x/x$ or $J_1(x)/x$ patterns, when both patterns have equal beamwidth and points within one-half beamwidth of the direction of maximum gain are compared. Of more importance here, the Gaussian function is analytically simpler to use.

The Gaussian power gain is

$$G(\theta) = G_o e^{-\ell n 2(2\theta/\theta_3)^2} \tag{7.19}$$

where θ_3 is the -3 dB beamwidth.

The error slope coefficient is then

$$k_s = 8\ell n 2\left(\frac{\theta_s}{\theta_3}\right) \approx 5.54\left(\frac{\theta_s}{\theta_3}\right) \tag{7.20}$$

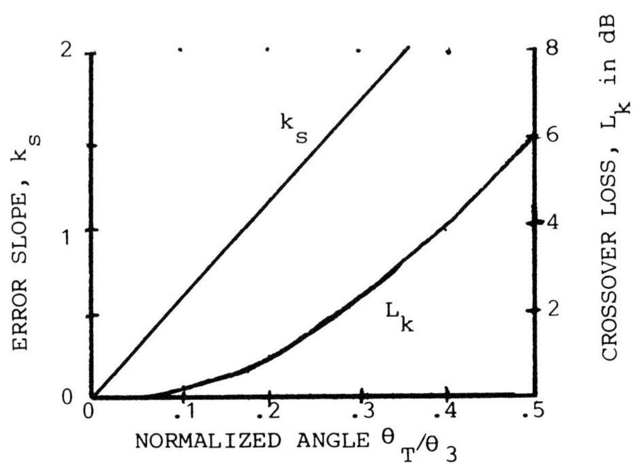

Fig. 7.14 CONSCAN error slope and crossover loss.

The crossover loss L_K is simply twice the difference between the peak gain (in dB) and squint angle gain as both the transmit and receive antennas are scanning. The crossover loss is essentially the signal power loss resulting from tracking the target at less than peak gain. For the Gaussian pattern, this is

$$L_k = 20 \log_{10} \frac{G(0)}{G(\theta_s)} \approx +24 \left(\frac{\theta_s}{\theta_3}\right)^2 \text{dB} \tag{7.21}$$

This error slope coefficient and crossover loss are plotted in Fig. 7.14 for $\theta_s \leq 0.4\theta_3$; the results are almost identical to those found in Barton [16, p. 269] for a $\sin x/x$ pattern over this restricted range.

The "optimum" squint angle can be assumed to be that which maximizes the ratio of the error slope coefficient to the square root of crossover loss. Such a maximization yields an optimum squint angle of $\theta_s/\theta_3 = 0.425$.

CONSCAN Summary

The operation of a conical scan radar can be analyzed with the aid of a relatively simple model that uses CW signals if the range gating and the gain control are functioning properly. Such a model can be modified to describe other types of radars and to analyze the effects of other signals on the radar. The azimuth channel of the proposed model is shown in Fig. 7.15. The voltage $V_s(t)$ represents the signal transmitted by the radar and $\sqrt{\sigma}$ the effect of the target cross section on the received voltage. The voltage gain variation with time of the CONSCAN antenna is $g(t)$, the power gain varies as $g^2(t) = G(t)$, and $g(t)$ is presumed to be a real function. The linear detector is shown in this model, but mathematical expediency dictates a square-law detector in later analysis.

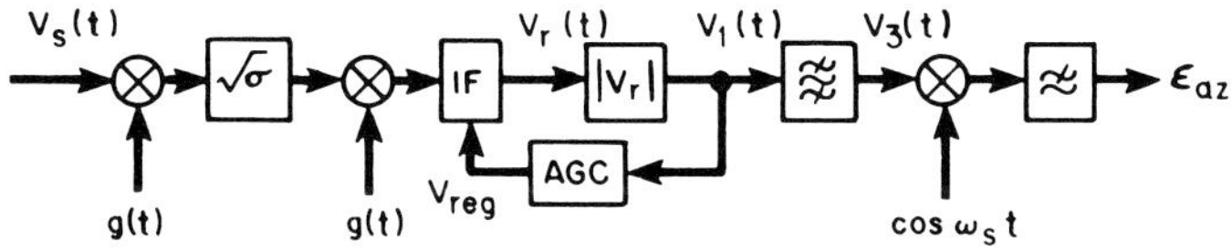

Fig. 7.15 CONSCAN radar model.

7.2 ANGLE TRACKING THROUGH SCAN WITH COMPENSATION

A scan-with-compensation (SWC) radar consists of a conical scan radar with an additional receive channel and the circuitry by which an auxiliary signal is subtracted from the conical scan (main) signal. An analysis model for such a radar appears in Fig. 7.16. Of course, subtraction of the auxiliary channel can be implemented at other locations in the model, depending on the availability of various radar components.

The primary advantage of a SWC radar, which cannot be appreciated until jamming is analyzed, is that the auxiliary channel (a mirror image of the main channel) can be used to cancel jamming common to both channels. For the present, it will simply be shown that Fig. 7.3 is still useful for analysis in the absence of jamming. The gain functions of the main and auxiliary antennas are assumed to be unbalanced as follows in order to determine the impact of any imbalance:

$$g_M(t) = g_{oM}\left(1 + \frac{k_{sM}\theta_T}{\theta_3}\cos\omega_s t\right)$$

$$= g_{oM}(1 + m_M \cos\omega_s t) \tag{7.22}$$

$$g_A(t) = g_{oA}\left(1 - \frac{k_{sA}\theta_T}{\theta_3}\cos\omega_s t\right)$$

$$= g_{oA}(1 - m_A \cos\omega_s t) \tag{7.23}$$

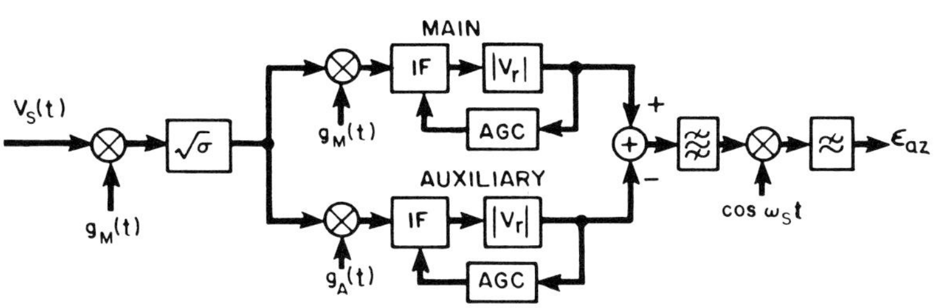

Fig. 7.16 Scan-with-compensation radar.

If both AGC circuits generate the same average value out of the IF amplifiers, then it is easy to show that the error voltage is of the proper sense for good tracking (i.e., positive for positive errors in angle) and is given by

$$\varepsilon_{az} = \frac{1}{2}\left[\frac{2m_M/(1+m_M^2)}{2} - \frac{m_M - m_A}{1-(m_M m_A/2)}\right] \tag{7.24}$$

Since the antenna gains have been linearized, validity requires m_M and m_A to be much smaller than 1, resulting in

$$\varepsilon_{az} = \tfrac{1}{2}(k_{SM} + k_{SA})\frac{\theta_T}{\theta_3} \tag{7.25}$$

The SWC radar receiver can thus be modeled by the single block in Fig. 7.3 with $K_D = (k_{SM} + k_{SA})/2\theta_3$.

7.3 ANGLE ESTIMATION BY MONOPULSE RADAR

A monopulse radar is capable of estimating the angular position of a target relative to the direction of the main beam of the antenna and the target range on a single pulse. Hence, the name monopulse. Many separate estimates are usually averaged together, typically through the radar tracking servos, as any estimate made from a single pulse may be grossly in error due to wave interference effects caused by extended target or ground scattering or just receiver noise. Still, a monopulse radar provides an estimate of the target position on a single pulse, whereas a scanning radar requires at least one scan (composed of many pulses), and usually more than one scan, to establish an AGC level prior to angular estimation.

Discussion of monopulse angle tracking will be approached here in a stepwise fashion (range tracking is similar in all radars). Consideration will first be given to tracking in one plane (either azimuth or elevation) by a variety of monopulse implementations, with the focus on one type of monopulse: amplitude sum-difference. The discussion will then follow the model used by Rhodes [25] to briefly examine the various categories of monopulse radar. Finally, this section will address monopulse angle tracking in two planes and analyze a two-channel monopulse.

Angle Estimation in One Plane

Sensing the Angle of Incidence. The two squinted antennas in Fig. 7.17 have phase centers separated by a distance d. The signals received from a source at an angle θ_T relative to the reference direction are

$$v_1 = Ag(\theta_S - \theta_T)\cos(\omega t + \varphi) \tag{7.26}$$

$$v_2 = Ag(\theta_S + \theta_T)\cos(\omega t - \varphi) \tag{7.27}$$

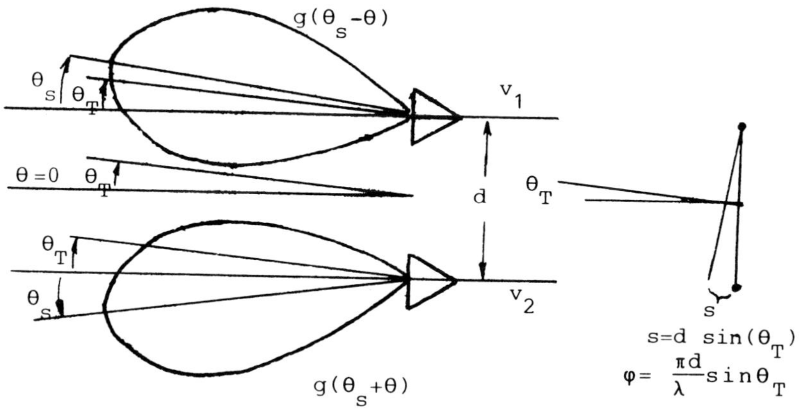

Fig. 7.17 Separated, squinted antennas.

where A and ω are the amplitude and radian frequency of the incident signal, respectively.

If the antenna separation is zero, then the phase centers of the antennas coincide and the angular offset of the target is contained in the amplitudes of the received voltages. Alternately, if the squint angle is zero, the gain functions are identical and the angle of the source from the reference direction is contained in the phase difference between the voltages v_1 and v_2.

Amplitude Sensing-Amplitude Detection. With the antenna separation set to zero, the amplitude detection circuit shown in Fig. 7.18 can be attached to the antennas to develop a voltage proportional to the angle between the target and the reference direction (i.e., the antenna boresight). In computation of the error voltage, the notation $|v_1|_{LP}$ means v_1 is envelope detected in the log IF amplifier and is then low-pass filtered. The error voltage is

$$\varepsilon = \log|v_1|_{LP} - \log|v_2|_{LP} = \log\left|\frac{v_1}{v_2}\right|_{LP} \tag{7.28}$$

Since the antenna signals are in phase,

$$v_1 = Ag(\theta_S - \theta_T)\cos\omega t \tag{7.29}$$

$$v_2 = Ag(\theta_S + \theta_T)\cos\omega t \tag{7.30}$$

For small angular errors, the variation in the antenna gain with the angle can be approximated as linear

$$|v_1|_{LP} = Ag(\theta_S - \theta_T) \approx g_o A(1 + k\theta_T) \tag{7.31}$$

$$|v_2|_{LP} = Ag_o(1 - k\theta_T) \tag{7.32}$$

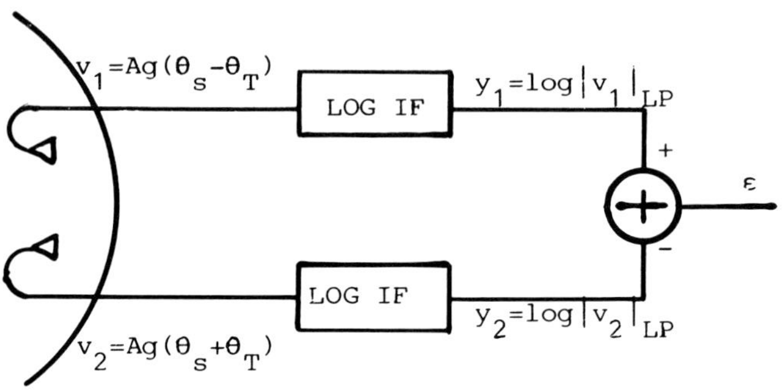

Fig. 7.18 Amplitude-amplitude monopulse.

resulting, ideally, in an error of

$$\varepsilon = \log \frac{1 + k\theta_T}{1 - k\theta_T} \approx \log(1 - 2k\theta_T) \approx \frac{2k}{2.3}\,\theta_T \tag{7.33}$$

Note that the signal has been normalized through subtraction of the logarithmic voltages.

Following the approach of Leonov and Fomichev [26, p. 253], one can show that any mismatch of the channel gains results in an angle bias error. Consider the same monopulse with unbalanced log IF gains shown in Fig. 7.19 and Gaussian antenna patterns. Assume the conditions $\theta_T \ll \theta_S < \theta_3/2$ and that the gain in the transmit pattern is constant with the angle. Relative to their maximum, the individual patterns are

$$g(\theta) = \text{Gaussian} = \sqrt{G_o e^{-2(\theta/\theta_3)^2 \ell n 2}} \tag{7.34}$$

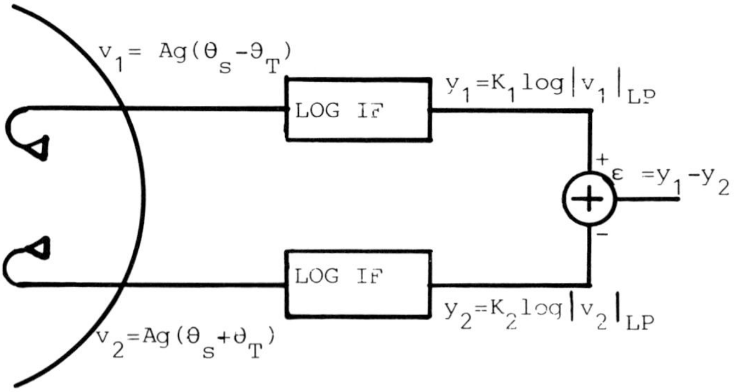

Fig. 7.19 Amplitude-amplitude monopulse with unequal gains.

The radar will point at an angle at which $\varepsilon = 0$,

$$\varepsilon = 0 = y_1 - y_2 = K_1 \ell n v_1 - K_2 \ell n v_2 = \ell n \frac{v_1^{K_1}}{v_2^{K_2}} \tag{7.35}$$

or where

$$v_1^{K_1} = v_2^{K_2} \tag{7.36}$$

Inserting Eq. (7.34) in Eq. (7.36) and ignoring terms θ_T^2 relative to θ_s^2 yields

$$\left\{A_o\sqrt{G_o}\,\exp\left[-2\ell n 2\left(\frac{\theta_s^2 - 2\theta_s\theta_T}{\theta_3^2}\right)\right]\right\}^{K_1}$$

$$= \left\{A_o\sqrt{G_o}\,\exp\left[-2\ell n 2\left(\frac{\theta_s^2 + 2\theta_s\theta_T}{\theta_3^2}\right)\right]\right\}^{K_2}$$

Combining comparable terms

$$\left\{A_o\sqrt{G_o}\,\exp\left[-2\ell n 2\left(\frac{\theta_s}{\theta_3}\right)^2\right]\right\}^{K_1 - K_2} = \left[\exp\left(-4\ell n 2\,\frac{\theta_S\theta_T}{\theta_3^2}\right)\right]^{K_2 + K_1}$$

and defining

$$g = \frac{K_1}{K_2} \qquad k = 4\left(\frac{\theta_S}{\theta_3}\right)\ell n 2$$

yields

$$\ell n[(A_o\sqrt{G_o}e^{-k\theta_s/2\theta_3})^{K_2(g-1)}] = -k\frac{\theta_T}{\theta_3}K_2(1+g)$$

and finally

$$\frac{\theta_T}{\theta_3} = \frac{-(g-1)}{(g+1)k}\ell n(A_o\sqrt{G_o}e^{-k\theta_s/2\theta_3}) \tag{7.37}$$

Thus, the boresight is driven off the target in order to make the error voltage zero.

Phase Sensing-Phase Detection. Similarly, separated antennas and a phase detector can be used to extract the source angle offset, as shown in Fig. 7.20. The received voltages are now

$$v_1 = A\cos(\omega t + \varphi) \qquad y_1 = \sqrt{2}\cos(\omega t + \varphi) \tag{7.38}$$

$$v_2 = A\cos(\omega t - \varphi)$$

$$y_2 = \sqrt{2}\cos\left(\omega t - \varphi - \frac{\pi}{2}\right) = \sqrt{2}\sin(\omega t - \varphi) \tag{7.39}$$

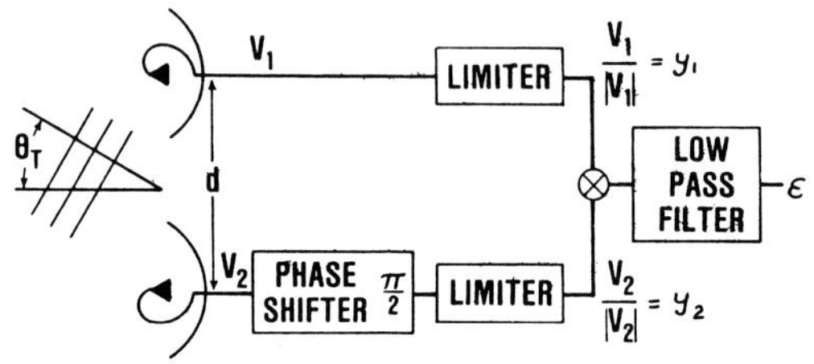

Fig. 7.20 Phase-phase monopulse.

and the error voltage can be easily computed as

$$y_3 = y_1 y_2 = \sin(2\omega t) + \sin(-2\varphi) \tag{7.40}$$

$$\varepsilon = (y_3)_{LP} = -\sin 2\varphi \approx -2\varphi \approx -kd\theta_T \tag{7.41}$$

Again, a bias error can exist in a less than ideal system. If the phase shifter (or limiter or anything else) introduces a fixed phase error ξ, then

$$y_1 = \sqrt{2}\cos(\omega t + \varphi) \tag{7.42}$$

$$y_2 = \sqrt{2}\cos\left(\omega t - \varphi - \frac{\pi}{2} - \xi\right) \tag{7.43}$$

The error voltage is computed to be

$$y_3 = 2\sin(\omega t - \varphi - \xi)\cos(\omega t + \varphi) \tag{7.44}$$

$$= \sin(2\omega t - \xi) + \sin(-2\varphi - \xi) \tag{7.45}$$

$$\varepsilon = (y_3)_{LP} = \sin(-2\varphi - \xi) \tag{7.46}$$

Again, the servo will stop where $\varepsilon = 0$ or when

$$2\varphi = kd\sin\theta_T = -\xi \tag{7.47}$$

and will result in a bias error

$$-\theta_T = \sin^{-1}\left(\frac{\xi}{kd}\right) \tag{7.48}$$

Sum-Difference Detection. The most common monopulse system, shown in Fig. 7.21, can be either amplitude or phase sensing. The error voltage will be derived here as a function of signal level and angular offset, following the approach of Shustov and Vakin [4]. If $\theta_s = 0$, $\psi_o = \pi/2$, and

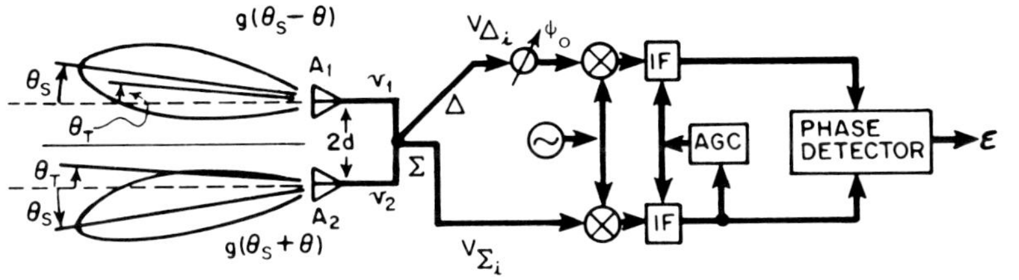

Fig. 7.21 Sum-difference monopulse.

$d \neq 0$, the radar is phase sensing and sum-difference detection; but if $\theta_s \neq 0$, $\psi_o = 0$, and $d = 0$, the radar is amplitude sensing with sum-difference detection. Shustov and Vakin considered reception only in a jamming environment in which jamming signals dominate over target variations caused by the transmit pattern. The target tracking case yields similar results near the track axis where the sum pattern is approximately constant. The received voltages are

$$\begin{aligned} v_{\substack{\Sigma i \\ \Delta i}} &= v_1 \pm v_2 \\ &= A[g(\theta_S - \theta_T)\cos(\omega t + \varphi) \pm g(\theta_S + \theta_T)\cos(\omega t - \varphi)] \end{aligned} \tag{7.49}$$

where $\varphi = kd\sin\theta_T$ $(k = 2\pi/\lambda)$.

To simplify the analysis, complex signal notation is employed. Let

$$v_1 = \mathrm{Re}\{V_1 e^{j\omega t}\} \quad V_1 = Ag(\theta_S - \theta_T)e^{j\varphi} = Ag_1 e^{j\varphi} \tag{7.50}$$

$$v_2 = \mathrm{Re}\{V_2 e^{j\omega t}\} \quad V_2 = Ag_2 e^{-j\varphi} \tag{7.51}$$

Then, out of the IF amplifiers, the signals are

$$V_{\Sigma o} = K_\Sigma(V_1 + V_2) = K_\Sigma A(g_1 e^{j\varphi} + g_2 e^{-j\varphi}) \tag{7.52}$$

$$V_{\Delta o} = e^{-j\psi'_o} K_\Delta(V_1 - V_2) = K_\Delta A(g_1 e^{j\varphi} - g_2 e^{-j\varphi})e^{-j\psi'_o} \tag{7.53}$$

where $\psi'_o = \psi_o + \varphi_\Sigma - \varphi_\Delta$, ψ_o is the difference channel phase shift, and $(\varphi_\Sigma - \varphi_\Delta)$ the phase shift difference between Σ and Δ channels. Let the IF channels be matched in phase, $\varphi_\Delta = \varphi_\Sigma$.

From the AGC analysis of the CONSCAN radar, the IF gains are related to the input sum voltage as

$$K_\Sigma = K_\Delta = \frac{K_o}{1 + \alpha k_g |v_{\Sigma_i}|} \tag{7.54}$$

but now the control voltage can be the instantaneous sum channel output.

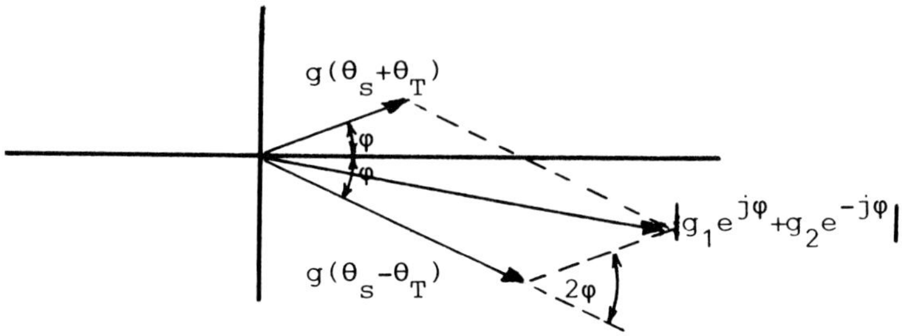

Fig. 7.22 Sum channel input voltage.

This sum input voltage can be visualized with the aid of Fig. 7.22. As

$$v_{\Sigma_i} = R_e\{A(g_1 e^{j\varphi} + g_2 e^{-j\varphi}) e^{j\omega_{IF}t}\} \tag{7.55}$$

then

$$|v_{\Sigma_i}| = A|g_1 e^{j\varphi} + g_2 e^{-j\varphi}| \tag{7.56}$$

where

$$g_{\substack{1\\2}} = g(\theta_S \mp \theta_T) \tag{7.57}$$

Finally

$$|v_{\Sigma_i}| = A\sqrt{g_1^2 + g_2^2 + 2g_1 g_2 \cos 2\varphi} \tag{7.58}$$

The phase detector output after low-pass filtering is the error voltage,

$$\varepsilon = \tfrac{1}{2}\,\mathrm{Re}\{V_{\Sigma_o} V^*_{\Delta_o}\} \tag{7.59}$$

$$\begin{aligned}\varepsilon &= \frac{A^2 K_\Sigma^2}{2}\,\mathrm{Re}\{(g_1 e^{j\varphi} + g_2 e^{-j\varphi})(g_1 e^{-j(\varphi - \psi_o)} - g_2 e^{j(\varphi + \psi_o)})\}\\ &= \frac{K_\Sigma^2 A^2}{2}[(g_1^2 - g_2^2)\cos\psi_o + 2g_1 g_2 \sin 2\varphi \sin\psi_o]\end{aligned} \tag{7.60}$$

Incorporating the IF gain from Eq. (7.54) results in

$$\varepsilon = \frac{K_o^2 A^2}{2}\,\frac{(g_1^2 - g_2^2)\cos\psi_o + 2g_1 g_2 \sin 2\varphi \sin\psi_o}{(1 + \alpha k_g A\sqrt{g_1^2 + g_2^2 + 2g_1 g_2 \cos 2\varphi})^2} \tag{7.61}$$

which is identical to Eq. (4.9) of Shustov and Vakin [4].

For the amplitude sensing monopulse, $\varphi = \psi_o = 0$ and the error voltage becomes

$$\varepsilon = \frac{\dfrac{K_o^2 A^2}{2}(g_1^2 - g_2^2)}{[1 + \alpha k_g A(g_1 + g_2)]^2} \tag{7.62}$$

Continuing to follow Shustov and Vakin, let

$$\mu = \alpha k_g A \qquad K_{pd} = \frac{K_o^2 A^2}{2\mu^2} = \frac{K_o^2}{2\alpha^2 k_g^2} \tag{7.63}$$

Then the error voltage is

$$\varepsilon = \frac{\mu^2 K_{pd}(g_1^2 - g_2^2)}{[1 + \mu(g_1 + g_2)]^2} \tag{7.64}$$

which, as the received signal increases, becomes

$$\varepsilon = K_{pd}\left(\frac{g_1 - g_2}{g_1 + g_2}\right) \tag{7.65}$$

Equation (7.64) is plotted in Shustov and Vakin [4, p. 183] and reproduced here as Fig. 7.23 for $K_{pd} = 1$ and for a $\sin x/x$ pattern.

In Fig. 7.23,

$$\theta_S = \frac{\theta_3}{3} \qquad \theta_3 = \frac{\lambda}{D} \tag{7.66}$$

where D is the aperture diameter and θ_3 the beamwidth of each horn. In Fig. 7.23, note the many stable tracking points and the width of the main tracking region in beamwidths $\Delta\,\theta_{max}$, which is approximately $2\theta_3$ for large signals.

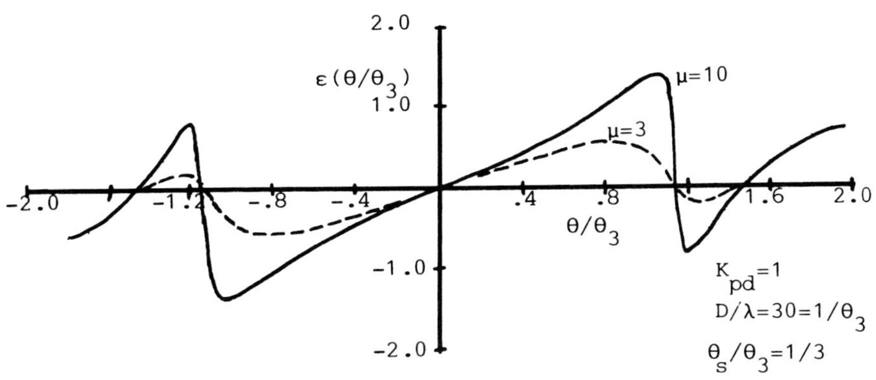

Fig. 7.23 Error voltage vs angular error for sin x/x pattern.

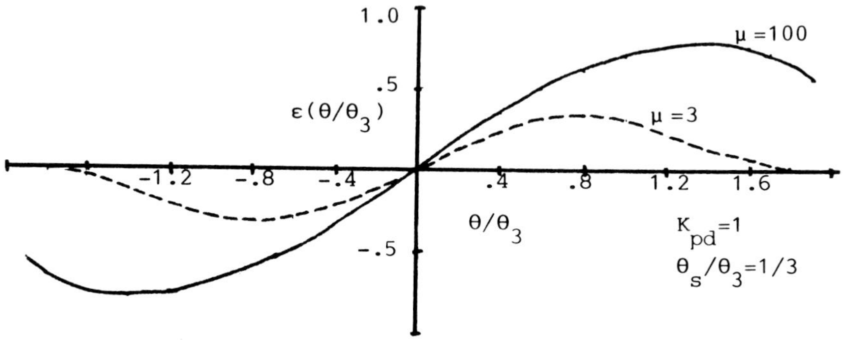

Fig. 7.24 Error voltage vs angular error for Gaussian pattern.

Figure 7.24 shows the error voltage vs angular error for a Gaussian pattern

$$g(\theta) = \sqrt{G_0}e^{-2\ell n 2(\theta/\theta_3)^2} = g_0 e^{-k_1(\theta/\theta_3)^2} \tag{7.67}$$

with $K_{pd} = 1$ and $k_1 = 1.3863$.

The two antenna voltage patterns are

$$g_1 = g_o \exp\left[-k_1\left(\frac{\theta_S}{\theta_3} - \frac{\theta}{\theta_3}\right)^2\right] \tag{7.68}$$

$$g_2 = g_o \exp\left[-k_1\left(\frac{\theta_S}{\theta_3} + \frac{\theta}{\theta_3}\right)^2\right] \tag{7.69}$$

The coefficient K_D, relating error voltage to angular error near the boresight, is important because it controls the speed of the servo response. This coefficient is given by

$$K_D = \frac{d\varepsilon}{d\theta}\bigg|_{\theta=0} \tag{7.70}$$

Using Eq. (7.64),

$$\frac{d\varepsilon}{d\theta}\bigg|_{\theta=0} = -\frac{4\mu^2 K_{pd} g(\theta_S) \dfrac{dg}{d\theta}\bigg|_{\theta_S}}{[1 + 2\mu g(\theta_S)]^2} \tag{7.71}$$

For the Gaussian pattern of Eq. (7.67), this coefficient, normalized by the

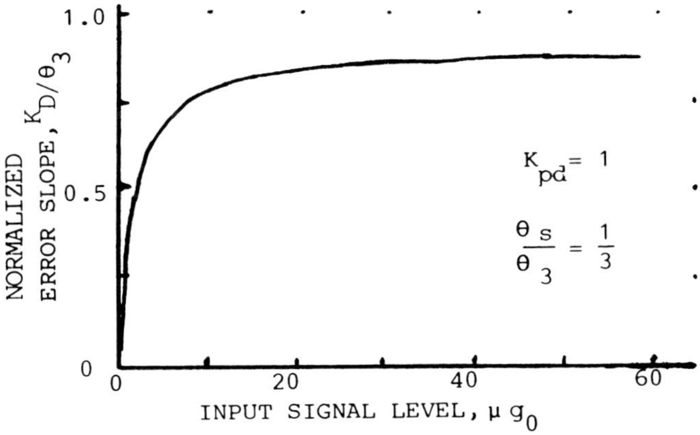

Fig. 7.25 Error slope coefficient vs signal level.

beamwidth, is

$$\frac{K_D}{\theta_3} = \frac{8\mu^2 K_{pd} g^2(\theta_S) 2\ell n 2 \left(\dfrac{\theta_S}{\theta_3}\right)}{[1 + 2\mu g(\theta_S)]^2} \tag{7.72}$$

which is plotted in Fig. 7.25.

Theory of Monopulse

Three different monopulse angle sensors have been described and other monopulse variations will be discussed in later sections. It is desirable to explain a fundamental theory of monopulse in order to gain a better understanding of the many variations. Such a theory was proposed by Rhodes [25] and later expanded by Leonov and Fomichev [26]. Rhodes postulated three principles of monopulse and organized all existing monopulse radars into six classifications. Leonov and Fomichev showed that Rhodes' postulates lead to nine classifications. Rhodes' postulates will be discussed using the previously described radars.

The first radar considered used amplitude sensing and amplitude detection. The error voltage resulting from an angular tracking error θ_T appeared as

$$\varepsilon = \ell n \left[\frac{g(\theta_S - \theta_T)}{g(\theta_S + \theta_T)}\right] = \ell n \left[\frac{g_a(\theta_T)}{g_b(\theta_T)}\right] \tag{7.73}$$

where $g_a(0) = g(\theta_S)$. The function $g(\theta)$ is the voltage gain of either antenna when it is not squinted. As shown in Fig. 7.26, $g_a(\theta)$ and $g_b(\theta)$ are the squinted pattern gains. If odd symmetry is required [i.e., $g_a(\theta) = g_b(-\theta)$] in

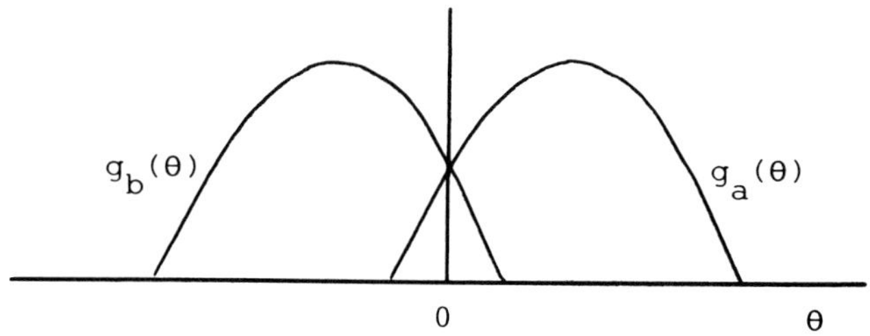

Fig. 7.26 Amplitude monopulse patterns.

the patterns about $\theta = 0$, then

$$\varepsilon = \ell n \frac{g_a(\theta)}{g_a(-\theta)} = \ell n \rho(\theta) \tag{7.74}$$

The error information appears in the form of an amplitude ratio $\rho(\theta)$, which can also be called a multiplicative ratio, $r_m(\theta)$ if

$$r_m(\theta) = \rho(\theta) = \frac{g_a(\theta)}{g_a(-\theta)} \tag{7.75}$$

since

$$r_m(\theta) r_m(-\theta) = \rho(\theta)\rho(-\theta) = 1 = \frac{g_a(\theta)}{g_a(-\theta)} \frac{g_a(-\theta)}{g_a(\theta)} \tag{7.76}$$

In the second or phase sensing radar, the error information can again be expressed as a ratio. Using complex signal notation and the antenna patterns shown in Fig. 7.27, the received voltages are

$$v_1 = A\cos(\omega t + \varphi_a) = g(\theta)\,\mathrm{Re}\{e^{j\omega t} e^{j\varphi_a(\theta)}\} \tag{7.77}$$

$$v_2 = g(\theta)\,\mathrm{Re}\{e^{j\omega t} e^{j\varphi_b(\theta)}\} \tag{7.78}$$

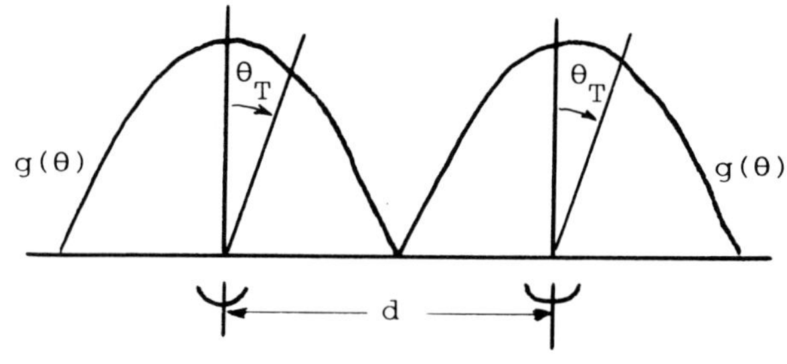

Fig. 7.27 Phase monopulse patterns.

A limiter in the receiver forces the amplitude of each channel to unity

$$y_1 = \mathrm{Re}\{e^{j(\omega t + \varphi_a)}\} \tag{7.79}$$

$$y_2 = \mathrm{Re}\{e^{j(\omega t + \varphi_b - \pi/2)}\} \tag{7.80}$$

and one channel is phase shifted by 90 deg.

If a phase detector followed by a low-pass filter is used to extract the error voltage, then the error voltage is

$$\begin{aligned} \varepsilon &= \mathrm{Re}\{y_1 y_2^*\} = \mathrm{Re}\{e^{j(\varphi_a - \varphi_b + \pi/2)}\} \\ &= \cos\left(\varphi_a - \varphi_b + \frac{\pi}{2}\right) = -\sin(\varphi_a - \varphi_b) \\ &= -\mathrm{Im}\left\{\frac{e^{j\varphi_a}}{e^{j\varphi_b}}\right\} \end{aligned} \tag{7.81}$$

Requiring $\varphi_a(\theta) = -\varphi_b(\theta) = -\varphi_a(-\theta)$, a multiplicative ratio $r_m(\theta)$ can be defined as

$$r_m(\theta) = \frac{e^{j\varphi_a(\theta)}}{e^{j\varphi_a(-\theta)}} = e^{j[\varphi_a(\theta) - \varphi_a(-\theta)]} = e^{j\Phi(\theta)} \tag{7.82}$$

since

$$r_m(\theta) r_m(-\theta) = 1 \tag{7.83}$$

With the above concepts, Rhodes' three postulates of monopulse can be stated meaningfully as

1. Angle information appears in the form of a ratio. For example, the multiplicative ratios formed by

(a) Amplitude sensing:

$$r_m(\theta) = \frac{g(\theta)}{g(-\theta)}$$

amplitude ratio

(b) Phase sensing:

$$r_m(\theta) = \frac{e^{j\varphi(\theta)}}{e^{j\varphi(-\theta)}}$$

phase ratio

2. This sensing ratio for positive error is the inverse of that for negative error,

$$r_m(\theta) = \frac{1}{r_m(-\theta)}$$

3. The angle error is derived (detected) from the sensing ratio as a real,

odd function:

(a) Amplitude

$$\varepsilon = \log r_m(\theta) = \log \rho(\theta)$$

(b) Phase

$$\varepsilon = -\operatorname{Im} r_m(\theta) = \sin \Phi(\theta)$$

Rhodes shows that the amplitude ratio $\rho(\theta)$ can be formed even if phase sensing is used. An example is shown in Fig. 7.28.

According to Rhodes, four forms of monopulse can be distinguished by the method of detection and the method of sensing. Rhodes called the form with amplitude detection and either amplitude or phase sensing class II and the form using phase detection with amplitude or phase sensing class III.

Only multiplicative ratios have been discussed. Rhodes showed that a more desirable additive ratio can be derived by sum-difference processing either amplitude or phase sensing antenna outputs. For example, the additive ratio for amplitude sensing antennas is

$$r_A(\theta) = \frac{g_a(\theta) - g_b(\theta)}{g_a(\theta) + g_b(\theta)} = \frac{g_a(\theta) - g_a(-\theta)}{g_a(\theta) + g_a(-\theta)} = \frac{g_{\text{odd}}(\theta)}{g_{\text{even}}(\theta)} \tag{7.84}$$

The three postulates of Rhodes are still satisfied if *inverse* is defined as the operation that produces the identity element (i.e., in matrice theory $AA^{-1} = I$) and the *identity* element is defined as one for multiplicative ratios and zero for additive ratios. Thus,

$$r_m(\theta) r_m(-\theta) = 1 \tag{7.85}$$

$$r_A(\theta) + r_A(-\theta) = 0 \tag{7.86}$$

This additive ratio can be derived from the multiplicative ratio $r_m(\theta)$, Eq. (7.75) as

$$r_A(\theta) = \frac{r_m(\theta) - 1}{r_m(\theta) + 1} \tag{7.87}$$

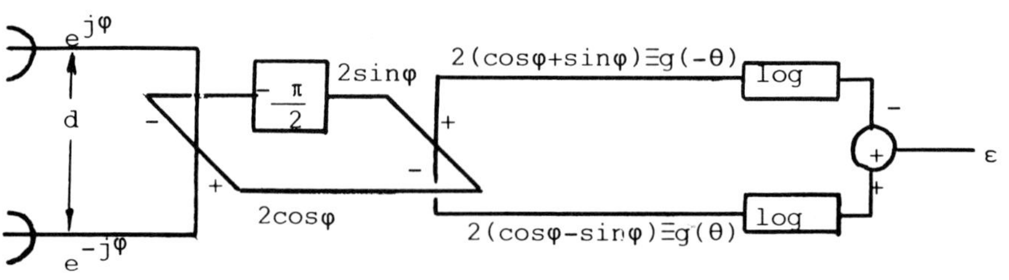

Fig. 7.28 Phase sensing and amplitude detection.

The target angle is embedded in this additive ratio and its recovery requires sum-difference detection (an example of this is shown in Fig. 7.21). Sum-difference processing is often preferred because no angle bias results from drift in the active IF components.

Six types of monopulse result, as defined by the method of angle sensing and the method of angle detection. These six are tabulated in Table 7.1, in which the first symbol relates to the method of sensing and the second symbol to the type of detection.

Table 7.1 Monopulse types

	Sensing		
Detection	Amplitude	Phase	Complex
Amplitude	A-A	P-A	C-A
Phase	A-P	P-P	C-P
Σ-D	A-ΣD	P-ΣD	C-ΣD

In considering the generally complex nature of the antenna pattern, Leonov and Fomichev [26] define a third method of angle sensing called complex. Such sensing is consistent with Rhodes' general theory. Since angle detection again can take three forms, Leonov and Fomichev consider nine classifications of monopulse, of which only four are widely used: amplitude-amplitude, phase-phase, amplitude sum-difference, and phase sum-difference. The notation in Table 7.1 follows that of Ref. 26 (p. 13).

Monopulse Angle Estimation in Two Planes. Measurement of the angle error in one plane is seldom sufficient for tracking. Normally, the angle error in two orthogonal directions is required. This measurement of error can be accomplished through the use of four antenna patterns. Each antenna pattern can be modeled near the track axis as a squinted Gaussian beam. Each one-way voltage gain function is given by

$$g(\theta) = g_o e^{-2\ell n 2(\theta/\theta_3)^2} \tag{7.88}$$

where θ is the total angle from the beam peak. For small angular displacements, θ can be computed from the azimuth and elevation displacements defined in Fig. 7.29 (a factor R_1, the range, is suppressed in all of the following). This figure presents the field of view as seen from the radar.

For a beam with its maximum in quadrant 1, displacement from the

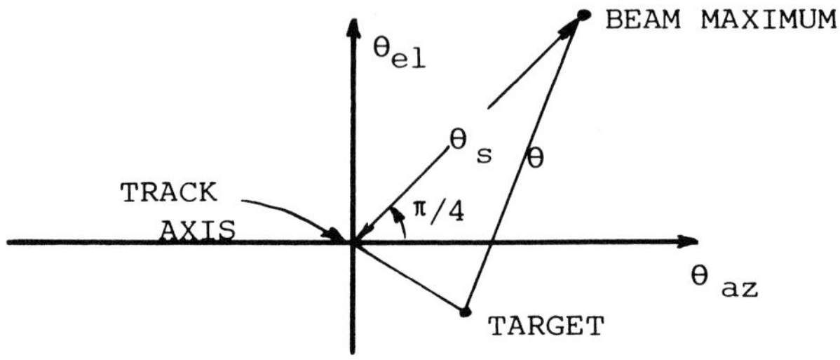

Fig. 7.29 Definition of angular displacement from track axis.

beam maximum can be written in terms of azimuth and elevation displacements as

$$\left(\frac{\theta_S}{\sqrt{2}} - \theta_{el}\right)^2 + \left(\frac{\theta_S}{\sqrt{2}} - \theta_{az}\right)^2$$

$$= \theta^2 \approx \frac{\theta_S^2}{2} - \sqrt{2}\theta_S\theta_{el} + \frac{\theta_S^2}{2} - \sqrt{2}\theta_S\theta_{az} \tag{7.89}$$

The quadrant 1 voltage gain is then,

$$g_1(\theta) = g[\theta(\theta_{az}, \theta_{el})] = g(\theta_{el}, \theta_{az})$$

$$= g_o \exp\left[-2\ell n 2\left(\frac{1}{\theta_3^2}\right)[\theta_S^2 - \sqrt{2}\theta_S(\theta_{el} + \theta_{az})]\right]$$

$$= g(\theta_S) \exp\left[+2\sqrt{2\ell n 2}\left(\frac{\theta_S}{\theta_3}\right)\left(\frac{\theta_{el} + \theta_{az}}{\theta_3}\right)\right] \tag{7.90}$$

If a new monopulse error slope coefficient is defined as

$$k_m = 2\sqrt{2\ell n 2}\,\frac{\theta_S}{\theta_3} \tag{7.91}$$

and if $k_m(\theta_{el} + \theta_{az})/\theta_3 \ll 1$, then the voltage gain for the beam in quadrant 1 is

$$g_1(\theta_{az}, \theta_{el}) = g(\theta_S)\left[1 + \frac{k_m}{\theta_3}(\theta_{az} + \theta_{el})\right] \tag{7.92}$$

The gain functions of the antenna patterns directed into the other quadrants can be found to be (under the assumption that all patterns are in phase)

$$g_1 = g_o\left[1 + \frac{k_m}{\theta_3}(\theta_{az} + \theta_{el})\right], \quad g_3 = g_o\left[1 - \frac{k_m}{\theta_3}(\theta_{el} + \theta_{az})\right]$$
$$g_2 = g_o\left[1 + \frac{k_m}{\theta_3}(\theta_{el} - \theta_{az})\right], \quad g_4 = g_o\left[1 + \frac{k_m}{\theta_3}(\theta_{az} - \theta_{el})\right] \tag{7.93}$$

The hybrid network of Fig. 7.30, when attached to the four antenna feeds, extracts signals related to the azimuth or elevation angles of the target from the track axis.

Using the gain functions of Eq. (7.93), the combining circuitry of Fig. 7.30, and a received signal voltage of amplitude A, the output sum and difference voltages are (in complex signal notation)

$$\sum = 2g_o A \tag{7.94}$$

$$\Delta_{el} = \frac{A}{2}\left[2g_o\left(1 + \frac{k_m}{\theta_3}\theta_{el}\right) - 2g_o\left(1 - \frac{k_m}{\theta_3}\theta_{el}\right)\right] = 2g_o\frac{k_m}{\theta_3}\theta_{el}A \tag{7.95a}$$

$$\Delta_{az} = \frac{A}{2}\left[2g_o\left(1 + \frac{k_m}{\theta_3}\theta_{az}\right) - 2g_o\left(1 - \frac{k_m}{\theta_3}\theta_{az}\right)\right] = 2g_o\frac{k_m}{\theta_3}\theta_{az}A \tag{7.95b}$$

These signals are near the radar's RF frequency. The difference signals are normalized in separate IF amplifiers by the common sum signal. Normalization can be accomplished in many ways, but a common approach was shown in the earlier section on tracking in one plane. The resultant low-frequency voltages from phase detectors comparing sum and difference channels are

$$\varepsilon_{az} = \frac{k_m}{2\theta_3}\theta_{az} \tag{7.96}$$

$$\varepsilon_{el} = \frac{k_m}{2\theta_3}\theta_{el} \tag{7.97}$$

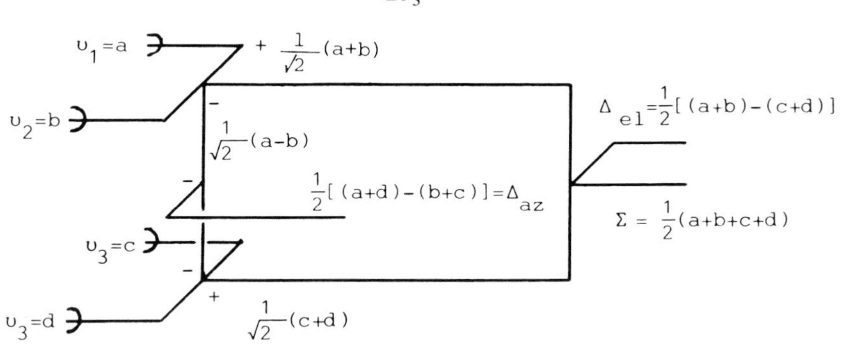

Fig. 7.30 Full monopulse hybrid network.

These error voltages can be used to drive two separate antenna servo loops similar to those shown in Fig. 7.3 to maintain the alignment of the track axis with the target.

Two-Channel Monopulse. Three IF channels are not essential for the recovery of target angular errors. The difference and sum channels can be combined (multiplexed) in many ways, resulting in a need for only two (sometimes one) IF channels. Two examples of two-channel monopulse radars follow.

In the first example (Fig. 7.31), the difference channels are combined in phase quadrature before amplification. Signal normalization is provided by the sum signal. If V_o is the amplitude of the AGC controlled IF voltage from the sum channel, then the difference channel signal is

$$\Delta(t) = \frac{V_o k_m}{\sqrt{2}\theta_3}(\theta_{az}\cos\omega_{IF}t + \theta_{el}\sin\omega_{IF}t) \tag{7.98}$$

With reference to Fig. 7.31, the error voltages are now

$$\varepsilon_{el} = \frac{V_o k_m}{2\sqrt{2}\theta_3}\theta_{el} \tag{7.99}$$

$$\varepsilon_{az} = \frac{V_o k_m}{2\sqrt{2}\theta_3}\theta_{az} \tag{7.100}$$

These error voltages are functions of any phase imbalance between the Σ and Δ channels.

A second two-channel monopulse receiver (shown in Fig. 7.32) combines the complex difference signal in quadrature with the sum signal. One benefit of such a combination is that both IF channels amplify a signal of almost equal amplitude, which usually makes it simpler to match the channels in phase or amplitude. Normalization is provided by the hard limiter in each channel. Near the tracking axis, it is assumed that $|\Sigma| \approx |\Sigma \pm j\Delta|$ and that the

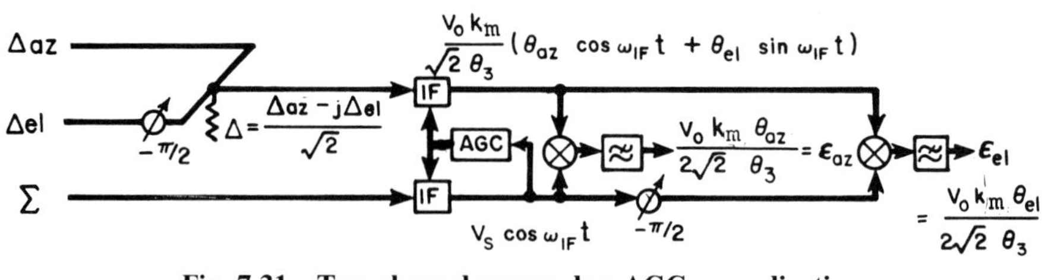

Fig. 7.31 Two-channel monopulse: AGC normalization.

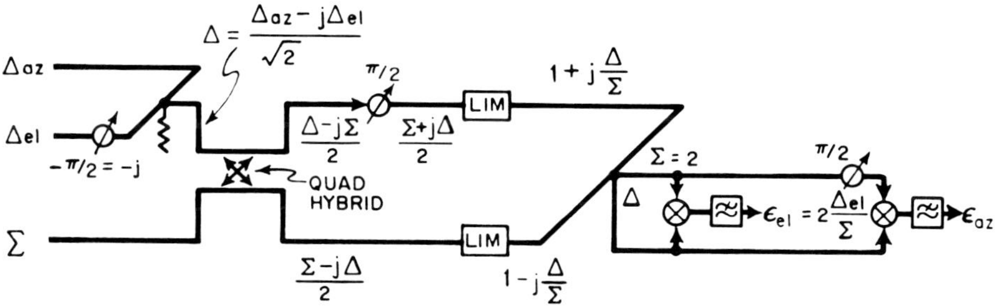

Fig. 7.32 Two-channel monopulse: limiter normalization.

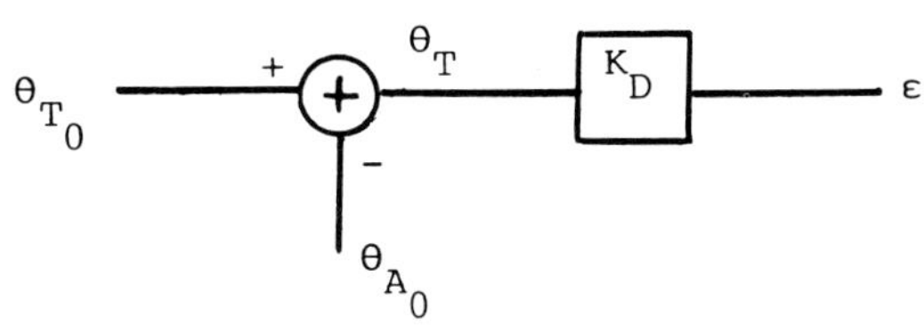

Fig. 7.33 Radar receiver model.

error voltages, as confirmed by inspection of Fig. 7.32, are

$$\varepsilon_{az} = \frac{k_m \theta_{az}}{\theta_3} \tag{7.101}$$

$$\varepsilon_{el} = \frac{k_m \theta_{el}}{\theta_3} \tag{7.102}$$

Monopulse Summary. In all cases, the radar receiver has been reduced to the very simple model shown in Fig. 7.33. The constant gain K_D depends on the antenna beamwidth, squint angle, and AGC normalization. The error voltage in a plane depends on the angular error in that same plane; i.e., the azimuth and elevation errors are uncoupled.

7.4 RANGE ESTIMATION

A typical range gate tracking circuit was discussed in Chapter 2 (Fig. 2.29). The portion of that circuit generating an error voltage proportional to the time difference between the centers of the range gate and the received video pulse is shown in Fig. 7.34. Let the gating function and the centered received signal be defined by

$$\begin{aligned} p(t) &= K \qquad 0 < t < \frac{\tau}{2} \\ &= 0 \qquad \text{else} \end{aligned} \tag{7.103}$$

$$v(t) = 1 \qquad -\frac{\tau}{2} < t < \frac{\tau}{2}$$

$$= 0 \qquad \text{else} \tag{7.104}$$

Then the error voltage due to a received signal delayed by $t_T < \tau/2$ is given by

$$\varepsilon(t_T) = -\int_{-\tau/2}^{0} p(-t)v(t - t_T)\,\mathrm{d}t + \int_{0}^{\tau/2} p(t)v(t - t_T)\,\mathrm{d}t$$

$$= -\int_{-\tau/2+t_T}^{0} K\,\mathrm{d}t + \int_{0}^{\tau/2} K\,\mathrm{d}t = -K\left[0 - \left(t_T - \frac{\tau}{2}\right) - \frac{\tau}{2}\right] = Kt_T$$

$$\varepsilon(t_T) = Kt_T \tag{7.105}$$

as can be seen by observing the overlapped gating pulses and received signal in Fig. 7.35.

For larger positive or negative time delays, the error voltage relation becomes that shown in Fig. 7.36. This error voltage can be expressed in terms of the actual and estimated range as t_T is equal to $2(R - \hat{R})/c$, where R is the

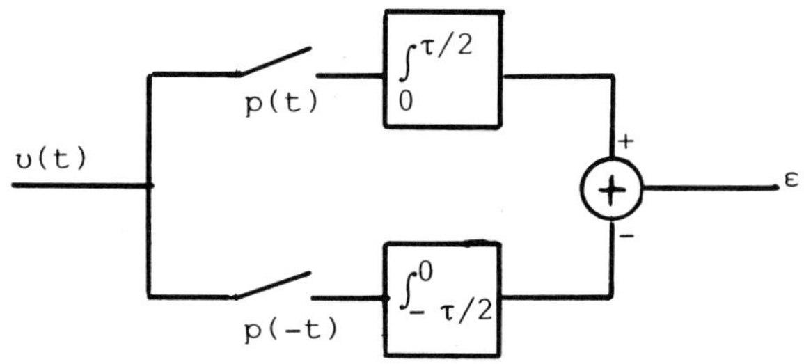

Fig. 7.34 Range gate model.

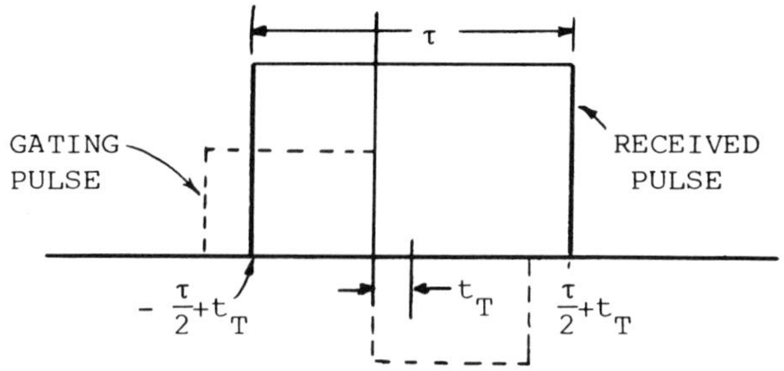

Fig. 7.35 Gating and received pulses.

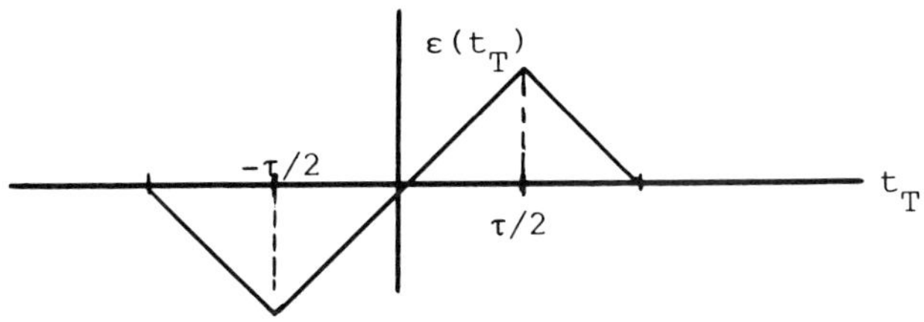

Fig. 7.36 Error voltage vs range error.

actual range and $\hat{R}$ the center of the gating pulses or estimated range. The error voltage relationship to the range tracking error expressed in Eq. (7.105) means that, for purposes of range tracking, the radar receiver can be replaced by an amplifier of constant gain similar to the one shown in Fig. 7.3 for angle tracking.

7.5 ANGLE SERVO SYSTEMS

Previous sections have shown that the antenna and receiver portions of all radars can be modeled as a summing junction followed by a constant gain, provided that the angular tracking error is small in relation to the beamwidth and provided that the radar receiver properly normalizes the signals sensed. The remaining element of the radar, the servo control system, uses the receiver output to reposition the antenna, thus maintaining the track axis aligned with the target line of sight. Modern TWS radars use a digitally implemented tracking loop. Characteristics of typical angle servo systems may limit radar system performance even in a nonjamming environment and they may have a serious impact on the type of jamming techniques that can reasonably be used against a radar. (A typical angle servo system is shown in Fig. 7.37.)

The error voltage in this system is amplified and applied to an electric or hydraulic motor that moves the antenna. The motor-drive/antenna portion of this servo can be modeled as a linear system if gear backlash, coulomb friction, and other realities are ignored. Analysis of the linear system can be performed with the aid of Laplace transforms.

The Laplace transform of an electric motor and load is still more complicated than desired for a first analysis. The transform can be simplified if only the dominant low-frequency poles and zeros are retained; however, such a model still provides valuable insight. Usually, roll-off in gain at higher

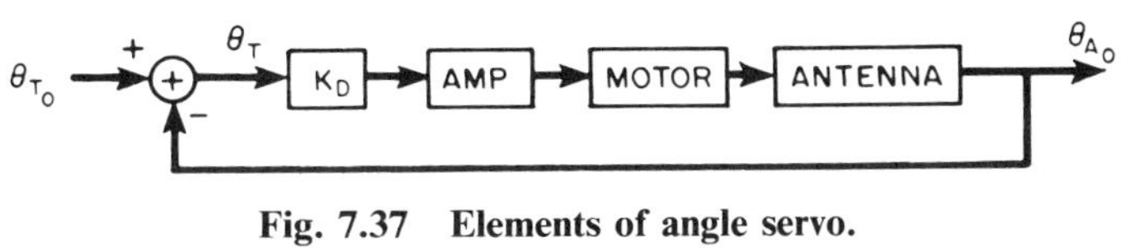

Fig. 7.37 Elements of angle servo.

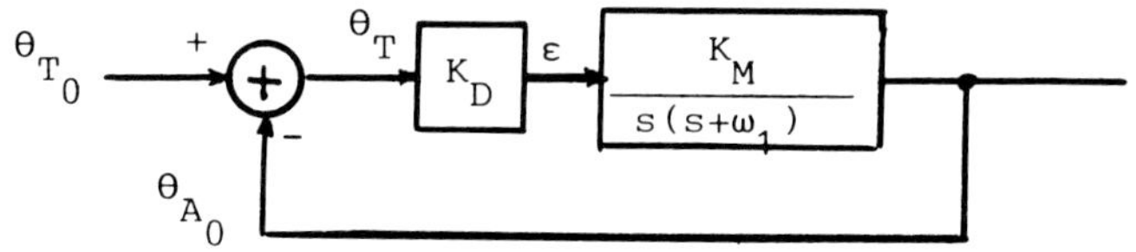

Fig. 7.38 Type I angle servo.

quencies coupled with the knowledge that previously ignored nonlinearities will dominate the response further justifies this simplification. The simplified servo is shown in Fig. 7.38 and is called a type I system [6, p. 236; 75]. The corner frequency ω_1 is fixed by the antenna inertia and motor torque. As a result, the designer has control of only one parameter—the gain K_M.

Once the angle tracking loop of the radar is specified, the response of the loop to all inputs is also specified. The problem then becomes one of selecting the inputs that provide the most insight. This selection is not a simple task, because the ultimate question is the performance of the loop in tracking a real target. Only the following approaches discussed in Chapter 6 will be applied here to gain an insight into the performance of the tracking loop: (1) absolute stability via a Bode plot or Nyquist diagram, (2) system step response, and (3) quasi-steady-state response. This discussion will also point out the significant relationships between the Bode plot and the Laplace transform.

Type I Servo

When the constants K_D and K_M in Fig. 7.38 are combined, the open-loop transfer function for a type I servo becomes

$$G(s) = \frac{K}{s(s+\omega_1)} \tag{7.106}$$

The closed-loop servo system is absolutely stable, but as K increases, the phase and gain margin will decrease. The closed-loop response is

$$\frac{\theta_{Ao}(s)}{\theta_{To}(s)} = H(s) = \frac{K}{s^2 + s\omega_1 + K} \tag{7.107}$$

The step response can be solved for or found in numerous texts. This response can be expressed in many ways, but the one used here is inspired by Locke [6, p. 242] and permits comparisons with other servos to be discussed later. The step response is

$$\theta_{A_o}(t) = 1 - \frac{e^{-\xi\omega_n t}}{\sqrt{1-\xi^2}}\cos(\omega_n\sqrt{1-\xi^2}\,t - \varphi) \tag{7.108}$$

where $\omega_n = \sqrt{K}$,

$$\varphi = \tan^{-1}\left(\frac{\omega_1/2\omega_n}{\sqrt{1-\xi^2}}\right) \tag{7.109}$$

and

$$\xi = \frac{\omega_1}{2\sqrt{K}} \tag{7.110}$$

This, of course, is a classical response in linear control theory. Since ω_1 is fixed by system elements that do not lend themselves to change, system step response for various values of K is of interest. Again, by a number of standards, a value of K yielding $\xi = 1/\sqrt{2}$ is optimum (i.e., minimum settling time or minimum integral of time times absolute error [76]). The risetime t_r of the type I step response, defined as the time from step input to peak response, is

$$t_r = \frac{\pi}{\omega_n\sqrt{1-\xi^2}} \tag{7.111}$$

and is shown in Fig. 7.39. The quasi-steady-state error can be expressed as

$$\theta_T = \frac{\theta_{To}}{K_o} + \frac{\dot{\theta}_{To}}{K_v} + \frac{\ddot{\theta}_{To}}{K_a} \tag{7.112}$$

where

$$K_o = \infty$$

$$K_v = \frac{K}{\omega_1}: \quad \text{velocity constant} \tag{7.113}$$

$$K_a = \frac{K^2}{\omega_1[(K/\omega_1) - \omega_1]}: \quad \text{acceleration constant} \tag{7.114}$$

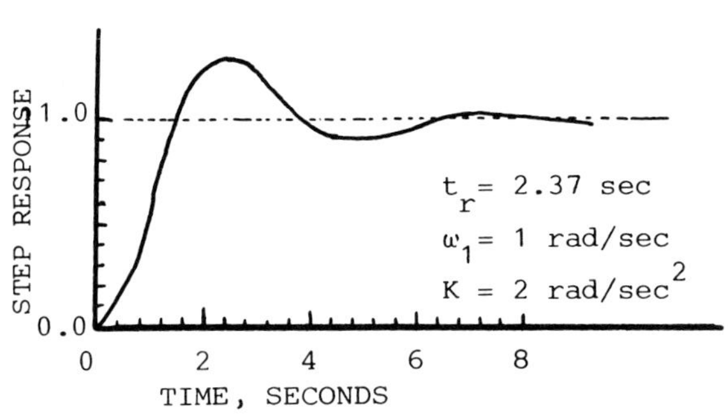

Fig. 7.39 Type I step response.

Equation (7.112) can be used in a pass-course problem to evaluate error buildup after an initial transient decay [6, Chap. 7; 16, Chap. 9; 77]. If a target flying a constant heading and velocity V_T passes a radar at time zero at a minimum range R_o, the position of the target is given by its range R_T and angle θ_{To} as specified in Fig. 7.40. The angular position, velocity, and acceleration are given by

$$\theta_{To} = \cos^{-1}\left(\frac{R_o}{R_T}\right) \tag{7.115}$$

$$\dot{\theta}_{To} = \frac{V_T}{R_o}\cos^2\theta_{To} \tag{7.116}$$

$$\ddot{\theta}_{To} = -\left(\frac{V_T}{R_o}\right)^2 \cos^2\theta_{To}\sin 2\theta_{To} \tag{7.117}$$

The maximum angular velocity and acceleration, which do not occur at the same time, are

$$\dot{\theta}_{To\,\max} = \frac{V_T}{R_o} \tag{7.118}$$

$$\ddot{\theta}_{To\,\max} = \frac{3\sqrt{3}}{8}\left(\frac{V_T}{R_o}\right)^2 \tag{7.119}$$

If $V_T = 500$ knots (257 m/s), $R_o = 1000$ yd (914 m), and the radar angle servo is specified by $\omega_1 = 10$ rad/s and $\xi = 0.707$, then

$$K_v = \frac{\omega_1}{2} = 5 \text{ rad/s} \tag{7.120}$$

$$K_a = \frac{\omega_1^2}{2} = -50 \text{ rad/s}^2 \tag{7.121}$$

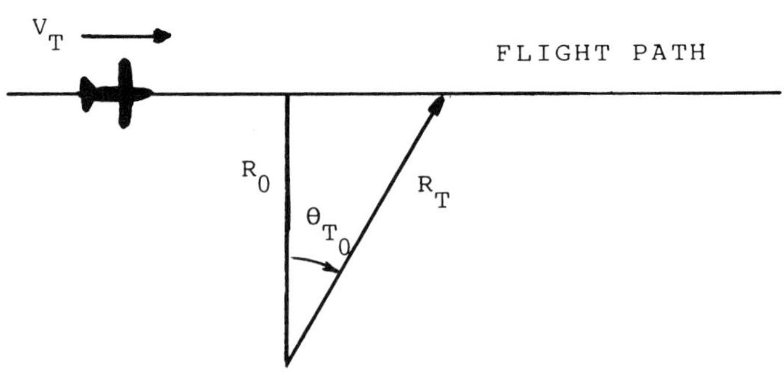

Fig. 7.40 Pass course geometry.

and, after including Eqs.(7.115 – 7.121), Eq. (7.112) becomes

$$\theta_T \approx \frac{2}{\omega_1}\frac{V_T}{R_o}\cos^2\theta_{To} + \frac{2}{\omega_1^2}\left(\frac{V_T}{R_o}\right)^2\cos^2\theta_{To}\sin 2\theta_{To} \tag{7.122}$$

$$= 0.056\cos^2\theta_{To} + 0.0016\cos^2\theta_{To}\sin 2\theta_{To}$$

Such a type I system has a large tracking error at target flyby (0.056 rad = 3.2 deg) requiring excessive antenna beamwidth to prevent losing the target.

Before leaving the type I system, its open-loop Bode plot, shown in Fig. 7.41, will be discussed. At very low frequencies, the open-loop gain decreases at 20 dB per decade as Eq. (7.123) implies,

$$|G(\omega)| \approx \frac{K}{\omega} \tag{7.123}$$

If the initial slope is continued past the corner frequency ω_1 to a frequency at which $|G(\omega)| = 1$, the point at which the frequency axis is intercepted, called ω_v, is the velocity constant. The actual gain function intercepts the frequency axis at ω_c, which is the crossover frequency and is related to ω_1 and ω_v as

$$\omega_c^2 = \omega_v\omega_1 \tag{7.124}$$

The gain plotted in Fig. 7.41 is for a highly underdamped and therefore nonideal system, but these notations will be useful in later more realistic systems. Finally, for the gain function in Fig. 7.41, the acceleration constant

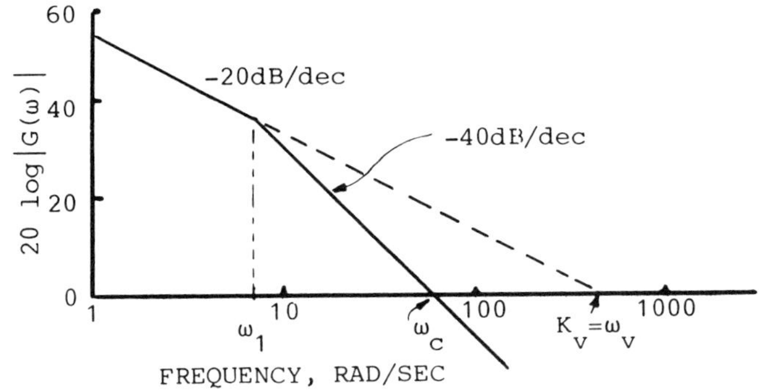

Fig. 7.41 Type I servo Bode plot.

and ω_c are related. If $\omega_v \gg \omega_1$, from Eq. (7.114)

$$K_a \approx K_v\omega_1 = \omega_v\omega_1 = \omega_c^2 \tag{7.125}$$

Improved Type I Servo

The type I servo error in angle for the pass-course problem can be reduced by the addition of a phase-lead network. Locke [6] calls the resultant servo system an improved type I. Of course, a pure phase-lead network is unrealizable, but the pole necessary for realizability can be at a high frequency where the system gain is small. Provided the drive signal is predominately a low-frequency signal, the effect of ignoring this pole should be negligible.

The open-loop transfer function is now

$$G(s) = \frac{K(s + \omega_2)}{s(s + \omega_1)\omega_2} \tag{7.126}$$

From the improved type I Bode plot, shown in Fig. 7.42, the velocity constant $K_v = \omega_v$ can again be read as the point where the $1/\omega$ slope intercepts unity gain on the frequency axis. The acceleration constant, $K_a = \omega_a^2$, is related approximately to the frequency at which the -40 dB/decade slope intercepts the frequency axis at unity gain.

The value of these constants can be evaluated exactly via the techniques of Chapter 6 or found in Barton [16, p. 298]. They are

$$K_v = \frac{K}{\omega_1} \tag{7.127}$$

$$K_a = \frac{K_v^2}{K_v\left(\dfrac{1}{\omega_1} - \dfrac{1}{\omega_2}\right) - 1} \tag{7.128}$$

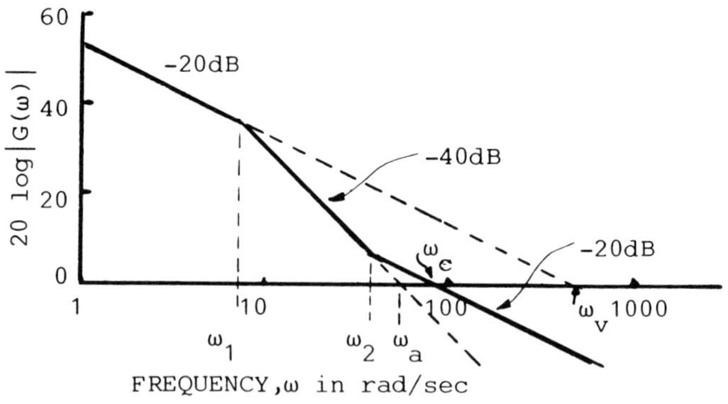

Fig. 7.42 Improved type I Bode plot.

If $\omega_2 \gg \omega_1$ and $\omega_v \gg \omega_1$, then the approximate formulation for the acceleration constant is

$$\omega_a^2 = K_a \approx K_v\omega_1 = \omega_v\omega_1 \tag{7.129}$$

which is given by the -40 dB/decade intercept as previously stated.

The step response is [6, p. 254]

$$\theta_{A_o}(t) = 1 - \frac{e^{-\xi\omega_n t}}{\sqrt{1-\xi^2}}\sqrt{1-\frac{\omega_1}{\omega_2}}\cos[\omega_n\sqrt{1-\xi^2}t - \varphi] \tag{7.130}$$

where

$$\varphi = \tan^{-1}\left[\frac{\dfrac{\omega_1}{\omega_n} - \dfrac{\omega_n}{\omega_2}}{2\sqrt{1-\xi^2}}\right] \tag{7.131}$$

$$\omega_n^2 = K \tag{7.132}$$

$$2\xi\omega_n = \frac{K}{\omega_2} + \omega_1 \tag{7.133}$$

The improved type I step response reduces to that of type I [Eq. (7.108)] as $\omega_2 \to \infty$. The advantage of this servo is that the damping factor ξ can be held to approximately one-half, even with very large gain K. This eliminates the ringing problem of type I when the gain is increased to reduce the risetime of the step response. If $\omega_n \approx \omega_2$ and $\omega_2 \gg \omega_1$, the damping factor is

$$\xi \approx \frac{\omega_n^2 + \omega_1\omega_n}{2\omega_n^2} \approx \frac{1}{2} \tag{7.134}$$

for all gain $K = \omega_n^2$.

The response time is now

$$t_r = \frac{\pi - \tan^{-1}\left[\dfrac{\left(\dfrac{\omega_1}{\omega_n} + \dfrac{\omega_n}{\omega_2}\right)}{2(1-\xi^2)^{1/2}}\right] + \tan^{-1}\left[\dfrac{\left(\dfrac{\omega_1}{\omega_n} - \dfrac{\omega_n}{\omega_2}\right)}{2(1-\xi^2)^{1/2}}\right]}{\omega_n(1-\xi^2)^{1/2}} \tag{7.135}$$

The improved type I servo is absolutely stable as the Nyquist diagrams in Fig. 7.43 show. However, the gain and phase margins are functions of gain.

If the higher-order pole of the phase-lead network had been retained, the Nyquist diagram would change to the dotted curves shown in Fig. 7.43. Again, a reduction in gain reduces the phase margin. For some servo systems, such a reduction in gain can actually result in instability, as pointed

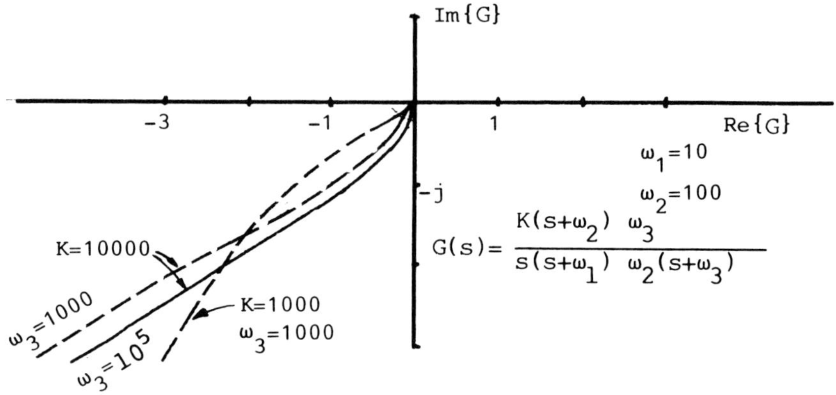

Fig. 7.43 Nyquist diagram of improved type I servo.

out by Shustov and Vakin [4, p. 176]. This point shall be addressed further in Chapter 10.

The improved type I servo functions much better than the type I in the same pass-course problem. Again, let $\omega_1 = 10$ rad/s as a result of the motor/antenna constants. If $\omega_n \approx 10\omega_1$, as suggested by Locke, then the generalized error coefficients are

$$K_v = \frac{K}{\omega_1} = 100\omega_1 = 1000 \text{ rad/s}$$

$$K_a \approx K_v\omega_1 \approx 10{,}000 \text{ rad/s}^2 \text{ from Eq. (7.129)}$$

$$= 112\omega_1^2 \qquad \text{from Eq. (7.128)}$$

Based on Eq. (7.112), the error in angular tracking as the target phase flies by is

$$\theta_T = 0.00028 \text{ rad} = 0.016 \text{ deg}$$

7.6 RANGE SERVO SYSTEMS

The early-gate/late-gate circuit was shown to develop a voltage proportional to the time difference between the common point of their sampling pulses and the center of the pulse being tracked. A servo can be used to adjust the time of sampling on the next received pulse. Such a servo, combined with the error sensing early/late gates, constitutes an automatic tracker in range. Such a circuit, able to control the time of sampling of the early and late gates, can enable another gate (range gate) to pass the desired received pulse from the detected IF output on to other circuits.

The open-loop transfer function of the range servo is not limited by mechanical constraints if it is constructed electronically; it normally has the form

$$G(s) = \frac{K(s + \omega_2)}{s^2\omega_2} \tag{7.136}$$

This transfer function, called type II, can be derived from Eq. (7.126) by letting $\omega_1 \to 0$. Similarly, from Eqs. (7.130) and (7.135), the step response and risetime are

$$\hat{R}_{To}(t) = 1 - \frac{1}{\sqrt{1 - \xi^2}} e^{-\xi\omega_n t} \cos[\omega_n\sqrt{1 - \xi^2}t + \cos^{-1}\sqrt{1 - \xi^2}] \tag{7.137}$$

with

$$\omega_n^2 = K, \qquad \xi\omega_n = \frac{K}{2\omega_2}$$

and

$$t_r = \frac{\pi - 2\cos^{-1}\sqrt{1 - \xi^2}}{\omega_n\sqrt{1 - \xi^2}} \tag{7.138}$$

A linear model for the range tracker is shown in Fig. 7.44. The actual range to the target is R_{To} and the estimated range is the servo-filtered output of the summing junction $\hat{R}_{To}$. The resultant error R_T can be found from Eq. (7.137) for a step change in target range.

The generalized error coefficients are now

$$K_v = \infty \tag{7.139}$$

$$K_a = K \tag{7.140}$$

$$\dot{K}_a = -K\omega_2 \tag{7.141}$$

The resultant error in range for the pass-course problem considered earlier is related to the required pulse width for that problem (i.e., the error must not exceed one-half the pulse width or the linear model fails). The error in

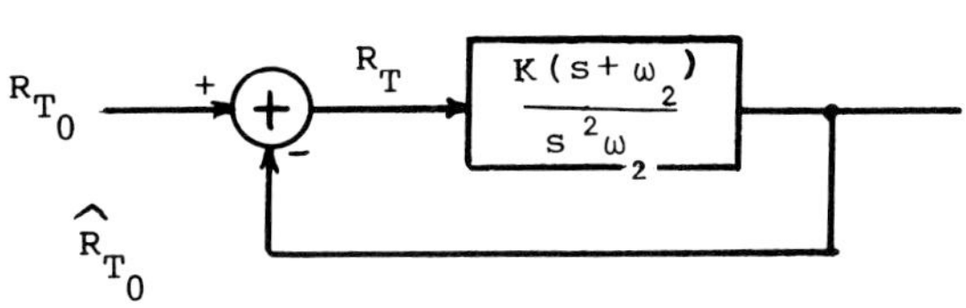

Fig. 7.44 Type II range servo.

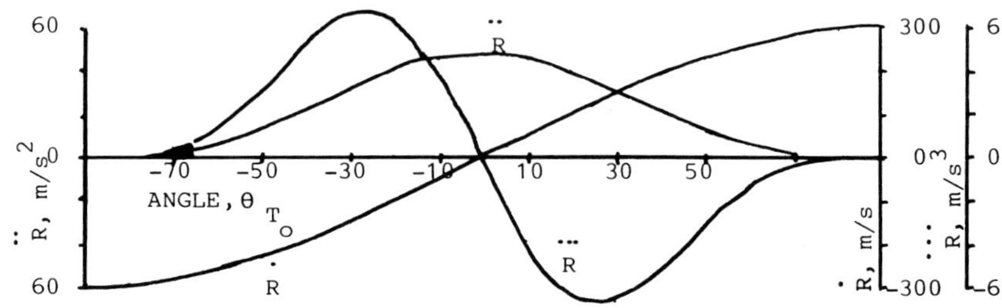

Fig. 7.45 Derivatives of range in pass course problem.

tracking is

$$R_T = \frac{1}{K}\left(\frac{d^2 R_{To}}{dt^2}\right) - \left(\frac{1}{K\omega_2}\right)\left(\frac{d^3 R_{To}}{dt^3}\right) \tag{7.142}$$

The derivatives of position are

$$\frac{d^2 R_{To}}{dt^2} = \ddot{R}_{To} = \frac{V_T^2}{R_o}\cos^3\theta_{To} \tag{7.143}$$

$$\frac{d^3 R_{To}}{dt^3} = -\frac{V_T^3}{R_o^2}3\cos^4\theta_{To}\sin\theta_{To} \tag{7.144}$$

and are plotted in Fig. 7.45.

If a 0.5 μs radar pulse width is assumed and an error in range tracking of less than one-tenth the resolution in range is required, then $R_T < 7.5$ m. If this total error in range is due to the acceleration term, the gain must satisfy

$$R_T = \frac{\ddot{R}_{To}}{K} < 7.5 \text{ m} \tag{7.145}$$

and from Eq. (7.145), $K > 9.73$ rad/s^2. If the damping factor is assumed to be 0.707 and $\omega_2 = \sqrt{K/2}$ is 2.19 rad/s, then the maximum range error due to the derivative of acceleration is

$$\frac{1}{\omega K}\frac{d^3 R_{To}}{dt^3} = 0.83 \text{ m} \tag{7.146}$$

This contribution to the error is small enough compared to 7.5 m to justify ignoring it. A range servo with $K = 10$ rad/s^2 and $\omega_2 = 2$ rad/s will track the pass-course target with the specified accuracy.

7.7 TRACKING ERRORS IN HIGH *S/N* ENVIRONMENT

The analysis thus far has been deterministic and must be expanded to include the effect of receiver noise. The effect of external noise is discussed in later chapters. The servo model with noise added in units of the measured variable is shown in Fig. 7.46. If the input signal is a constant, it can arbitrarily be set to zero and the servo loop is driven by the noise alone. The loop output is a direct measure of the error in tracking, as clearly demonstrated in Fig. 7.47. It can be assumed that the receiver noise is zero mean and possesses a white power spectrum of density (two sided) N_A. Then, the variance of the measured variable is given by

$$\sigma_x^2 = \frac{N_A}{2\pi} \int_{-\infty}^{\infty} |H(\omega)|^2 \, \mathrm{d}\omega \tag{7.147}$$

where $|H(\omega)|$ is the magnitude of the Fourier transform of the closed-loop transfer function. It is common practice to define a single-sided noise bandwidth β_n for low-pass filters with Fourier transform $H(\omega)$ by [63, p. 4-35]

$$2\beta_n = \frac{1}{2\pi} \int_{-\infty}^{\infty} |H(\omega)|^2 \, \mathrm{d}\omega \tag{7.148}$$

provided $|H(0)| = 1$. The variance of the error is then

$$\sigma_x^2 = 2N_A\beta_n \tag{7.149}$$

The density of the noise power is expressed in equivalent units of the measured quantity x. It is necessary to develop this density in terms of the IF signal-to-noise ratio.

The IF signal can be expressed as the sum of a signal term and a noise term

$$V_r(t) = V_o[1 + \tilde{m}\cos(\omega_s t - \tilde{\varphi})]\cos\omega_{IF} t + \sum_{n=-f_r/2\Delta f}^{f_r/2\Delta f} \alpha_n \cos[(\omega_{IF} + n\Delta\omega)t + \tilde{\psi}_n] \tag{7.150}$$

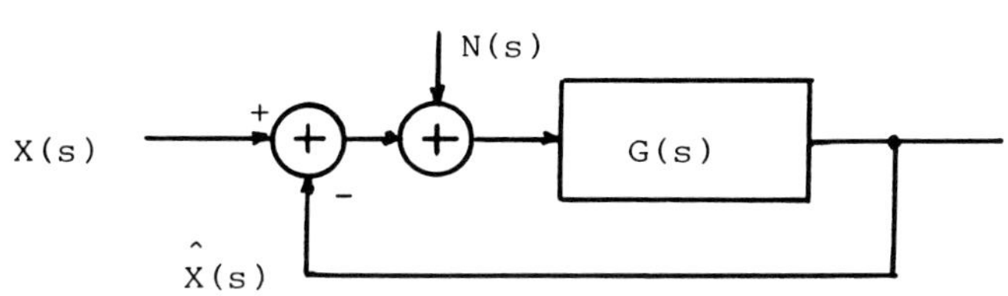

Fig. 7.46 Servo loop with noise.

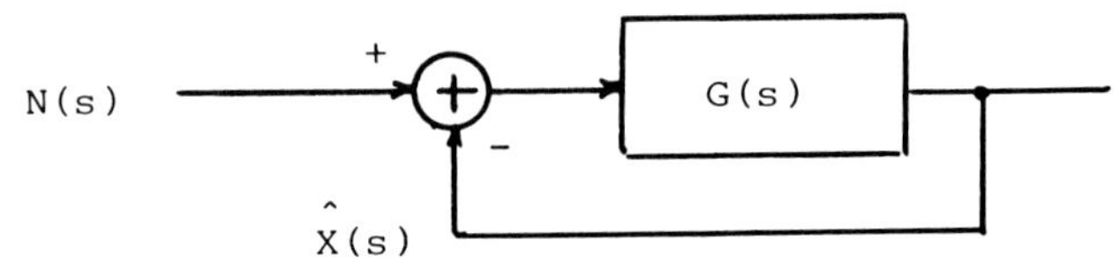

Fig. 7.47 Noisy loop without drive signal.

The continuous noise spectrum has been modeled as a discrete set of sinusoids of equal amplitude and statistically independent phases. The following relationships are assumed:

$$E\{\tilde{\psi}_n\} = 0 \qquad \alpha_n^2 = \alpha_m^2 \quad \text{all } n, m \tag{7.151}$$

$$E\{\tilde{\psi}_n \tilde{\psi}_m\} = 0 \tag{7.152}$$

$$p(\psi_n) = \frac{1}{2\pi}, \quad |\psi| < \pi$$

$$= 0 \qquad \text{else} \tag{7.153}$$

The noise is assumed to occupy a bandwidth f_r centered on the carrier. As shown by Barton [16, p. 276], the IF signal-to-noise ratio computed with this model is identical to the signal-to-noise ratio of a pulsed radar if f_r is equated to the radar pulse-repetition frequency. The total noise power is $\Sigma\alpha_n^2/2 = N_o f_r$, where the noise power density N_o is defined in Fig. 7.48. The IF signal-to-noise ratio is defined as

$$\left(\frac{S}{N}\right)_{IF} = \frac{V_o^2}{\Sigma\alpha_n^2} = \frac{V_o^2 \Delta f}{\alpha_n^2 f_r} = \frac{V_o^2}{2N_o f_r} \tag{7.154}$$

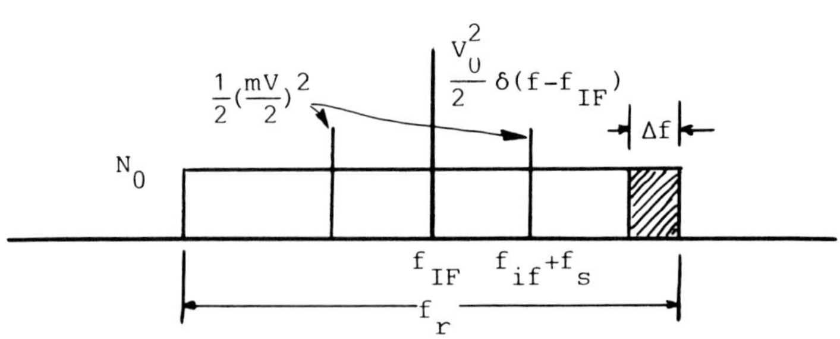

Fig. 7.48 CW CONSCAN radar spectrum.

The noise that competes with the signal term, after conversion to baseband, consists of the two terms offset by the scan frequency above and below the carrier. This assumes that the scan filter has a width Δf,

$$V_3(t) = \frac{V_o\tilde{m}}{2}\cos(\omega_s t - \tilde{\varphi}) + \frac{\alpha_n}{2}\cos(n\Delta\omega t + \tilde{\psi}_n)$$

$$-\frac{\alpha_{-n}}{2}\cos(n\Delta\omega t - \tilde{\psi}_{-n}) \tag{7.155}$$

The error voltage driving the servo with the $\cos\omega_s t$ reference can be expressed as

$$\varepsilon = \frac{V_o\tilde{m}}{4}\cos\tilde{\varphi} + \frac{\alpha_{+n}}{4}\cos\tilde{\psi}_{+n} - \frac{\alpha_{-n}}{4}\cos\tilde{\psi}_{-n} \tag{7.156}$$

If the target is on the track axis, it does not contribute to the error voltage. The error voltage then present is due entirely to the receiver noise, but it can be equated to an equivalent target offset as [i.e., $\tilde{m}\cos\tilde{\varphi} = (k_s\tilde{\theta}_T/\theta_3)\cos\tilde{\varphi} = (k_s/\theta_3)\tilde{\theta}_x$]

$$V_o\tilde{m}\cos\tilde{\varphi} = \alpha_n\cos\tilde{\psi}_n - \alpha_{-n}\cos\tilde{\psi}_{-n} \tag{7.157}$$

The expected error is zero and the mean square error voltage is

$$V_o^2\overline{(\tilde{m}\cos\tilde{\varphi})^2} = V_o^2\frac{k_s^2}{\theta_3^2}\overline{\theta_x^2} = \alpha_n^2 \tag{7.158}$$

Combining Eqs. (7.154) and (7.158) yields

$$\overline{\theta_x^2} = \frac{\theta_3^2\Delta f}{k_s^2(S/N)_{IF}f_r} \tag{7.159}$$

as the equivalent angular noise power available over a bandwidth of Δf. The servo loop is driven with an angular noise power density of

$$N_A = \frac{\overline{\theta_x^2}}{\Delta f} = \frac{\theta_3^2}{k_s^2 f_r(S/N)_{IF}} \tag{7.160}$$

resulting in a tracking error variance via Eq. (7.149) of

$$\sigma_x^2 = \frac{\theta_3^2}{k_s^2\dfrac{f_r}{2\beta_n}\left(\dfrac{S}{N}\right)_{IF}} \tag{7.161}$$

Equation (7.161) shows that the measurement error resulting from system noise has a mean square value proportional to β_n, the noise bandwidth of the closed-loop servo system [see Eq. (7.148)]. Thus, the narrower the servo bandwidth, the more likely it is that the estimated range or angle nears the actual range or angle. However, the mean square error given in Eq. (7.161) is applicable only to the case having an unknown constant value of the measured variable. When the input to the filter (servo) is time varying rather than constant, the filter with the narrowest bandwidth does not develop an output that best approximates the input. In more general terms, what filter will process the available data to yield the best estimate of some desired parameter of the input signal (e.g., its current value)?

In Chapter 6, system performance was measured in terms of stability, steady-state error, and the transient response to specific, deterministic input signals. The input signals considered had step, linear, and parabolic variations in time. The filter transient response resulting from a specific input was given in terms of risetime, overshoot, and settling time. Such measurements are important, because with them it can be determined when transient errors are large enough to nullify the linearity assumptions basic to the models. Even for deterministic inputs, other performance criteria have been used. Shinners [76, p. 93] discusses the minimization of the integral of the time-times-squared error as a criterion that penalizes the initial transient error less (and any steady-state error more) than the usual mean square error criterion. Another aspect of the approach taken in Chapter 6 and earlier sections of this chapter is that it amounted to the determination of the performance of a particular linear filter when given the input signal and performance criteria. An alternate approach is to specify the input signal and performance criteria and then to search for the "best" filter. The filter search can be restricted to those filters with one or two integrators, the gain of a known filter, etc. Which filter is best?

There is no one "optimum" filter for all situations! The following four factors, however, must be specified [97, p. 219] before the best filter for a given situation can be determined:

(1) Input to the filter. For example, is the signal a random process; is it correlated with the noise; what is its power spectrum?

(2) Output of the filter. For example, is an estimate of the current or future value of the input signal desired?

(3) Method of measuring the filter performance. For example, minimizing the mean square error is a popular criterion, but not the only one.

(4) Types of filter. Only linear, time-invariant filters are considered here, but others are possible.

What then are some examples of optimum filters when a time-varying input is corrupted by noise? Consider first the mean square error in angle due to receiver noise as given by Eq. (7.161). Barton [16, p. 307] adds the steady-state error in the angle resulting from the acceleration of a target to the mean square error of Eq. (7.161). Both errors are formulated for a specific filter and the error due to acceleration is inversely proportional to servo bandwidth. Barton then considers the optimum servo bandwidth to be the one that minimizes the total mean square error.

Gardner [98, p. 138] provides the results of an analysis by Jaffe and Rechtin [103]. In this analysis, a specific time-varying signal plus white Gaussian noise are applied to the input of a filter. The deterministic time signals considered are steps, ramps, and parabolic ramps. Again, the criterion of goodness is the minimum mean square error, but Jaffe and Rechtin derive the optimum linear filter, as opposed to evaluating the performance of a specified filter as a number of the parameters of the filter are varied.

Davenport and Root [97] derive the optimum linear, time-invariant filter when both inputs to the filter are sample functions of random processes. The inputs are characterized only by their power spectrum and the mean square error is the criterion of goodness. The result is called a Weiner filter.

As a final example, Garnell and East [99] provide a simple derivation of the Kalman filter. The input to the filter consists of the sum of two random processes representing the signal and noise. The noise is assumed to be white and Gaussian. The goodness criterion is again minimizing the mean square error and the form of the filter is specified by the dynamics of the system creating the signal portion of the input. For stationary inputs, the Kalman and Weiner filters are identical.

All these examples of optimization minimize the mean square error and, except for the Kalman filter, result in filters that are time-invariant.

As an example of optimizing a filter, the approach followed by Barton [16] will be applied to derive the optimum gain of the type II range servo of Fig. 7.44. The servo error in range due to target acceleration is given by Eq. (7.145). The mean square error due to receiver noise is given by Eq. (7.147), where

$$H(\omega) = \frac{(K/\omega_2)(s + \omega_2)}{s_2 + (K/\omega_2)s + K} \tag{7.162}$$

The integral in Eq. (7.147) can be evaluated with the aid of tables contained in Shinners [76, p. 320]. The resulting noise induced mean square error in range is,

$$\sigma_R^2 = N_A \frac{(K/\omega_2) + \omega_2}{2} = 2N_A\beta_n \tag{7.163}$$

Again, N_A is the amplitude of the two-sided noise spectrum in units of range at the servo input. It can be evaluated in terms of the IF signal-to-noise ratio as follows. Let the amplitude of the received signal be A. Using Fig. 7.36, an error in range of R will result in an error voltage of v, such that

$$v = 2R\frac{A}{\Delta R} \tag{7.164}$$

where $\Delta R = c\tau/2$ is the range resolution of the radar.

Equivalently, a noise voltage will be balanced by offsetting the correct range measurement in the same ratio. From Eq. (7.164), the mean square error voltage resulting from a large number of received pulses is

$$\overline{v^2} = \left(\frac{2A}{\Delta R}\right)^2 \overline{R^2}$$

or equivalently

$$\overline{R^2} = \frac{\Delta R^2}{2[A^2/\overline{v^2}]} = \frac{\Delta R^2}{2(S/N)_1} \tag{7.165}$$

This "noise" power can be considered distributed uniformly over a bandwidth equal to the PRF of the radar, resulting in the noise spectrum

$$N_A = \frac{\overline{R^2}}{f_{PRF}} = \frac{\Delta R^2}{2(S/N)_1 f_{PRF}} \tag{7.166}$$

Combining Eqs. (7.163), (7.166), and (7.145) yields total mean square error

$$\sigma_R^2 = \frac{\Delta R^2 \left(\dfrac{K}{\omega_2} + \omega_2\right)}{4(S/N)_1 f_{PRF}} + \left(\frac{\ddot{R}_{To}}{K}\right)^2 \tag{7.167}$$

Differentiating Eq. (7.167) with respect to K yields an optimum gain of

$$K_{\mathrm{OPT}} = \left[\frac{8\ddot{R}_{To}\omega_2(S/N)_1 f_{PRF}}{\Delta R^2}\right]^{\frac{1}{3}} \tag{7.168}$$

Problems

7.1 The AGC circuit of Fig. 7.9 is modified as shown below to incorporate a delay voltage V_d.

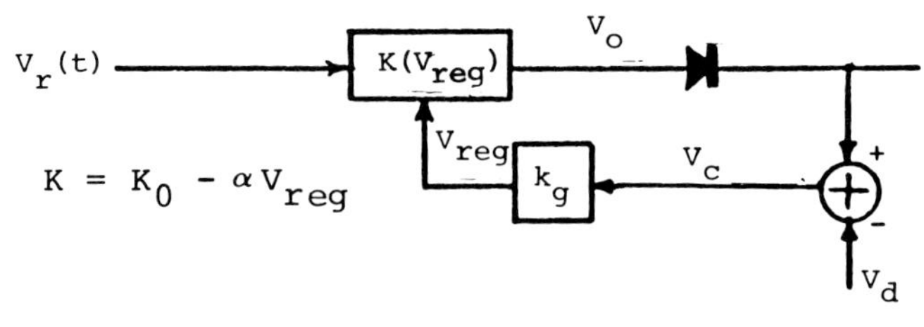

Assuming the control voltage is

$$V_{\text{reg}} = k_g(V_o - V_d) \qquad V_o > V_d$$

$$V_{\text{reg}} = 0 \qquad V_o < V_d$$

show that the amplifier gain depends on the amplitude of the received voltage as

$$K = K_o \qquad V_r < V_d/K_o$$

$$K = \frac{K_o + \alpha k_g V_d}{1 + \alpha k_g V_r} \qquad V_r > V_d/K_o$$

7.2 Because the transient response of an AGC circuit is very important, the interested reader may find the following useful in later chapters.

Consider the AGC circuit of Figs. 7.9 and 7.10 with $k = 1$, and assume the impulse response of the AGC filter to be

$$h(t) = (1/T)e^{-t/T} \qquad t > 0$$

$$= 0 \qquad t < 0$$

Assume the input to the receiver to be ($V_a < V_b$)

$$V_r(t) = V_a \qquad t < 0$$

$$V_r(t) = V_b \qquad t > 0$$

Show that for $t > 0$, the control voltage is governed by

$$V_{\text{reg}}(t) = V_o(o)e^{-t/T} + \int_o^t V_o(\tau)h(t - \tau)\, d\tau$$

that the regulation voltage is

$$V_{\text{reg}}(t) = \frac{K_o V_b}{1 + \alpha V_b}[1 - e^{-(1 + \alpha V_b)t/T}] + \frac{K_o V_a}{1 + \alpha V_o} e^{-(1 + \alpha V_b)t/T}$$

and that the output voltage is

$$V_o(t) = \frac{K_o V_b}{1 + \alpha V_b} + \alpha V_b \left[\frac{K_o V_b}{1 + \alpha V_b} - \frac{K_o V_a}{1 + \alpha V_a}\right] e^{-(1 + \alpha V_b)t/T}$$

7.3 A lobe-on-receive-only (LORO) radar illuminates a target with a constant antenna gain and receives the energy scattered by a target with an antenna that points sequentially in a finite number of directions. If the receive antenna scanned continuously [conical-scan-on-receive-only (COSRO)], the envelope detector output is given by Eq. (7.4), with a factor of two inserted to account for the constant illumination, as

$$V_1(t) = V_o\left[1 - \frac{k_s\theta_T}{2\theta_3}\cos(\omega_s t - \varphi_T)\right]$$

Assuming Fig. 7.15 models a LORO radar when the leftmost $g(t) = 1$ and the second $g(t)$ is

$$g(t_i) = \sqrt{G(\theta_s)}\left[1 + \frac{k_s\theta_T}{2\theta_3}\cos(\omega_s t_i - \varphi_T)\right], \quad t_i = \frac{i\pi}{2\omega_s}, \quad i = 0, 1, 2, 3, \ldots$$

show that the voltage out of the scanning filter is

$$V_3(t) = \frac{\sqrt{2}}{\pi} V_o \frac{k_s\theta_T}{\theta_3}\cos\left(\omega_s t - \varphi_T - \frac{\pi}{4}\right)$$

The amplitude of this voltage is slightly less than one-half the amplitude of the equivalent signal of a CONSCAN radar and it is shifted in phase by 45 deg. (A CONSCAN radar radiating only four pulses per scan period will also exhibit a 45 deg phase shift when compared to a CONSCAN radar with, say, 100 pulses per scan period.)

7.4 A LORO and COSRO radar is often implemented through electronic switching or scanning. A monopulse antenna is used to form the single illumination beam and multiple receive beams. A microwave resolver or switching assembly is employed to combine the outputs of the receive beams to simulate a scanning beam or to step the position of the receive beam. As soon as the position of the receive beam is controlled electronically, new and more flexible scanning patterns can be introduced to reduce the tracking errors caused by known modulations imposed on the received signal by sources external to the radar. For example, Felsenthal [100] discusses an approach which periodically reverses the scan direction of a CONSCAN radar. Pifer [101] introduces a design employing the pseudorandom selection of the antenna scan rate, executing a pattern of scan frequency variation in a time that is short in comparison with the response time of the antenna servo. Finally, Clayton [102] describes an approach that pseudorandomly positions the receive pattern. Assume the following figure adequately describes a radar as discussed by Clayton,

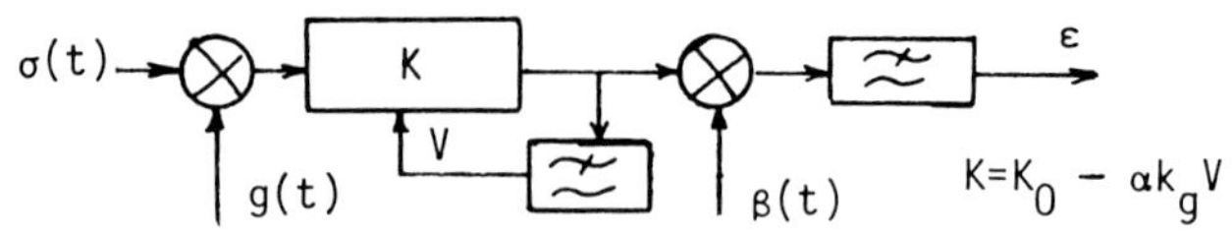

Define the voltage gain of the receive antenna to be

$$g(t) = g_o \left[1 + \frac{k_s \theta_T}{2\theta_3} \beta(t) \right]$$

where $\beta(t)$ randomly switches between the values ± 1.

Define the time average as

$$\overline{\beta(t)} = \lim_{T \to \infty} \frac{1}{2T} \int_{-T}^{T} \beta(t) \, dt$$

and assume $\overline{\beta(t)} = 0$, $\overline{\beta^2(t)} = 1$.

The received signal is composed of a constant and a time-varying part such that

$$\sigma(t) = \sigma_o + \sigma_1(t), \qquad \overline{\sigma_1(t)} = 0, \qquad \overline{\sigma_1(t)\beta(t)} = 0$$

Assume the regulation and error voltages are given by

$$V = K\overline{\sigma(t)g(t)} \qquad \text{and} \qquad \varepsilon = K\overline{\sigma(t)g(t)\beta(t)}$$

Show that the error voltage is

$$\varepsilon = \frac{K_o}{\alpha k_g} \frac{k_s \theta_T}{2\theta_3}$$

7.5 Consider the two-channel, amplitude sensing, sum-difference detection monopulse shown in the following figure.

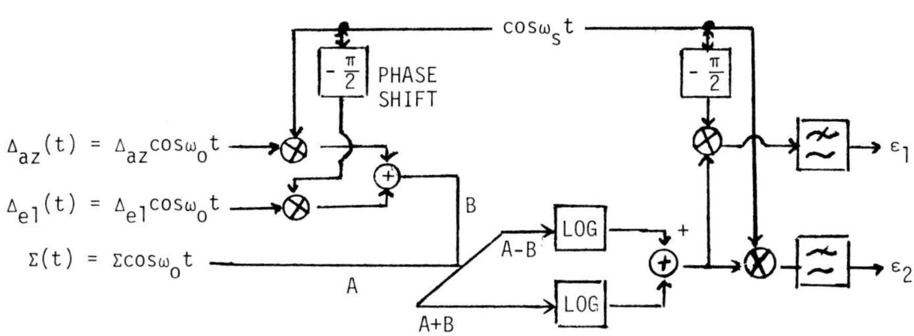

Given that $|\Delta_{az}| \ll |\Sigma|$, and $|\Delta_{el}| \ll |\Sigma|$, derive formulas giving the dependence of ε_1 and ε_2 on Δ_{az}, Δ_{el}, and Σ.

Note $\omega_s \sim 100$ Hz and $\omega_o \sim 10$ GHz. Log amplifiers output a voltage proportional to the $\log_{10}$ of the *amplitude* of the *video* input. Finally, ε_1 and ε_2 are filtered to have a bandwidth of several hertz.

7.6 Assume a radar angle tracking loop has the following *open-loop* transfer function

$$\frac{\theta_A(s)}{\theta_T(s)} = G(s) = \frac{2.5(s+1.4)}{s(s+0.02)} \qquad \begin{aligned} \omega_1 &= 0.02 \\ \omega_3 &= 1.4 \\ K &= (2.5)(1.4) \end{aligned}$$

a) Show that the closed-loop response of the above radar to a unit step input is

$$\theta_A(t) = 1 - \left(\frac{1-\omega_1/\omega_3}{1-\xi^2}\right)^{1/2} e^{-\xi\omega_n t}$$

$$\times \cos\left[\omega_n\sqrt{1-\xi^2}\,t - \tan^{-1}\left(\frac{(\omega_1/\omega_n)-(\omega_n/\omega_3)}{2\sqrt{1-\xi^2}}\right)\right]$$

where

$$\xi = \frac{K+\omega_1\omega_3}{2\omega_3\omega_n} \qquad \text{and} \qquad \omega_n^2 = K$$

b) Derive the following "steady-state" error coefficients

$$K_o = \infty \qquad K_v = \frac{K}{\omega_1} \qquad K_a = \frac{\left(\dfrac{K}{\omega_1}\right)^2}{\dfrac{K}{\omega_1}\left(\dfrac{1}{\omega_1}-\dfrac{1}{\omega_3}\right)-1}$$

using, for example, the power series expansion.

c) Discuss the effects of a reduction of gain K on the stability/shape of step response.

7.7 A radar angle servo has the following open-loop transfer function:

$$G(s) = \frac{K\omega_1(s+\omega_2)^2\omega_3}{s(s+\omega_1)^2\omega_2^2(s+\omega_3)}$$

where $\omega_1 = 0.5$ rad/s, $\omega_2 = 5$ rad/s, $\omega_3 = 100$ rad/s, and $K = 500$.

a) Sketch the Bode plot in amplitude and phase.
b) Sketch the associated Nyquist diagram.
c) Is the system stable?
d) If the gain K is reduced to 5, is the system stable?

7.8 What are K_v and K_a for the closed-loop servo of Problem 7.7? Derive only as a *last* resort!!

7.9 The radar servo of Problem 7.7 is used to track a 1000 m/s target. If the target minimum range is 2000 m, what is the maximum angular error due to the target angular velocity and angular acceleration?

8 ACTIVE ECM VS ANGLE SCANNING SYSTEMS

An earlier chapter classified active ECM techniques by the modulation applied to either a sinusoidal RF carrier or to a statistical characterization of a noise-like signal. The effect of a specific ECM technique on a specific radar was stated without underlying rationale. This chapter provides previously missing rationale for a number of ECM/radar encounters and analytical tools that can be used to support the assertions made in other encounters. The effects of the selected ECM techniques on the radar receivers are determined analytically. The principal effects sought are a shift in angle from the target at which the error voltage out of the radar receiver is zero or a change in the value of gain to be associated with the radar receiver. The presentation order of the ECM techniques proceeds from the simplest analysis to the most complex.

8.1 AMPLITUDE MODULATED REPEATER VS COSRO RADAR

This analysis follows that of Shustov and Vakin [4, Sec. 3.3], changing only the symbology to one more familiar to the expected reader. The analysis is of "first order"; thus some of the conclusions drawn here and in Ref. 4 are invalid because of the assumed linearities. Nevertheless, this section should provide useful insight and aid in understanding more detailed analyses.

Assume an incident target return

$$v_s(t) = V_s e^{j\omega_{IF} t} \tag{8.1}$$

and a jamming signal given by

$$v_j(t) = V_j[1 + m_j\cos(\omega_j t - \varphi_j)]e^{j\omega_{IF} t} \tag{8.2}$$

The signal reflected from the target has an envelope of constant amplitude V_s, as the illuminating antenna does not scan when the target is near the tracking axis, and the target cross section is assumed constant for simplicity. The jamming signal is amplitude modulated by a tone at ω_j and it has a constant carrier amplitude V_j.

Both signals are expressed as if at IF frequency, ignoring the unimportant conversion from RF to IF. The jamming and radar signals are assumed coherent and in phase, simplifying the analysis considerably but yielding results consistent with a more complex analysis to follow. The radar model

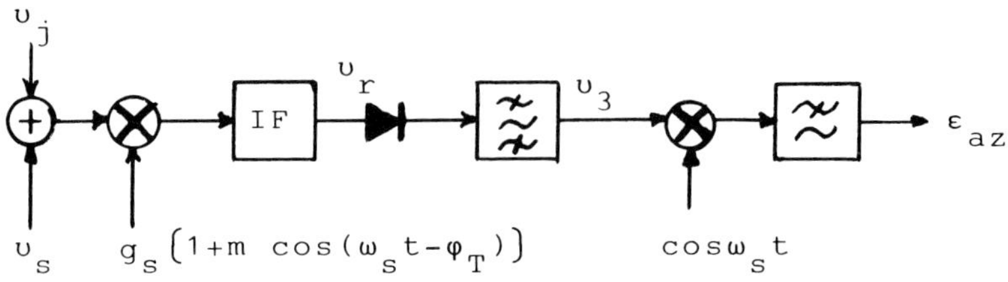

Fig. 8.1 COSRO radar model for AM jamming analysis.

in Fig. 7.15 is applied to this problem in Fig. 8.1. The modulation introduced by the receive antenna is assumed linearly dependent on the tracking error, $m = k_s\theta_T/\theta_3$, which is valid only for $(\theta_T/\theta_S)^2 \ll 1$. The phase φ_T associated with the antenna modulation depends on the relative locations of the scan axis and target position as shown in Fig. 7.5.

The signal at the IF amplifier input port is

$$g_s[v_s(t) + v_j(t)][1 + m\cos(\omega_s t - \varphi_T)] \tag{8.3}$$

If the IF amplifier gain is K, the detector input is

$$\begin{aligned} v_r(t) &= g_s K[V_s + V_j + V_j m_j \cos(\omega_j t - \varphi_j)] \\ &\quad \times [1 + m\cos(\omega_s t - \varphi_T)]e^{j\omega_{IF}t} \end{aligned} \tag{8.4}$$

The envelope detector output is given by (with $K' = Kg_s$)

$$\begin{aligned} v_d(t) &= |V_r(t)|_{LP} \\ &= K'[V_s + V_j + V_j m_j \cos(\omega_j t - \varphi_j) + (V_s + V_j)m\cos(\omega_s t - \varphi_T) \\ &\quad + V_j m_j m \cos(\omega_j t - \varphi_j)\cos(\omega_s t - \varphi_T)] \end{aligned} \tag{8.5}$$

The only terms in Eq. (8.5) to pass the scan filter are those near the scan frequency of the receive antenna. If the frequency of the jammer modulation exactly equals the radar scan frequency, the input to the angle error detector is

$$v_3(t) = K'V_j m_j \cos(\omega_s t - \varphi_j) + K'(V_s + V_j)m\cos(\omega_s t - \varphi_T) \tag{8.6}$$

resulting in the following error voltages:

$$\varepsilon_{az} = \frac{K'V_j m_j}{2}\cos\varphi_j + \frac{K'(V_s + V_j)m}{2}\cos\varphi_T \tag{8.7}$$

$$\varepsilon_{el} = \frac{K'V_j m_j}{2}\sin\varphi_j + \frac{K'(V_s + V_j)m}{2}\sin\varphi_T \tag{8.8}$$

It is apparent from Eqs. (8.7) and (8.8) that the jammer-induced signal is identical to the signal generated by a target at an angle φ_j from the time reference and with an offset from the scan axis encoded in the amplitude of the error signal. (The induced and real targets are shown in Fig. 8.2a.) The measured offset from the track axis of the induced target, as well as the offset of the real target, depends on the response of the radar's AGC system. If the AGC loop rapidly forces the average detected output in Eq. (8.5) to equal unity, then

$$K'\left\{V_s + V_j\left[1 + \frac{m_j m}{2}\cos(\varphi_T - \varphi_j)\right]\right\} = 1 \tag{8.9}$$

Initially, assuming $m \ll 1$, the error voltages are

$$\varepsilon_{az} = \frac{V_j m_j}{2(V_s + V_j)}\cos\varphi_j + \frac{m}{2}\cos\varphi_T \tag{8.10}$$

$$\varepsilon_{el} = \frac{V_j m_j}{2(V_s + V_j)}\sin\varphi_j + \frac{m}{2}\sin\varphi_T \tag{8.11}$$

resulting in an induced target at an angle given by

$$\frac{k_s\theta_j}{2\theta_3} = \frac{V_j m_j}{2(V_s + V_j)} \tag{8.12}$$

$$\theta_j = \frac{\theta_3 m_j V_j}{k_s(V_s + V_j)} < \theta_3 \tag{8.13}$$

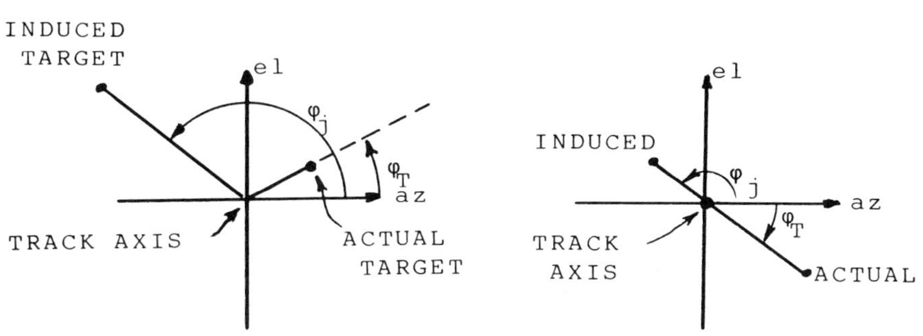

a) Initial track axis position.

b) Final track axis position.

Fig. 8.2 Relocation of track axis by jamming.

The scan axis will move under the influence of the error voltages given in Eq. (8.10) until *both* the azimuth and elevation error voltages are zero. This zero condition occurs only when $\varphi_j = \varphi_T \pm \pi$ and when

$$\frac{V_j m_j}{2(V_j + V_s)} = \frac{m}{2} = \frac{k_s \theta_T}{2\theta_3} \tag{8.14}$$

(The new track axis location is shown in Fig. 8.2b.) Based on Eq. (8.14), the normalized tracking error is

$$\frac{\theta_T}{\theta_3} = \frac{m_j}{k_s} \frac{V_j}{V_j + V_s} \tag{8.15}$$

If a jamming-to-signal ratio is defined by

$$\frac{J}{S} = \frac{V_j^2}{V_s^2} \tag{8.16}$$

and a Gaussian antenna pattern is assumed for the COSRO receive beam with

$$k_s = 4\ell n 2(\theta_s/\theta_3) \tag{8.17}$$

then the normalized track error is

$$\frac{\theta_T}{\theta_3} = \frac{m_j}{4\ell n 2(\theta_s/\theta_3)} \frac{\sqrt{J/S}}{1 + \sqrt{J/S}} \tag{8.18}$$

This result is plotted in Fig. 8.3 for specified parameters. For a jammer of infinite power, if $\theta_s/\theta_3 = 1/2$ and as $m_j \le 1$

$$\frac{\theta_T}{\theta_3} \le \frac{1}{4\ell n 2(\theta_s/\theta_3)} = \frac{1}{2\ell n 2} \approx 0.7 \tag{8.19}$$

as concluded by Shustov and Vakin [4, p. 152]. However, the error slope is not a constant over this large a variation in target angle and thus invalidates Eq. (8.19). Harmonics of the receive antenna modulating function will result in signals out of the scan filter that must be accounted for.

If the jammer modulation frequency does not exactly equal the scan frequency of the radar, the error voltages are easily found from Eq. (8.5) to be

$$\varepsilon_{az} = \frac{V_j m_j}{2(V_j + V_s)} \cos[(\omega_s - \omega_j)t + \varphi_j] + \frac{m}{2} \cos\varphi_T \tag{8.20}$$

$$\varepsilon_{el} = \frac{V_j m_j}{2(V_j + V_s)} \sin[(\omega_s - \omega_j)t + \varphi_j] + \frac{m}{2} \sin\varphi_T \tag{8.21}$$

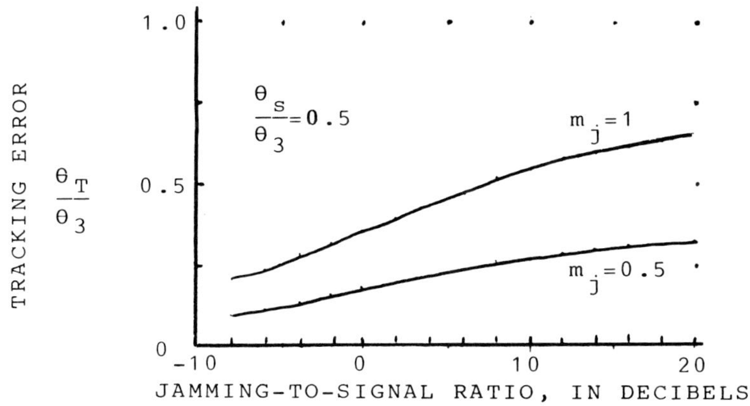

Fig. 8.3 Angle error of COSRO radar resulting from AM repeater jamming.

where the normalized signal of Eq. (8.9) is assumed. Again, the track axis will move until the error voltage is zero. If the change in phase due to the difference in modulation frequencies is small over the time period required for the servo to reposition itself, the track axis will move in a circle about the real target as shown in Fig. 8.4a.

Finally, since the actual scan frequency is unknown, the jammer will probably vary its modulation frequency over a region that contains the radar scanning frequency. When the jammer frequency of modulation differs from the modulation frequency of the radar scan mechanism by more than one-half the bandwidth of the scan filter, the jammer-induced error will go to

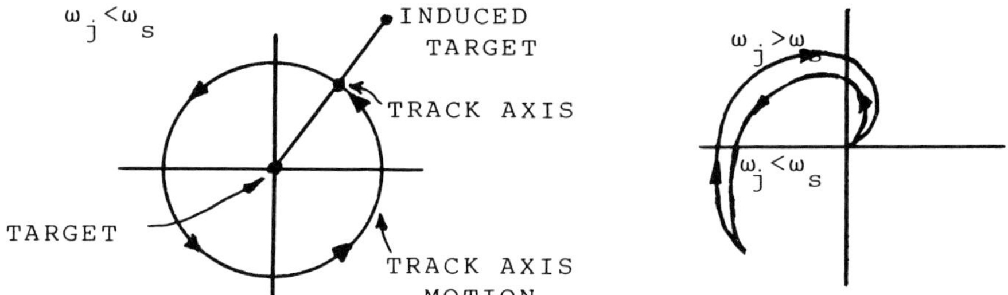

a) Constant frequency difference. **b) Varying frequency difference.**

Fig. 8.4 Track axis motion.

zero. In fact, the strong jammer signal provides better tracking than the skin return. The motion of the track axis as the jammer sweeps slowly in frequency is shown in Fig. 8.4b.

Effect of Random Phase of Jammer Carrier

If the carrier frequency of the jammer is within the IF bandwidth of the radar but not exactly at the radar carrier frequency, then the relative phase between the target return and the jammer signal changes rapidly. The detector output of a pulsed radar should appear pulse-like, although a large difference in frequency could produce "rounded" pulses. These pulses will fluctuate from pulse to pulse about an average value, which is all that will survive the scan filter and finally drive the servo. This same average value results when the frequencies of the jammer and the radar carrier are equal, but the relative phase is a random variable $\tilde{\psi}$ such that

$$\begin{aligned} p(\psi) &= \frac{1}{2\pi} \qquad |\psi| < \pi \\ &= 0 \qquad \text{else} \end{aligned} \tag{8.22}$$

Under these conditions, the instantaneous detector output is

$$|\tilde{V}_r(t)| = Kg_s[1 + m\cos(\omega_S t - \varphi_T)]\,|V_s + V_j(t)e^{j\tilde{\psi}}| \tag{8.23}$$

where

$$V_j(t) = V_j[1 + m_j\cos(\omega_j t - \varphi_j)] \tag{8.24}$$

With $Kg_s[1 + m\cos(\omega_S t - \varphi_T)] = A(t)$, the detector output is

$$\begin{aligned} V_d(t) &= E\{|\tilde{V}_r(t)|\} \\ &= \frac{A(t)}{\pi}\int_0^{\pi} \sqrt{V_s^2 + V_j^2(t) + 2V_sV_j(t)\cos\psi}\ d\psi \end{aligned} \tag{8.25}$$

In interpreting Eq. (8.25), one must remember that $V_j(t)$ and $A(t)$ are essentially a constant over the interval of averaging, which is on the order of a pulse width.

Equation (8.25) can be cast as an elliptic integral of the second kind [79, as p. 266].

$$V_d(t) = \frac{2A(t)}{\pi}[V_s + V_j(t)]\int_0^{\pi/2}\left(1 - \frac{4\sqrt{J/S}}{(1+\sqrt{J/S})^2}\sin^2\varphi\right)^{1/2} d\varphi \tag{8.26}$$

Table 8.1 Evaluation of integral in Eq. (8.26)

J/S, dB	Integral	J/S, dB	Integral
0	1.0	20	1.43
1	1.01	30	1.52
5	1.08	40	1.56
10	1.2	∞	$\pi/2$

where $\varphi = \psi/2$ and J/S is defined by

$$\frac{J}{S} = \left[\frac{V_j(t)}{V_s}\right]^2 \tag{8.27}$$

Table 8.1 shows the result of the integration for various J/S ratios. From Table 8.1, one can reasonably assume the value of the integral to be a constant in the range of 1.3 provided the jamming-to-signal ratio in Eq. (8.27) is at least unity. The detector output is then

$$V_d(t) \approx \frac{2.6}{\pi} Kg_s[1 + m\cos(\omega_S t - \varphi_T)] \times \{V_s + V_j[1 + m_j\cos(\omega_j t - \varphi_j)]\} \tag{8.28}$$

which is, within a factor of 1.2, equal to Eq. (8.5). Thus, the results of the previous subsection remain valid whenever the minimum jammer power equals or exceeds the target return, even if the frequency of the jammer carrier is not equal to the frequency of the radar.

Rate of Increase of Angular Error

Nothing has been said of the time required for the angular error induced in the radar to reach the value given in Eq. (8.18). Assuming perfect target tracking prior to the onset of jamming, the jamming results in the generation of a step-angular input to the servo. This step is not equivalent to target motion; it represents the instantaneous appearance of a new target. Thus, the motion of the radar toward the false target decreases the error voltage generated by the false target, but such motion simultaneously increases the error voltage from the real target.

The size of the equivalent angular step depends on the average IF output voltage, which varies with AGC control voltage. The AGC control voltage varies with the size of the angular tracking error according to Eq. (8.9) and the servo loop is nonlinear. However, the nonlinearity is not large and, with some approximation, the linear servo loop response can still be developed.

If one assumes perfect radar tracking, initially, and that the jammer-induced error is in azimuth (i.e., $\varphi_j = \pi$), the error voltage [from Eqs. (8.7) and (8.9)] is

$$\varepsilon_{az} = \frac{-V_j m_j + (V_s + V_j)m}{2\left[V_s + V_j\left(1 - \frac{mm_j}{2}\right)\right]} \tag{8.29}$$

If the jammer carrier is 100% modulated and very much greater in power than the target return ($V_j \gg V_s$), then the error voltage is

$$\varepsilon_{az} \approx \frac{m - 1}{2\left(1 - \frac{m}{2}\right)} \tag{8.30}$$

which is approximately linear for small m as shown in Fig. 8.5. Because of this linear relationship, the servo loop can be analyzed as a linear system subjected to both a new return displaced from the real target by $\theta_j = \theta_3/k_s$ and the original target. (The appropriate servo model is shown in Fig. 8.6.) The transfer function of the servo response to jamming is

$$\theta_{Ao}(s) = \frac{2G(s)}{1 + 2G(s)} \frac{1}{2k_s} \frac{\theta_3}{s} \tag{8.31}$$

which equates to the response of the original unjammed servo with twice the open-loop gain to a step input of $\theta_3/2k_s$. Using Eq. (7.135) the response time of an improved type I servo with the parameters $\omega_1 = 10$ rad/s, $\omega_2 = 100$ rad/s, and $K = 10{,}000$ rad/s^2 is $t_r = 16$ ms, which is 64% of the unjammed servo response (i.e., the jammed servo is quicker). The servo-controlled scan axis moves quickly toward the jammer-induced target and, after an 18% overshoot, settles rapidly on the false target.

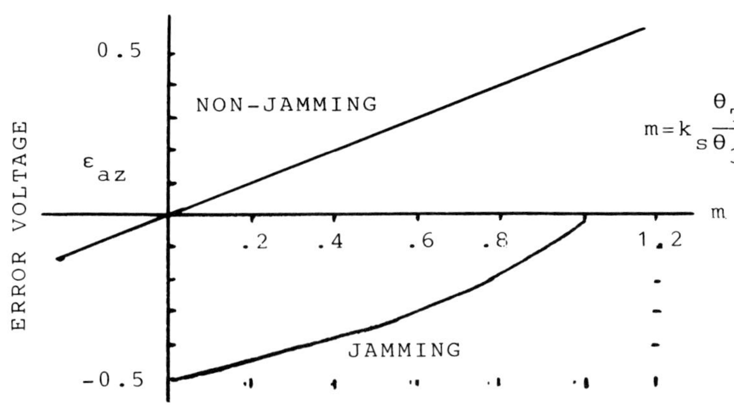

Fig. 8.5 Error voltage vs angle error under jamming.

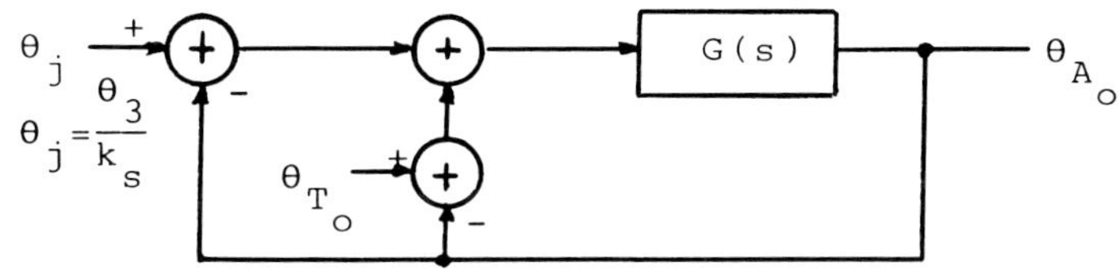

Fig. 8.6 Model of CONSCAN radar servo subjected to AM jamming.

8.2 AMPLITUDE MODULATION THROUGH INVERSE GAIN

When the threat reveals its scan frequency, in simple CONSCAN for example, the jammer can use the radar's illumination to control the modulation frequency of the jammer. This approach permits simultaneous jamming of multiple pulsed radars, provided different radar pulses are not time coincident in the final power amplifier of the jammer.

The circuit in Fig. 8.7 is assumed to adequately represent a CONSCAN radar receiver and the signals present at its input. A square law detector is assumed in this model because it simplifies the analysis. The variation in voltage gain of the antenna with time is

$$g_1(t) = g_s(1 + m\cos\omega_s t) \tag{8.32}$$

where g_s is the voltage gain on the radar scan axis. The IF amplifier acts simply as a constant gain.

The inverse-gain-jammer amplitude modulates its output at the radar scan rate. But rather than store various scan rates, it generates them internally from the received radar scan modulation. If the target signal is

$$v_s(t) = V_s(1 + m\cos\omega_s t)\cos\omega_{IF} t \tag{8.33}$$

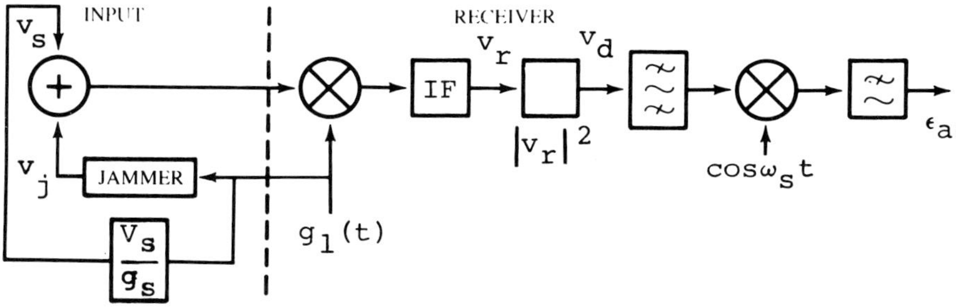

Fig. 8.7 Inverse gain vs CONSCAN.

then the jammer signal will be

$$\tilde{v}_j(t) = V_j(1 - \beta m \cos\omega_s t)\cos(\omega_{IF} t + \tilde{\psi}) \tag{8.34}$$

where $\tilde{\psi}$ is distributed as in Eq. (8.22). The modulation of the jammer amplitude is shifted by 180 deg from the modulation of the radar return or inverted; thus, the name inverse gain.

The voltage out of the IF amplifier is

$$\tilde{v}_r(t) = K[v_s(t) + \tilde{v}_j(t)]g_1(t)$$

and the voltage from the detector

$$v_d(t) = E\{|\tilde{v}_r(t)|^2\} = K^2 g_1^2(t) E\{|v_s(t) + \tilde{v}_j(t)|^2\} \tag{8.35}$$

Expanding Eq. (8.35) with Eqs. (8.22), (8.33), and (8.34), and assuming that, initially, m^2 and $\beta^2 m^2$ are much less than one and can thus be ignored, results in

$$\begin{aligned} v_d(t) = K^2 g_s^2 (1 + 2m\cos\omega_s t)[(V_j^2 + V_s^2) \\ + 2m(V_s^2 - \beta V_j^2)\cos\omega_s t] \end{aligned} \tag{8.36}$$

Only the term at the scan rate ω_s passes the scan filter, resulting in an error voltage

$$\varepsilon_{az} = K^2 g_s^2 [2V_s^2 + (1 - \beta)V_j^2] \frac{k_s \theta_T}{\theta_3} \tag{8.37}$$

before considering AGC effects and recognizing k_s to be for a one-way pattern. If AGC normalization is considered, the error voltage becomes

$$\varepsilon_{az} = \left(1 + \frac{V_s^2 - \beta V_j^2}{V_s^2 + V_j^2}\right)\frac{k_s}{\theta_3}\theta_T \tag{8.38}$$

Now we define a jamming-to-signal ratio by Eq. (8.16). The error voltage of the radar receiver is related to the tracking error in the nonjamming case through

$$\varepsilon_{az} = 2k_s \frac{\theta_T}{\theta_3} = K_D \theta_T \tag{8.39}$$

in which k_s is defined for the one-way radiation pattern. With jamming, the error voltage relates to the angular tracking error through

$$\varepsilon_{az} = \left[\frac{2 + (J/S)(1 - \beta)}{1 + (J/S)}\right]\frac{k_s \theta_T}{\theta_3} \tag{8.40}$$

which determines the radar response.

The constant K_D represents the radar receiver in the radar model shown in Fig. 7.3. This constant was shown to be a function of the antenna pattern through the antenna beamwidth and error slope coefficient. It is now shown to be a function of certain jammer parameters as given in Eq. (8.40). The value of K_D in the unjammed case is designated K_{Do}. The ratio K_D/K_{Do}, plotted in Fig. 8.8, is

$$\frac{K_D}{K_{Do}} = \frac{2 + (J/S)(1 - \beta)}{2[1 + (J/S)]} \tag{8.41}$$

If the jammer generates an unmodulated carrier (i.e., $\beta = 0$), radar sensitivity to angle errors drops by a factor of two as the J/S ratio becomes infinite. This is reasonable because only the scanning pattern of the receive antenna modulates the receiver input; the modulated target return is overpowered by jamming. If the jammer modulates in phase ($\beta < 0$), radar sensitivity to angle errors can actually increase if the magnitude of β exceeds unity. This will result in a radar servo response often referred to as "tight"—that is, servo motion is more rapid than normal and the overshoots are larger.

Of most interest is the response when inverse modulation is employed. When $\beta < 1$, it reduces the gain in the radar-servo loop and, as to be discussed in Chapter 10, can cause some servo systems to become unstable. When $\beta > 1$, a J/S ratio of

$$\frac{J}{S} \geq \frac{2}{\beta - 1} \tag{8.42}$$

results in a negative loop gain or a servo loop with positive feedback, causing all of the servo loops to become unstable.

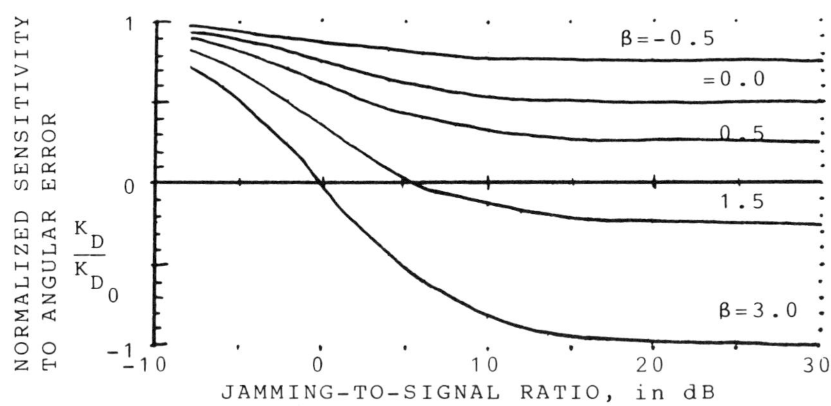

Fig. 8.8 Normalized angle error sensitivity of radar.

Given an unstable loop, the loop gain governs the rate of error increase. With an infinite J/S ratio, the ratio of jammed to unjammed loop gains is, from Eq. (8.41),

$$\frac{K_D}{K_{D_o}} = -\frac{\beta - 1}{2} \tag{8.43}$$

If the inverse gain parameter β is only slightly greater than one, any instability will increase slowly. Larger values of β will reduce the time required to achieve larger errors. For example, consider the improved type I servo in Chapter 7 with the following parameters: $K = 10{,}000$ rad/s^2, $\omega_1 = 20$ rad/s, and $\omega_2 = 100$ rad/s. If $\beta = 3$, the unstable servo loop gain is of the same magnitude as the gain of the unjammed loop. Assuming an initial tracking error of 0.01 θ_3, the error will exceed a beamwidth in 32 ms. If $\beta = 1.2$, the same initial tracking error will have grown to only 0.05 θ_3 in 70 ms.

8.3 INVERSE GAIN VS SCAN WITH COMPENSATION

One can show that the inverse gain jammer in the previous section will be ineffective against the analytical SWC radar modeled in Fig. 8.9. If the jammer is activated after radar tracking has begun, one can assume that the modulation depth m is small. The various signals are as follows. The target return and jammer signals are the same as those shown in Eqs. (8.33) and (8.34). The scan patterns of the main and auxiliary antennas are

$$\begin{aligned} g_m(t) &= g_s(1 + m\cos\omega_s t) \\ g_A(t) &= g_s(1 - m\cos\omega_s t) \end{aligned} \tag{8.44}$$

After performing the operations indicated in Fig. 8.9, the signal from the main channel scan filter is

$$v_{3_m} = \frac{2m[2V_s^2 + (1-\beta)V_j^2]\cos\omega_s t}{V_s^2(1+3m^2) + V_j^2\left[1 + \dfrac{m^2}{2}(1+\beta^2)\right] - 2\beta m^2 V_j^2} \tag{8.45}$$

and the signal in the auxiliary channel is

$$v_{3_A} = \frac{-2m[V_j^2(1+\beta)]\cos\omega_s t}{V_s^2(1-m^2) + V_j^2\left[1 + \dfrac{m^2}{2}(1+\beta^2)\right] + 2\beta m^2 V_j^2} \tag{8.46}$$

In Eqs. (8.45) and (8.46), the average voltage in each channel has been equated to one through separate AGC systems and terms of order m^3 and higher have been dropped.

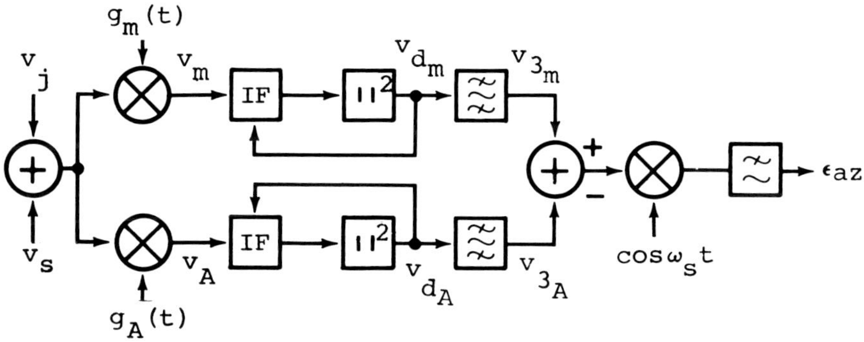

Fig. 8.9 Model of inverse gain vs SWC radar.

Performing the subtraction and allowing the J/S ratio to become infinite results in the error voltage

$$\varepsilon_{az} = \frac{2m\left[1 + \dfrac{m^2}{2}(1+\beta^2) - 2\beta^2 m^2\right]}{\left[1 + \dfrac{m^2}{2}(1+\beta^2)\right]^2 - 4\beta^2 m^4} \approx 2k_s \frac{\theta_T}{\theta_3} \tag{8.47}$$

Terms of order m^2 and higher have now been ignored and again k_s is defined as for a one-way antenna pattern. [Identical results are obtained in this example if terms of order m^2 and higher are ignored outright, beginning with Eqs. (8.45) and (8.46). In general, however, care must be exercised in ignoring terms in m^2 and higher as they may be multiplied by a large J/S ratio.] Of most importance in Eq. (8.47), the sense of the error—its sign—is correct. Second, the error voltage is zero only at zero angular error. A slight change in gain occurs, but it has insignificant impact. Thus, the SWC radar, as modeled, easily tracks the AM jamming signal.

This one example does not exhaust all possible encounters between SWC radars and inverse gain jammers. An example of another encounter assumes an imbalance in the antenna patterns of the radar as follows:

$$g_m(t) = g_s(1 + m\cos\omega_s t) \tag{8.48a}$$

$$g_A(t) = g_{s_A}(1 - m_A\cos\omega_s t) \tag{8.48b}$$

If terms involving m^2 and higher are ignored, the detected main and auxiliary signals out of the square law detectors are, with $m_A = m(1+\delta)$,

$$\begin{aligned} v_{d_M} &= g_s^2(V_s^2 + V_j^2) + g_s^2[V_s^2 4m + V_j^2(1-\beta)2m]\cos\omega_s t \\ v_{d_A} &= g_{s_A}^2(V_s^2 + V_j^2) - g_{s_A}^2[V_s^2 2m\delta + V_j^2(1+\delta+\beta)2m]\cos\omega_s t \end{aligned} \tag{8.49}$$

In Eq. (8.49), only the constant term, which determines the AGC control voltage, and the scan-frequency term are given since the other terms are not important.

If the AGC system again adjusts the IF amplifier gain so that the average of the detected IF output is unity, the error voltage will be

$$\varepsilon_{az} = (2+\delta)m = (2+\delta)\,\frac{k_s\theta_T}{\theta_3} \tag{8.50}$$

Once again, the jamming produces no tracking error on the part of the radar. This does not mean that error generation is impossible in a more accurate model. The terms dropped in this analysis because of the assumed smallness of the depth of modulation may be significant if target maneuvers cause a tracking error.

8.4 AN ANALYSIS TECHNIQUE FOR LARGER ANGULAR ERRORS

All of the preceding analyses linearized the variation of antenna gain with the angle from the scan axis. Such a linearization is obviously invalid when large tracking errors either exist or result from jamming. Similarly, the jammer has been modulated by a single tone and a methodology valid for square-wave modulation is required. Both of these deficiencies can be corrected by expanding the time functions defining the antenna gain and jammer signal variations into a Fourier series. This is possible only when both time signals are periodic; thus, the model developed is a stationary one. If the result of the analysis is an error voltage that will increase the angular error and thus change the antenna gain function, the validity of the model requires that the antenna motion is "slow" in relation to the antenna scan rate ω_s.

As in earlier sections, the likely pulsed nature of the radar is unimportant since it is assumed that the jamming pulses overlap the target returns. The jamming signal is now given by

$$v_j(t) = \sum_{n=0}^{\infty} V_{jn}\cos(n\omega_s t - \psi_n)e^{j\omega_{IF}t} \tag{8.51}$$

where, for simplicity, the jammer period is set equal to the scan frequency of the radar. The target return is given by Eq. (8.1). If a COSRO radar is modeled, only the variation in gain of the receive antenna needs to be considered, but it is still a function of the size of tracking error. Shown in Fig. 8.10 is one cycle of the variation in gain of a Gaussian antenna pattern for three different tracking errors. This functional dependency on tracking error can be incorporated by writing the gain function as

$$g(t) = \sum_{m=0}^{\infty} g_m(\theta_T)\cos(m\omega_s t - \phi_m) \tag{8.52}$$

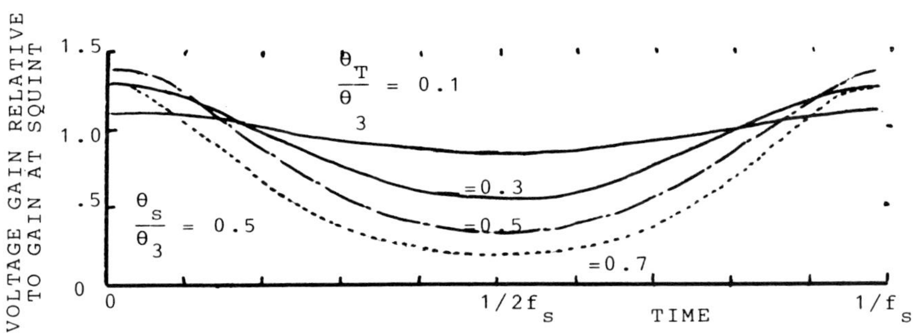

Fig. 8.10 Time variation of gain of Gaussian pattern.

Now any even function can be expanded as

$$f(t) = \sum_{n=0}^{\infty} a_n \cos n\omega_s t$$

and a time shift of τ results in

$$g(t) = f(t - \tau) = \sum_{n=0}^{\infty} a_n \cos\left(n\omega_s t - n2\pi \frac{\tau}{T}\right)$$

$$= \sum_{n=0}^{\infty} a_n \cos(n\omega_s t - n\varphi_1) \qquad (8.53)$$

The relationship between the phase shifts of the harmonics shown in Eq. (8.53), $\varphi_m = m\varphi_1$, can be used to advantage.

Using Fig. 8.1 as the radar model (except for the antenna gain function), the voltage at the envelope detector output is

$$v_d(t) = K|v_j(t) + v_s(t)|g(t)$$

$$= K\left[\sum_{n=0}^{\infty} V_{jn}\cos(n\omega_s t - \psi_n) + V_s\right]\sum_{m=0}^{\infty} g_m \cos(m\omega_s t - \varphi_m) \qquad (8.54)$$

Equation (8.54) is valid provided the jammer is modulated at no more than 100% and modulations that would vary the phase of the jammer RF carrier relative to the phase of the target in a deterministic manner are not used (i.e., double-sideband suppressed carrier). Additionally, Eq. (8.54) will be in error if the modulation of the antenna or jammer has significant harmonic content near the pulse-repetition frequency of a pulsed radar.

The detector output can be rewritten as a constant term plus a term at the scan frequency plus all higher harmonics. This representation is

$$v_d(t) = g_o(V_s + V_{j_o}) + \sum_{n=1}^{\infty} \frac{g_n V_{jn}}{2} \cos(\varphi_n - \psi_n)$$

$$+ V_s g_1 \cos(\omega_s t - \varphi_1) + \frac{g_o V_{j1}}{2} \cos(\omega_s t - \psi_1) + \frac{g_1 V_{jo}}{2} \cos(\omega_s t - \varphi_1)$$

$$+ \sum_{n=1}^{\infty} \frac{g_n V_{j,n-1}}{2} \cos(\omega_s t - \varphi_n + \psi_{n-1}) + \sum_{n=1}^{\infty} \frac{g_{n-1} V_{jn}}{2} \cos(\omega_s t + \varphi_{n-1} - \psi_n)$$

$$+ \text{higher-order terms} \tag{8.55}$$

The azimuth error voltage is then proportional to

$$\varepsilon_{az} = V_s g_1 \cos\varphi_1 + \frac{g_o V_{j1}}{2} \cos\psi_1 + \frac{g_1 V_{jo}}{2} \cos\varphi_1$$

$$+ \sum_{n=1}^{\infty} \frac{g_n V_{j,n-1}}{2} \cos(\varphi_n - \psi_{n-1}) + \sum_{n=1}^{\infty} \frac{g_{n-1} V_{jn}}{2} \cos(\varphi_{n-1} - \psi_n) \tag{8.56}$$

Repeater Modulation by a Single Frequency

If one recalls that the terms g_n are functions of the tracking error θ_T, it would be generally impossible to solve Eq. (8.56) for the tracking error at which Eq. (8.56) and the corresponding elevation error are zero. However, for very small tracking errors and single-tone jammers, it was found in Sec. 8.1 that $\psi_1 = \varphi_1 \pm \pi$. If one can assume that a single-tone jammer is operating and that $\psi_1 = \varphi_1 + \pi$, Eq. (8.56) yields

$$\varepsilon_{az} = (V_s g_1 - g_o V_{j1} + g_1 V_{jo}) \cos\varphi_1 - \frac{g_2 V_{j1}}{2} \cos(\varphi_2 - \varphi_1) \tag{8.57}$$

But $\varphi_2 = 2\varphi_1$ and the error voltages become

$$\varepsilon_{az} = \left[(V_s + V_{jo}) g_1 - V_{j1}\left(g_o + \frac{g_2}{2} \right) \right] \cos\varphi_1 \tag{8.58}$$

$$\varepsilon_{el} = \left[(V_s + V_{jo}) g_1 - V_{j1}\left(g_o + \frac{g_2}{2} \right) \right] \sin\varphi_1 \tag{8.59}$$

requiring the following term to be zero in order that both error voltages be zero:

$$(V_s + V_{jo}) g_1 - V_{j1}\left(g_o + \frac{g_2}{2} \right) = 0 \tag{8.60}$$

The solution to Eq. (8.60) is the tracking error. Equation (8.60) can be expressed as

$$\frac{2g_1(\theta_T)}{2g_o(\theta_T) + g_2(\theta_T)} = \frac{V_{j1}}{V_s + V_{jo}} = \frac{m_j}{\sqrt{(S/J) + 1}} \tag{8.61}$$

if the jamming-to-signal ratio is again defined as $(V_{jo}/V_s)^2$.

If the radar antenna is modeled as Gaussian, a functional expression exists for the left side of Eq. (8.61). The antenna gain function is $g(t) = g[\theta(t)]$ where, from Fig. 7.5,

$$\theta^2(t) = \theta_s^2 + \theta_T^2 - 2\theta_s\theta_T\cos\varphi_T \tag{8.62}$$

The expansion for $g(t)$ uses the formulation in Skolnik [80]

$$g(t) = g[\theta(t)] = g_g \exp\left[\frac{-2\ell n 2}{\theta_3^2}\theta^2(t)\right]$$

$$= g_g \exp\left[\frac{-2\ell n 2}{\theta_3^2}(\theta_3^2 + \theta_T^2)\right]\exp\left[\frac{4\ell n 2\theta_s\theta_T}{\theta_3^2}\cos(\omega_s t - \varphi_1)\right] \tag{8.63}$$

Using $k = 4(\theta_s/\theta_3)\ell n 2$ and the identity from Abramowitz and Stegun [81],

$$e^{z\cos\psi} = I_o(z) + 2\sum_{n=1}^{\infty} I_n(z)\cos n\psi \tag{8.64}$$

where $I_n(z)$ is the modified Bessel function, Eq. (8.63) becomes

$$g(t) = g_g \exp\left[\frac{2\ell n 2(\theta_s^2 + \theta_T^2)}{\theta_3^2}\right]$$

$$\times\left[I_o\left(\frac{k\theta_T}{\theta_3}\right) + 2\sum_{n=1}^{\infty} I_n\left(\frac{k\theta_T}{\theta_3}\right)\cos n(\omega_s t - \varphi_1)\right] \tag{8.65}$$

from which the series expansion terms are

$$g_o = g_g \exp\left[\frac{-2\ell n 2(\theta_s^2 + \theta_T^2)}{\theta_3^2}\right] I_o\left(\frac{k\theta_T}{\theta_3}\right) \tag{8.66}$$

$$g_n = 2g_g \exp\left[\frac{-2\ell n 2(\theta_s^2 + \theta_T^2)}{\theta_3^2}\right] I_n\left(\frac{k\theta_T}{\theta_3}\right)$$

$$\varphi_n = n\varphi_1 \tag{8.67}$$

Using Eq. (8.66) in Eq. (8.61) and equating $x = k\theta_T/\theta_3$ results in

$$\frac{2I_1(x)}{I_o(x) + I_2(x)} = \frac{m_j}{1 + \sqrt{S/J}} = \frac{I_1(x)}{(d/dx)I_1(x)} \tag{8.68}$$

where the last relationship used the identity [81, Sec. 9.8]

$$I_{n-1}(x) + I_{n+1}(x) = 2\frac{d}{dx}I_n(x) \tag{8.69}$$

For a specified value of J/S, the value of x can be incremented until equality holds in Eq. (8.68), at which point x yields the tracking error. Solutions of

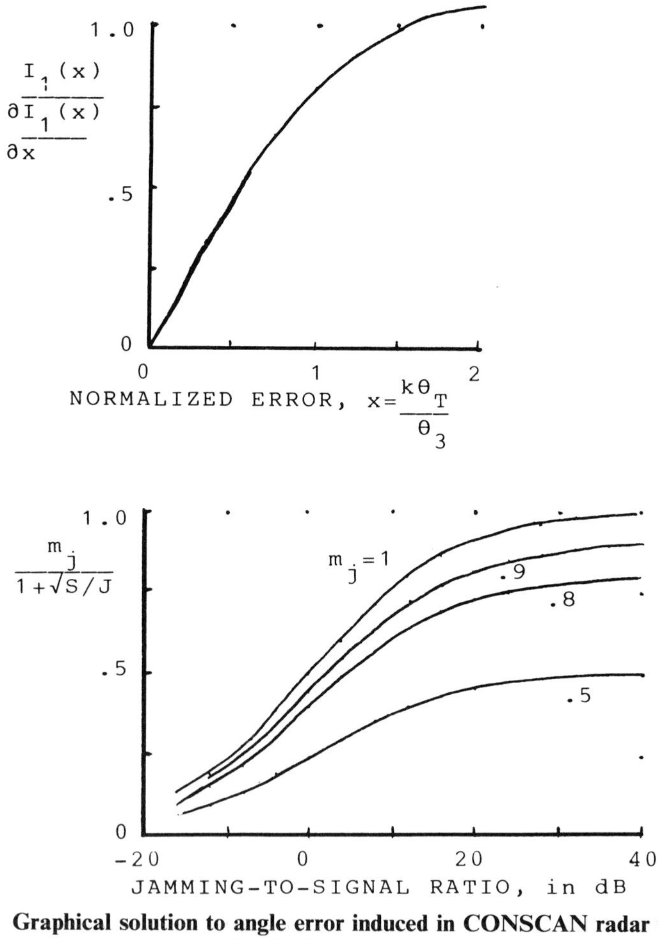

Fig. 8.11 Graphical solution to angle error induced in CONSCAN radar with Gaussian pattern by AM jamming of single tone.

Eq. (8.68) are presented in Fig. 8.11 for various depths of modulation. As $m_j \le 1$, there is always a stable tracking point at even when the J/S ratio becomes infinite. If the squint angle is $\theta_3/2$, the maximum angular error is $1.12\ \theta_3$. Of course, the Gaussian approximation of a typical antenna pattern begins to degrade rapidly when this size of angular offset is achieved. The implication of this analysis is that large angular errors are achieved and only more detailed consideration of the actual antenna patterns will permit prediction of the result of jamming.

Validity of Linear Servo Model

In addition to determining the angular tracking error, if any, at which the error voltage is zero, it is necessary to assess the magnitude of the error voltage resulting from a particular angular error and the linearity of any relationship between these two quantities. This latter information determines the validity of using linear radar servo models.

All the necessary information is contained in Eqs. (8.55) and (8.66). Given a single-tone jammer, the constant and scan frequency terms out of the detector are

$$v_d(t) = \left[g_o(V_s + V_{jo}) - \frac{g_1 V_{j1}}{2}\right] + \left[g_1(V_s + V_{jo}) - V_{j1}\left(g_o + \frac{g_2}{2}\right)\right]\cos(\omega_s t - \varphi_1) \tag{8.70}$$

Using the expansion coefficients defined in Eq. (8.66) yields a normalized azimuth error voltage of

$$\varepsilon_{az} = \frac{2I_1(x)(V_s + V_{jo}) - V_{j1}[I_o(x) + I_2(x)]}{I_o(x)(V_s + V_{jo}) - I_1(x)V_{j1}} \tag{8.71}$$

or with a J/S ratio defined by $(V_{jo}/V_s)^2$

$$\varepsilon_{az}(x) = \frac{2I_1(x)(1 + \sqrt{S/J}) - m_j[I_o(x) + I_2(x)]}{I_o(x)(1 + \sqrt{S/J}) - m_j I_1(x)} \tag{8.72}$$

This error voltage is plotted in Fig. 8.12 for various J/S ratios and jammer modulation levels. At all J/S ratios, a linear model is valid and a linear approximation to Eq. (8.72) can be derived from Fig. 8.12. Also, using identities in Ref. 81, Eq. (8.71) can be expressed as

$$\varepsilon_{az}(x) = 2\,\frac{I_1(x)\left(1 + \sqrt{\dfrac{S}{J}}\right) - m_j \dfrac{\mathrm{d}}{\mathrm{d}x} I_1(x)}{I_o(x)\left(1 + \sqrt{\dfrac{S}{J}}\right) - m_j \dfrac{\mathrm{d}}{\mathrm{d}x} I_o(x)} \tag{8.73}$$

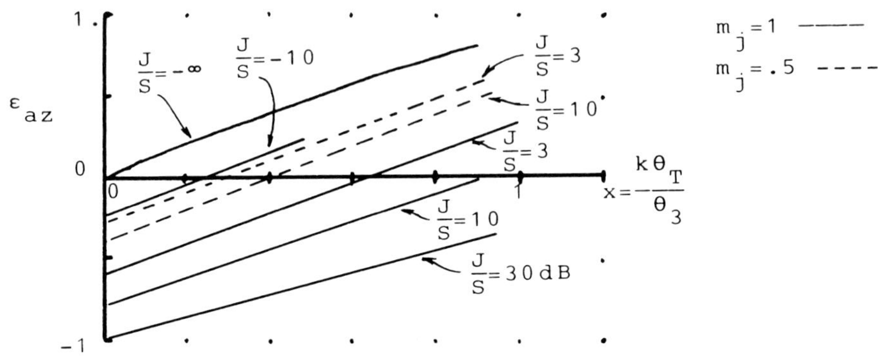

Fig. 8.12 Error voltage in CONSCAN radar due to single-tone AM repeater jammer.

Square-Wave Repeater Modulation

If a square-wave modulation is applied to the repeated signal and if the square-wave period is equal to the scan period, then Eq. (8.51) can still represent the jamming waveform. Let the jammer modulation be as shown in Fig. 8.13. The jammer coefficients are given by

$$V_{jo} = V_j/2 \tag{8.74}$$

$$V_{jn} = (-1)^{(n-1)/2}\frac{2V_j}{n\pi} \qquad n \geq 1 \text{ and odd}$$

$$= 0 \qquad n \text{ even} \tag{8.75}$$

All phase terms ψ_n are zero.

Upon initiation of jamming, the antenna scan axis will reposition itself to lie between the true target and the fictitious target induced by the jamming. From this new location, the phases of the jamming and target modulations will appear changed. Their relative phase, however, will be such that $\psi_1 = \varphi_1 + \pi$. Using this key relative phase relationship, Eq. (8.56) for the error voltage becomes

$$\varepsilon_{az} = V_s g_1 - V_{j1} g_o + V_{jo} g_1 - \sum_{\substack{n=2 \\ \text{even}}}^{\infty} \frac{g_n V_{j,n-1}}{2} - \sum_{\substack{n=3 \\ \text{odd}}}^{\infty} \frac{V_{jn} g_{n-1}}{2} \tag{8.76}$$

After applying the identity in Eq. (8.69), using Eq. (8.66), and equating the common term $g_g \exp[-2\ell n 2(\theta_s^2 + \theta_T^2)/\theta_3^2]$ to K, this error voltage becomes

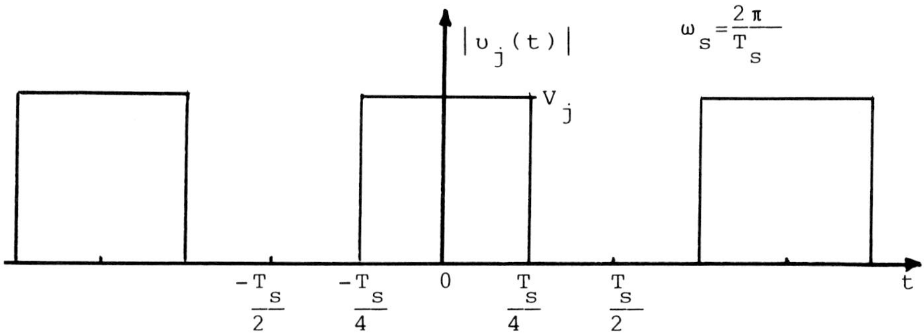

Fig. 8.13 Jamming modulation, square wave.

[with $I'_n(x) = (d/dx)I_n(x)$],

$$\varepsilon_{az} = 2K\left[(V_s + V_{jo})I_1(x) - \sum_{\substack{n=1\\ \text{odd}}}^{\infty} V_{jn}I'_n(x)\right] \tag{8.77}$$

The tracking axis will stabilize at a point so that $\varepsilon_{az} = 0$. Inserting Eq. (8.74) into Eq.(8.77) and setting $\varepsilon_{az} = 0$ results in

$$\left(V_s + \frac{V_j}{2}\right) I_1(x) = \sum_{\substack{n=1\\ \text{odd}}}^{\infty} (-1)^{(n-1)/2} \frac{2V_j}{n\pi} I'_n(x) \tag{8.78}$$

If a J/S ratio is defined by $J/S = (V_j/V_s)^2$, Eq. (8.78) becomes

$$\frac{I_1(x)}{\sum\limits_{\substack{n=1\\ \text{odd}}} (-1)^{(n-1)/2} \dfrac{I'_n(x)}{n}} = \frac{\dfrac{4}{\pi}\sqrt{\dfrac{J}{S}}}{2 + \sqrt{\dfrac{J}{S}}} \tag{8.79}$$

The tracking error can be found by incrementing x until equality holds in Eq. (8.79). (A graphical solution is presented in Fig. 8.14.) The maximum error is given by $x = 1.804$ as the J/S ratio becomes unbounded. Again, a stable tracking point is reached, but, at such large angles, the Gaussian pattern is not a good model. The error is larger than the error developed by the single-tone jammer because the ratio of the error producing term at the scan frequency to the target enhancing term at zero frequency is larger for a square wave.

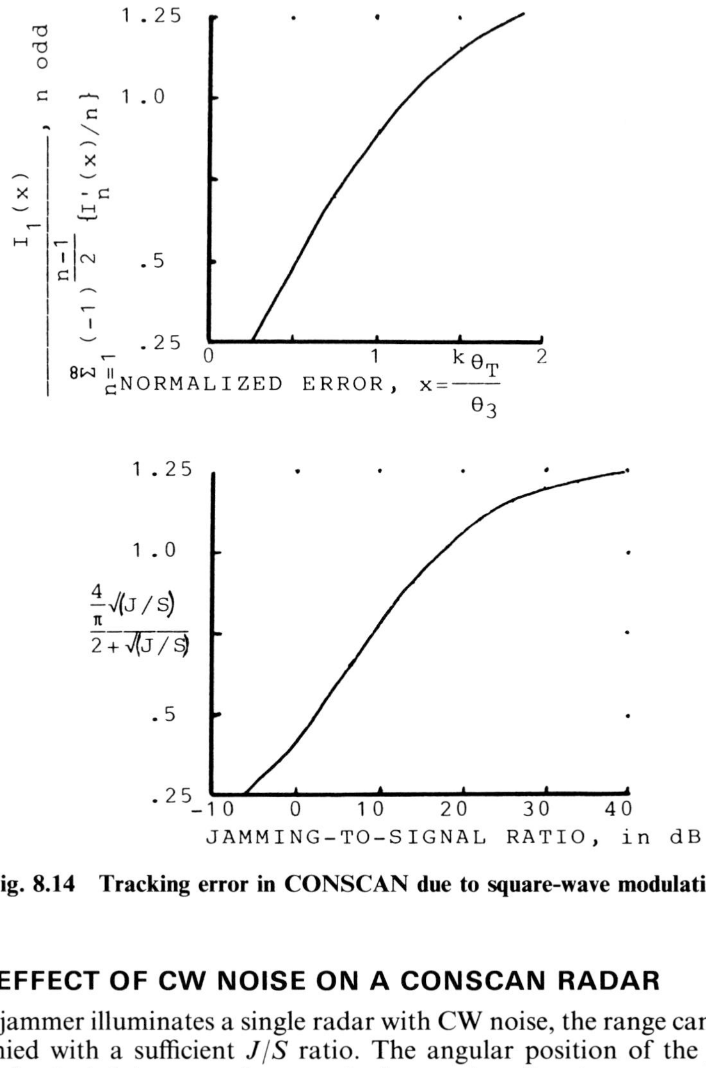

Fig. 8.14 Tracking error in CONSCAN due to square-wave modulation.

8.5 EFFECT OF CW NOISE ON A CONSCAN RADAR

If a jammer illuminates a single radar with CW noise, the range can always be denied with a sufficient J/S ratio. The angular position of the jammer cannot be denied, however, because the jammer's radiated energy reveals the jammer's location just as the sun's light reveals the sun's position in the sky.

Denial of Range of Target

At the outport port of the IF amplifier, let the target return be given by Eq. (8.1) and the jammer signal be

$$\tilde{v}_j(t) = \tilde{V}_j(t)e^{j[\omega_{IF}t + \tilde{\varphi}(t)]} \tag{8.80}$$

where $\tilde{v}_j(t)$ is defined as a narrow-band Gaussian process. Then, at any instant of time, the amplitude $\tilde{V}_j(t)$ follows a Rayleigh probability distribution [82, p. 101] and the phase $\tilde{\varphi}$ is distributed as in Eq. (8.22). If $E\{\tilde{V}_j^2\} = 2\sigma^2$, the Rayleigh distribution is

$$p(V_j) = \frac{V_j}{\sigma^2} \exp\left(-\frac{V_j^2}{2\sigma^2}\right) \qquad V_j \geq 0 \tag{8.81}$$

and $E\{\tilde{V}_j\} = \sigma\sqrt{\pi/2}$. The jamming-to-signal ratio is defined as

$$\frac{J}{S} = \frac{E\{\tilde{v}_j^2\}}{E\{\tilde{v}_s^2\}} = \frac{2\sigma^2}{V_s^2} \tag{8.82}$$

At the output port of an envelope detector, the voltage $\tilde{v}_d$ will follow a Ricean distribution [82, p. 105] with the following characteristics when the target is present:

$$p(v_d) = \frac{v_d}{\sigma^2} \exp\left(-\frac{v_d^2 + V_s^2}{2\sigma^2}\right) I_o\left(\frac{V_s v_d}{\sigma^2}\right)$$

$$E\{\tilde{v}_d\} \approx \sqrt{\frac{\pi}{2}}\,\sigma\left[1 + \frac{V_s^2}{4\sigma^2} - \frac{1}{16}\left(\frac{V_s^2}{2\sigma^2}\right)^2\right]$$

$$E\{\tilde{v}_d^2\} = 2\sigma^2\left[1 + \frac{V_s^2}{2\sigma^2}\right] \tag{8.83}$$

When the return from the target is not at the detector output, the detected voltage results from jamming alone, $v_d = v_j$, and the voltage distribution of the detector output is Rayleigh with

$$E\{\tilde{v}_d\} = \sqrt{\frac{\pi}{2}}\,\sigma \tag{8.84}$$

$$E\{\tilde{v}_d^2\} = 2\sigma^2 \tag{8.85}$$

On a typical A-scope, many separate traces will be integrated. The presentation should consist of a wide trace with mean and variance determined by Eqs. (8.83) and (8.84). Determination of target location is impossible if no observable variation in trace statistics occurs near the target return. Assuming that an operator uses only mean and variance as discriminants, the target will not be detectable above some J/S ratio. From Eqs. (8.83) and (8.84), these moments without a target are

$$E\{\tilde{v}_d\} = \sqrt{\frac{\pi}{2}}\,\sigma \tag{8.86}$$

$$V_{ar}\{\tilde{v}_d\} = \left(2 - \frac{\pi}{2}\right)\sigma^2 \tag{8.87}$$

Table 8.2 Normalized mean and variance of detected output with CW jamming

J/S, dB	Mean	Variance
0	1.438	1.085
3	1.235	1.271
10	1.049	1.091
13	1.025	1.048
20	1.005	1.010

and, with a target, they become

$$E\{\tilde{v}_d\} = \sqrt{\frac{\pi}{2}}\,\sigma\left[1 + \frac{1}{2}\left(\frac{S}{J}\right) - \frac{1}{16}\left(\frac{S}{J}\right)^2\right] \tag{8.88}$$

$$V_{ar}\{\tilde{v}_d^2\} \approx \left(2 - \frac{\pi}{2}\right)\sigma^2\left(1 + \frac{S}{J}\right) - \frac{\pi\sigma^2}{8}\left(\frac{S}{J}\right)^2 \tag{8.89}$$

The expectation and square root of the variance in Eqs. (8.88) and (8.89), normalized by the mean in Eq. (8.86) are plotted in Fig. 8.15 for various J/S ratios and tabulated in Table 8.2. In Fig. 8.15, the average detected output of jamming by itself is arbitrarily set equal to one. From Table 8.2 and Fig. 8.15, the target is apparently obscured on the A-scope at a J/S ratio between 10 and 15 dB.

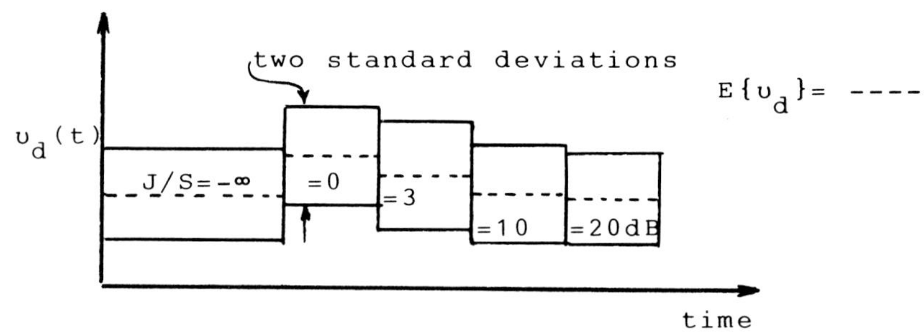

Fig. 8.15 Normalized mean and variance of detected output with CW jamming.

CONSCAN Angle Estimation in CW NOISE

Barring saturation of the output voltage of the receiver in the radar, the CW noise jammer does not deny angular information. If the radar scan axis points directly toward the jammer, the incident jammer signal is received at

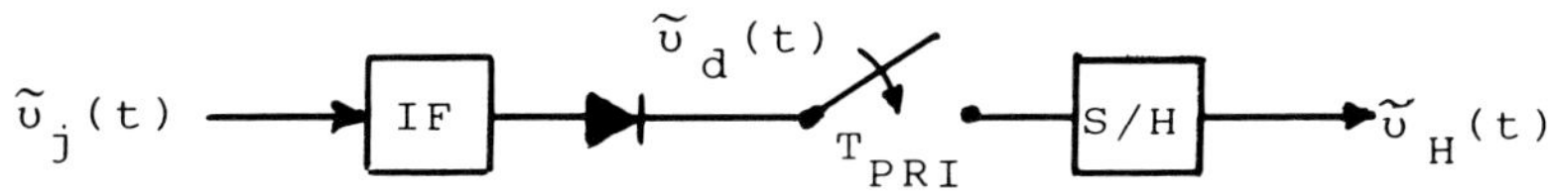

Fig. 8.16 Model of CONSCAN radar for noise analysis.

a constant antenna gain. If the J/S ratio is large or if the range gate no longer samples the receiver output at the time the target return is present, the voltage from the envelope detector will be Rayleigh distributed as in Eq. (8.81). This gated detector output is passed through a sample-and-hold (S&H) circuit, whose output is again Rayleigh. (A model of the radar receiver is shown in Fig. 8.16.) outputs of the S&H circuit (see Fig. 8.17) are independent for separations in time exceeding the pulse-repetition interval of the radar because samples separated by this much are caused by independent jammer/receiver outputs. The first and second moments of the output of the S&H circuit are

$$E\{\tilde{v}_H\} = \sqrt{\frac{\pi}{2}}\,\sigma \tag{8.90}$$

$$E\{\tilde{v}_H^2\} = 2\sigma^2 \tag{8.91}$$

where σ^2 can be related to the jamming power if desired.

This output can be represented as a constant plus a zero mean value noise component

$$\tilde{v}_H(t) = E\{\tilde{v}_H\} + \tilde{n}(t) \tag{8.92}$$

where $V_{ar}\{\tilde{v}_H\} = E\{\tilde{n}^2\} = [2 - (\pi/2)]\sigma^2$ and $E\{\tilde{n}\} = 0$. Each pulse from the S&H circuit possesses a $(\sin x/x)^2$ energy spectrum that has zeros at integral

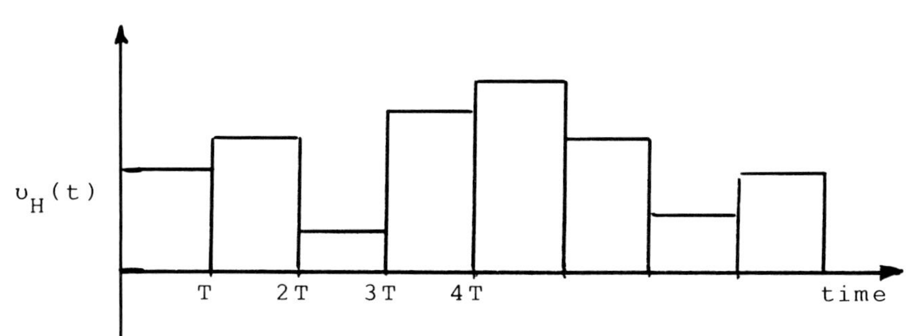

Fig. 8.17 Typical sample-and-hold (S&H) output.

multiples of the pulse-repetition frequency of the radar and a magnitude proportional to the square of the pulse amplitude. The power spectrum associated with the zero mean noise component is

$$S_n(\omega) = \frac{E\{\tilde{n}^2\}}{T} T^2 \left|\frac{\sin \omega T/2}{\omega T/2}\right|^2 = \left(2 - \frac{\pi}{2}\right)\sigma^2 T \left|\frac{\sin \omega T/2}{\omega T/2}\right|^2 \tag{8.93}$$

The constant output has a power spectrum of

$$S_c(\omega) = \pi^2\sigma^2\delta(\omega) \tag{8.94}$$

and the total power spectrum presented to the error detector is shown in Fig. 8.18.

The noise power near the scan frequency will induce a tracking error that, in turn, will generate a self-correcting error voltage. This phenomenon is the same as that discussed in Sec. 7.7, except for the correlation of the positive and negative frequency portions of the noise spectrum. Because of this correlation, only the noise and signal terms at positive frequencies need to be considered. Assuming $\omega_s \ll 2\pi/T$, the noise passing a filter of width Δf is,

$$\left(2 - \frac{\pi}{2}\right)\sigma^2 T\Delta f = \text{noise power} \tag{8.95}$$

The competing signal is

$$E\{\tilde{v}_H\}\frac{m}{2}\cos(\omega_s t - \tilde{\varphi}) \tag{8.96}$$

As in Eq. (7.158), the assumption is that the track axis moves until the error voltage is zero. The mean square tracking signal must then equal the noise power, or

$$\frac{\pi\sigma^2}{8} E\{\tilde{m}^2\} = \left(2 - \frac{\pi}{2}\right)\sigma^2 T\Delta f \tag{8.97}$$

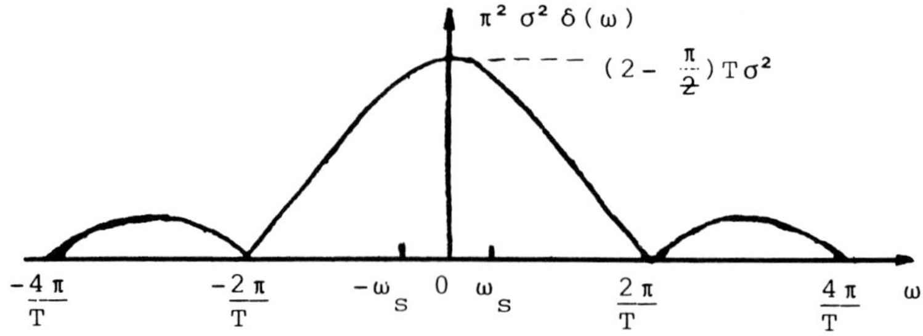

Fig. 8.18 Spectrum of error detector input due to CW noise jamming.

As $m = k_s(\theta_T/\theta_3)$, the mean square total angular error includes azimuth plus elevation and is

$$E\left\{\left(\frac{\tilde{\theta}_T}{\theta_3}\right)^2\right\} = \frac{8\left(2 - \frac{\pi}{2}\right)T\Delta f}{\pi k_s^2} \tag{8.98}$$

Noting that $T = 1/f_r$, where f_r is the pulse-repetition frequency of the radar and replacing Δf by twice the servo noise bandwidth, $2\beta_n$, results in

$$\frac{\sigma_\theta^2}{\theta_3^2} = \frac{4\left(\frac{4}{\pi} - 1\right)}{k_s^2} \frac{2\beta_n}{f_r} \tag{8.99}$$

This should be compared to Eq. (7.161) repeated below and multiplied by two to transform the error in one plane to the total error (i.e., $2\sigma_x^2 = \sigma_\theta^2$)

$$\frac{\sigma_\theta^2}{\theta_3^2} = \frac{2}{k_s^2\left(\frac{S}{N}\right)_{IF}\left(\frac{f_r}{2\beta_n}\right)} \tag{7.161}$$

The constant average power of a CW noise jammer results in a tracking error equal to the tracking error when the S/N ratio at the IF frequency is

$$\left(\frac{S}{N}\right)_{IF} = \frac{1}{2\left(\frac{4}{\pi} - 1\right)} \approx 1.8 \rightarrow 2.6 \text{ dB} \tag{8.100}$$

As an example of the significance of the tracking error given by Eq. (8.99), assume the angle error of a radar is distributed as a Gaussian random variable and that it assumes a new, independent value every $1/2\beta_n$ seconds. Assume further that tracking of the target by the radar stops if the angle tracking error, during any time interval of duration $1/2\beta_n$ exceeds one-half the radar one-way beamwidth θ_3. Then, the probability of a break in tracking during any interval of duration $1/2\beta_n$ is

$$P_b = 1 - \int_{-\theta_{3/2}}^{\theta_{3/2}} p(\theta)\, \mathrm{d}\theta = \operatorname{erfc}\left(\frac{\theta_3}{2\sqrt{2}\sigma_\theta}\right) \tag{8.101*}$$

Over a period of observation, consisting of N intervals each of length $1/2\beta_n$, the probability of breaklock during at least one interval is

$$P_B = 1 - (1 - P_b)^N = 1 - \left[2\int_o^{\theta_{3/2}} p(\theta)\, \mathrm{d}\theta\right]^N \tag{8.102}$$

If an aircraft flies at 250 m/s over a 10 km course, it will be observed for 40 s.

*erfc(z) is the complementary error function as defined in Ref. 81 (p. 297). For large values of z,

$$\operatorname{erfc}(z) \approx \frac{\exp(-z^2)[(1/z) - (1/2z^2)]}{\sqrt{\pi}}$$

If the one-sided servo bandwidth of the radar is 5 Hz, then there are also 400 separate chances for a breaklock. From Eq. (8.99), however, if the pulse repetition frequency is 2 kHz, the variance of the Gaussian distribution is

$$\sigma_\theta^2 = 0.0024\,\theta_3^2 \tag{8.103}$$

Since one-half beamwidth equals 10 standard deviations, the probability of breaklock is essentially zero.

8.6 THE EFFECTS OF AM NOISE ON A CONSCAN RADAR

The radiated power of a noise jammer can be varied at the scan rate of a conical scan radar, causing errors in the estimation of target position. The magnitude of error in estimating target position will be limited by the same causes of the error limitation affecting the repeater jammer: the balancing effect of a return from the target or jammer-radiated power that does not appear at the scan frequency of the radar. The computation of the error induced is conducted again because the noise-like nature of the jammer signal does complicate the analysis.

The assumption is that a sufficient J/S ratio is maintained at all levels of jammer power to prevent determination of target range. The voltage out of the IF receiver, which is sampled and passed on to the circuits used to estimate angular error, is due solely to jamming, and the follow-on circuits effectively operate in an infinite J/S ratio. The jamming signal present at the output of the sample-and-hold circuit can be expressed

$$\tilde{v}_m(t) = \tilde{v}_H(t)(1 + m_j\cos\omega_s t) \tag{8.104}$$

where $\tilde{v}_H(t)$ is defined in Fig. 8.17 and m_j is the jammer depth of modulation.

The autocorrelation function presented to the error detector circuits is

$$\begin{aligned} R_M(\tau) &= E\{\tilde{v}_m(t)\tilde{v}_m(t-\tau)\} \\ &= E\{\tilde{v}_H(t)\tilde{v}_H(t-\tau)\}\left[1 + \frac{m_j^2}{2}\cos\omega_s\tau\right] \end{aligned} \tag{8.105}$$

resulting in a power spectrum given by

$$S_M(\omega) = S_H(\omega) * \left\{\delta(\omega) + \frac{m_j^2}{4}\left[\delta(\omega - \omega_s) + \delta(\omega + \omega_s)\right]\right\} \tag{8.106}$$

The portion of this power spectrum presented in Fig. 8.19 shows the contribution of the various terms of Eq. (8.106) near the scan frequency. As the bandwidth of the scanning filter is much less than the pulse-repetition frequency,

$$\left(2 - \frac{\pi}{2}\right)\sigma^2(1 + m_j^2)\frac{B_s}{f_r} \ll \pi\frac{\sigma^2 m_j^2}{8} \tag{8.107}$$

or the continuous noise spectrum contributes negligibly and can be ignored.

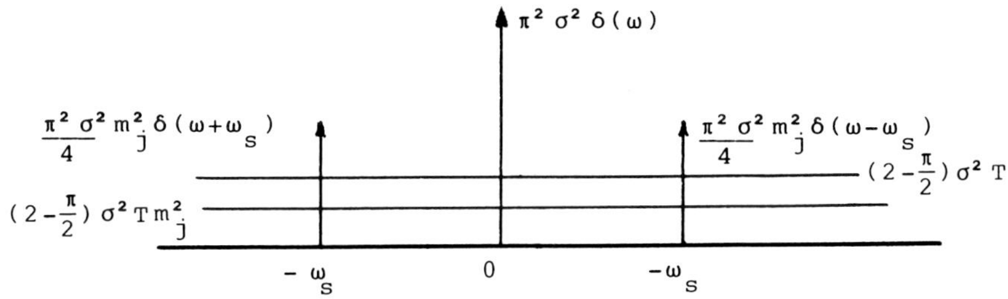

Fig. 8.19 Power spectrum of AM noise jammer at error detector input.

The resulting signal is identical to that covered in Sec. 8.1 and the angular error is given by Eq. (8.15) as

$$\frac{\theta_T}{\theta_3} = \frac{m_j}{k_s} \tag{8.108}$$

Problems

8.1 Consider a COSRO radar tracking two targets S_1 and S_2, as shown in the top drawing below. An appropriate model of the radar is shown in the lower figure.

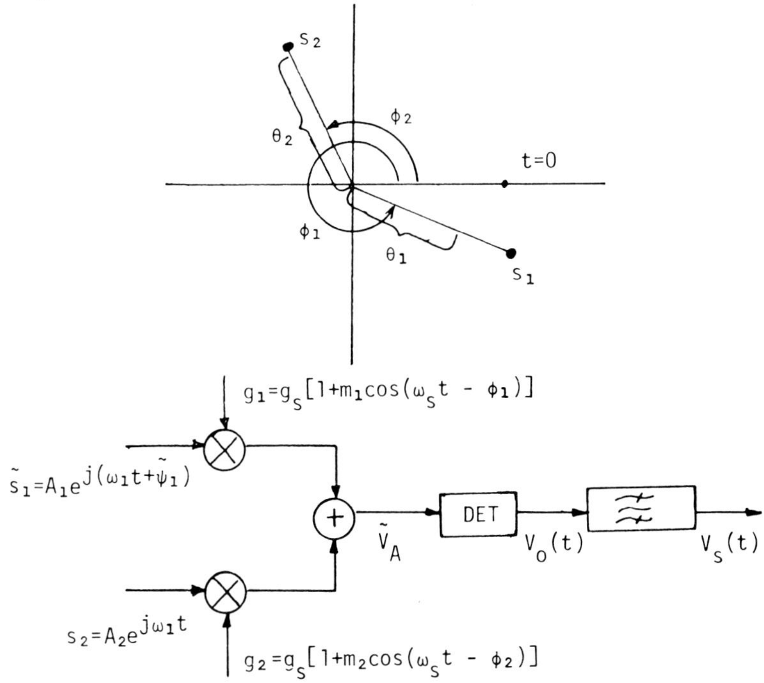

The phase $\tilde{\psi}_1$ is a random variable, uniform on 0–2π. The convention $V(t) = \text{Re}\{Ve^{j\omega t}\}$ has been used. The tilde denotes a random variable and E the expected value.

a) Show that $V_o(t) \equiv E\{|\tilde{V}_A(t)|^2\} = A_2^2 g_2^2(t) + A_1^2 g_1^2(t)$.

b) For m_1 and m_2 much less than one, show that the error voltage will be zero when

$$\varphi_2 = \varphi_1 \pm \pi \qquad A_2^2 m_2 = A_1^2 m_1$$

c) Defining $\Delta\theta \equiv \theta_1 + \theta_2$, show that the tracking error relative to S_1 is

$$\frac{\theta_1}{\Delta\theta} = \frac{A_2^2}{A_1^2 + A_2^2}$$

8.2 Building on Problem 7.4, consider the random-lobing radar and jammer defined in the following sketch and notation:

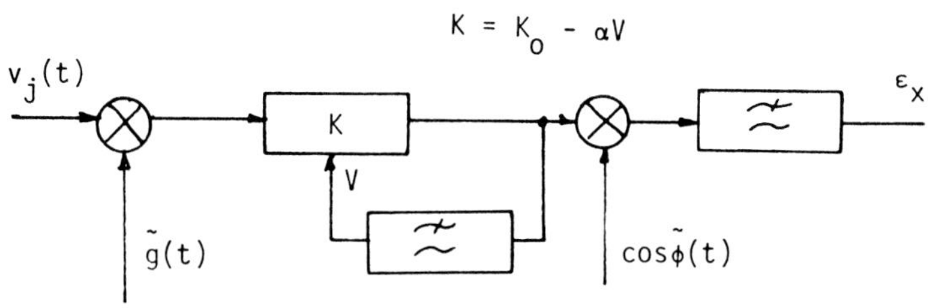

$$\tilde{g}(t) = g_o\left(1 + \frac{k_s\theta_T}{2\theta_s}\cos(\tilde{\varphi}(t) - \varphi_T)\right)$$

$$\tilde{\varphi}(t) = \sum_{n=-\infty}^{\infty} \tilde{\varphi}_n P_{T/4}\left(t - n\frac{T}{4}\right)$$

where $P_{T/4}(t)$ is a unit pulse of duration $T/4$.

The probability density on $\tilde{\varphi}_n$ is

$$p(\varphi_n) = \frac{1}{4}\delta(\varphi_n) + \frac{1}{4}\delta\left(\varphi_n - \frac{\pi}{2}\right)$$

$$+ \frac{1}{4}\delta\left(\varphi_n + \frac{\pi}{2}\right) + \frac{1}{8}\delta(\varphi_n - \pi) + \frac{1}{8}\delta(\varphi_n + \pi)$$

The low-pass filters develop an output voltage that is the time average of their input, which, for the terms containing a random phase, we equate to the expectations of such a term.

If the jamming input is amplitude modulated as $v_j(t) = V_j(1 + m_j \cos\omega_s t)$, show that the error voltage in the azimuth plane is

$$\varepsilon_x = \frac{K_o k_s \theta_T}{\alpha 4\theta_s} \cos\varphi_T$$

8.3 It was claimed, in Sec. 3.3 (in the subsection on repeater polarization modulation), that effective jammer operation against a scanning radar with a nutating feed requires that the orthogonally polarized signal from a jammer exceed the nominally polarized target return by a ratio greater than that of the radar antenna's copolarized peak gain to its cross-polarized peak gain. A simplified analysis supporting this statement begins with observation of the antenna radiation pattern and the locus of points within this pattern outlined by a target tracked with a one-half beamwidth error in a plane at 45 deg to the horizon, as shown in the sketch below. Only one of the cross-polarized lobes is shown and the pattern is much simplified.

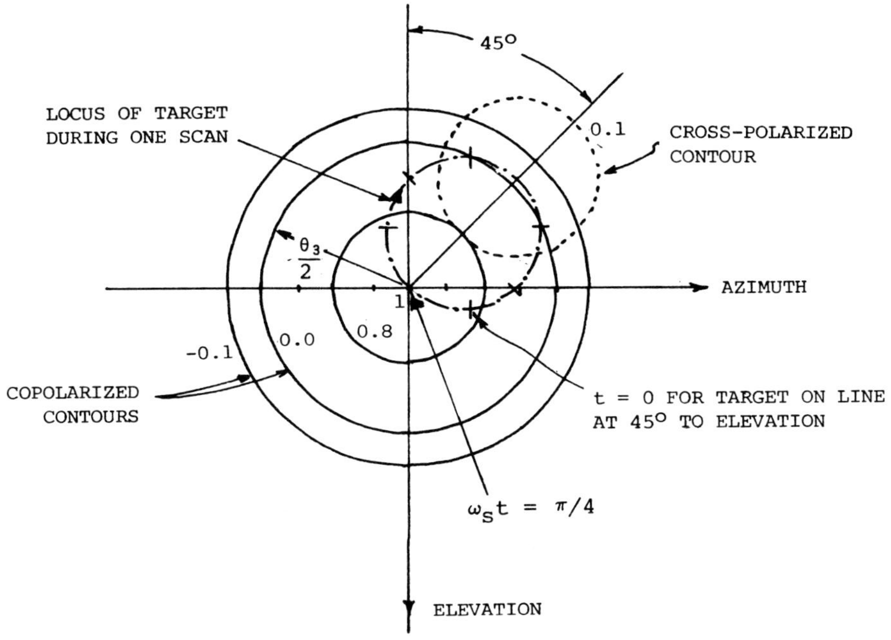

The time variation of the magnitude of the copolarized and cross-polarized fields, as seen at the target, are shown below,

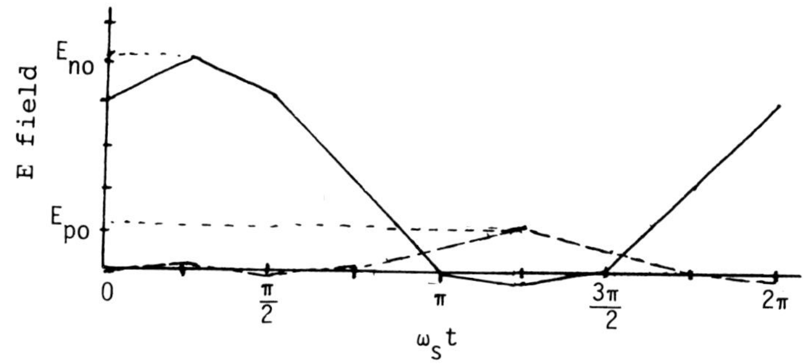

where $E_{no}(E_{po})$ is the peak copolarized (cross-polarized) field strength. The amplitude of the radiated fields can be approximated as

$$E_n(t) = \frac{E_{no}}{2}\left[1 + \sin\left(\omega_s t + \frac{\pi}{4}\right)\right]$$

$$E_p(t) = \frac{E_{po}}{2}\left[1 - \sin\left(\omega_s t + \frac{\pi}{4}\right)\right]$$

If the incident field is $\bar{E}_{in} = E_J\hat{x} + E_T\hat{y}$, the voltage out of the antenna will have a modulation given by

$$v_{in} = E_T E_{no} + E_J E_{po} + (E_T E_{no} - E_J E_{po})\sin\left(\omega_s t + \frac{\pi}{4}\right)$$

The average signal out of the receiver will be normalized by the AGC and the resulting azimuth error voltage will be proportional to

$$\varepsilon_x \propto \frac{E_T E_{no} - E_J E_{po}}{E_T E_{no} + E_J E_{po}}$$

For radar purity defined as $E_{no}/E_{po} = \gamma$ and a jamming-to-signal ratio defined by $J/S = (E_J/E_T)^2$, this error voltage becomes

$$\varepsilon_x \propto \frac{\gamma - \sqrt{J/S}}{\gamma + \sqrt{J/S}}$$

Thus, when $E_J/E_T > E_{no}/E_{po}$, the sign of the error voltage is incorrect.

The actual motion of the tracking axis of the radar requires the above analysis be repeated until both azimuth and elevation error voltages are zero.

Show that the J/S requirement for an initial tracking error of $\theta_T \ll \theta_3$ in the first quadrant is given by

$$\frac{J}{S} > \left(\frac{3\pi k_s}{8}\right)^2 \left(\frac{E_{no}}{E_{po}}\right)^2$$

where the cross-polarized field is assumed to be as shown below.

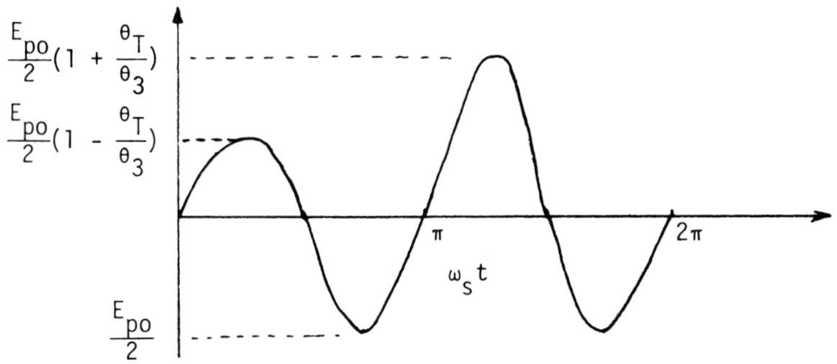

and is defined by

$$E_p(t) = \frac{E_{po}}{2}\sin 2\omega_s t\left\{1 - \frac{\theta_T}{\theta_3}\left[p(t) - p\left(t - \frac{\pi}{2\omega_s}\right)\right]\right\}$$

and $p(t)$ is a unit pulse of duration $T_{S/4}$.

9
ACTIVE ECM VS ANGLE MEASUREMENT IN MONOPULSE

A monopulse system estimates the target location at an angle from the boresight of its antenna through a comparison of two or more signals received simultaneously from the target. The variation of incident signal parameters in time has no effect on the estimation process unless the physical realization of the radar results in a channel imbalance in amplitude or phase [4, p. 177; 26]. Because of the inability of a single-point source to create significant errors in the monopulse system, Shustov and Vakin [4] concentrate on the effects created by two-point sources. The error created by two-point sources is the subject of the first part of this chapter. The second part explores the effect on an angle estimate of the polarization of the incident wavefront.

The half-power beamwidth θ_3 of the pattern of each of the elements making up the monopulse feed is used throughout this chapter for normalization. This is not the same as the beamwidth of the resulting sum-radiation pattern.

9.1 MULTIPLE-SOURCE JAMMING OF MONOPULSE

Shustov and Vakin developed the relationship between the error voltage of a monopulse radar and the characteristics of signals incident on the radar from two unresolved sources. The two sources could be related coherently or incoherently, with coherent sources producing signals having a constant phase relationship at the radar. This section reproduces their development and builds on the sections concerned with monopulse radar in Chapter 7. The discussion is limited to amplitude sensing with sum-difference detection as it is the most common.

Figure 9.1 shows the problem geometry. The sum-radiation pattern and two sources, S_1 of amplitude A_1 and S_2 of amplitude A_2, are shown. The feed antenna patterns comprising the sum and difference patterns are shown in Fig. 9.2. Note that the angle θ_1 can be interpreted as the error in tracking source 1. If the two sources are of the same RF frequency ω_o but have an arbitrary phase relationship ψ, the complex signal representation of the antenna output voltages are

$$V_1 = A_1 g(\theta_1 - \theta_s) + A_2 g(\theta_1 - \Delta\theta - \theta_s)e^{j\psi} \tag{9.1}$$

$$V_2 = A_1 g(\theta_1 + \theta_s) + A_2 g(\theta_1 - \Delta\theta + \theta_s)e^{j\psi} \tag{9.2}$$

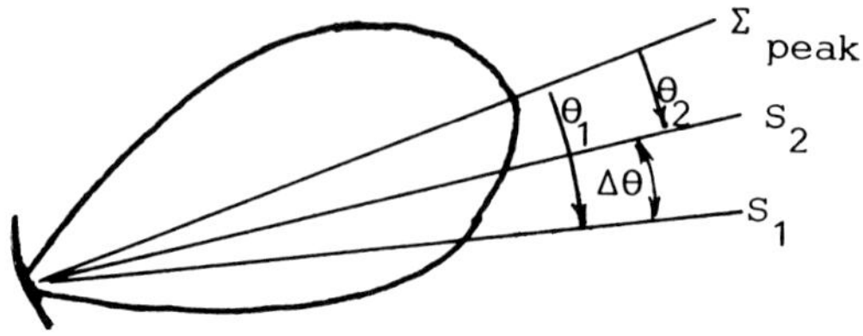

Fig. 9.1 Two-source jamming geometry.

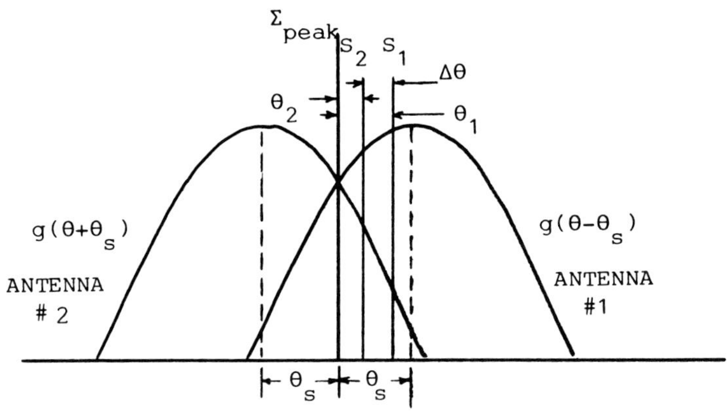

Fig. 9.2 Monopulse feed antenna patterns.

Assuming phase-matched channels, the IF amplifier output voltages will be

$$V_{\Sigma_o} = K_\Sigma(V_1 + V_2) \tag{9.3}$$

$$V_{\Delta_o} = K_\Delta(V_1 - V_2) \tag{9.4}$$

The output voltage of the phase detector driving the angle servo is

$$\varepsilon_{az} = \tfrac{1}{2}\mathrm{Re}\{V_{\Sigma_o}V^*_{\Delta_o}\} = \frac{K_\Sigma K_\Delta}{2}(|V_1|^2 - |V_2|^2) \tag{9.5}$$

Two Incoherent Sources

If the relative phase between the two jamming sources changes rapidly in time compared to the impulse response time of the radar servo or the filter following the phase detector, the average output voltage of the phase detector drives the servo. For example, the relative phase between sequential pairs of pulses arriving at a pulsed radar may be independent. With pulse-repetition intervals of less than a millisecond and servo response times of a hundred milliseconds, many pairs of pulses influence any given servo output.

A uniform distribution for the RF phase [Eq. (8.22)] adequately describes two noise sources or two constant-frequency sources with a difference in frequency much larger than the filter bandwidths previously mentioned (i.e., servo). The average error voltage resulting from a uniform phase distribution is

$$\varepsilon_{az} = \frac{K_\Sigma K_\Delta}{2} E\{|V_1|^2 - |V_2|^2\}$$

$$= \frac{K_\Sigma K_\Delta}{2} \{A_1^2[g^2(\theta_1 - \theta_s) - g^2(\theta_1 + \theta_s)]$$

$$+ A_2^2[g^2(\theta_1 - \Delta\theta - \theta_s) - g^2(\theta_1 - \Delta\theta + \theta_s)]\} \qquad (9.6)$$

The voltage gain of the IF amplifier is found from Eq. (7.54) to be

$$K_\Sigma = K_\Delta = \frac{K_o}{1 + \alpha k_g E\{|v_{\Sigma_i}|\}} \qquad (9.7)$$

where

$$|v_{\Sigma_i}| = |A_1[g(\theta_1 - \theta_s) + g(\theta_1 + \theta_s)]$$

$$+ A_2[g(\theta_1 - \Delta\theta - \theta_s) + g(\theta_1 - \Delta\theta + \theta_s)]e^{j\psi}| \qquad (9.8)$$

Following the argument in Sec. 8.1, the expected magnitude of the sum-channel input voltage is (the multiplier $2.6/\pi$ is appropriate for $A_1 = A_2$; if $A_1 \gg A_2$ or $A_2 \gg A_1$, this multiplier becomes 1)

$$E\{|v_{\Sigma_i}|\}$$

$$= \frac{2.6}{\pi} \{A_1[g(\theta_1 - \theta_s) + g(\theta_1 + \theta_s)] + A_2[g(\theta_2 - \theta_s) + g(\theta_2 + \theta_s)]\} \qquad (9.9)$$

If both sources are close enough to the track axis to justify a linear approximation to the radiation pattern of the feed antennas, then the sums of the gains in Eq. (9.9) result in constants equal to twice the gain of the feed pattern at the squint angle from the direction of maximum gain,

$$g(\theta_1 - \theta_s) + g(\theta_1 + \theta_s) = 2g(\theta_s) \qquad (9.10)$$

Finally, if the ratio of the amplitudes of the two sources is defined as $\beta = A_2/A_1$ and the constant gain in Eq. (9.10) is assumed, the error voltage becomes

$$\varepsilon_{az} = \left(\frac{K_o^2}{2\alpha^2 k_g^2}\right) \times \frac{\pi^2\{g^2(\theta_1 - \theta_s) - g^2(\theta_1 + \theta_s) + \beta^2[g^2(\theta_1 - \Delta\theta - \theta_s) - g^2(\theta_1 - \Delta\theta + \theta_s)]\}}{(2.6)^2 g^2(\theta_s)(1+\beta)^2} \tag{9.11}$$

where Eq. (7.63) can be used to define

$$K_{pd} = \frac{K_o^2}{2\alpha^2 k_g^2} \tag{9.12}$$

The position of stable tracking can be found from the zeros of Eq. (9.6). Dynamic effects generally require consideration of the gain variation of the servo loop. However, for small source separations and small errors in angle tracking, it follows from Eq. (9.11) that this gain does not vary. Under such conditions, it is valid to approximate the radiation patterns of the feeds as

$$g(\theta \pm \theta_s) = g_o \exp\left[-2\ell n 2\left(\frac{\theta \pm \theta_s}{\theta_3}\right)^2\right] \approx g(\theta_s)\left(1 \mp k\frac{\theta}{\theta_3}\right) \tag{9.13}$$

where $k = 4\ell n 2$. Then the error voltage becomes

$$\varepsilon_{az} \approx K_{pd}\frac{\pi^2 k}{(1.3)^2(1+\beta)^2}\left(\frac{\theta_1}{\theta_3} + \beta^2\frac{\theta_1 - \Delta\theta}{\theta_3}\right) \tag{9.14}$$

This error voltage is zero when

$$\theta_1 = \frac{\beta^2}{1+\beta^2}\Delta\theta \tag{9.15}$$

which is just the power centroid discussed in Chapter 3 and where θ_1 is the error in tracking source 1.

When the two sources are of equal power, Eq. (9.6) can be used to determine the maximum source separation before the radar selects and tracks one of the sources, i.e., resolves the two sources. The error voltage from Eq. (9.6) is plotted in Fig. 9.3 for various source separations and a Gaussian radiation pattern. For small separations, the radar tracks halfway between the two sources as given by Eq. (9.15). As the source separation increases, the slope of the curve relating the error voltage to the angular error decreases and, when $\Delta\theta \approx 0.9\,\theta_3$, this slope becomes zero. For other element radiation patterns or squint angles, the source separation resulting in a zero slope will differ. At this source separation, the tracking of the radar will be very

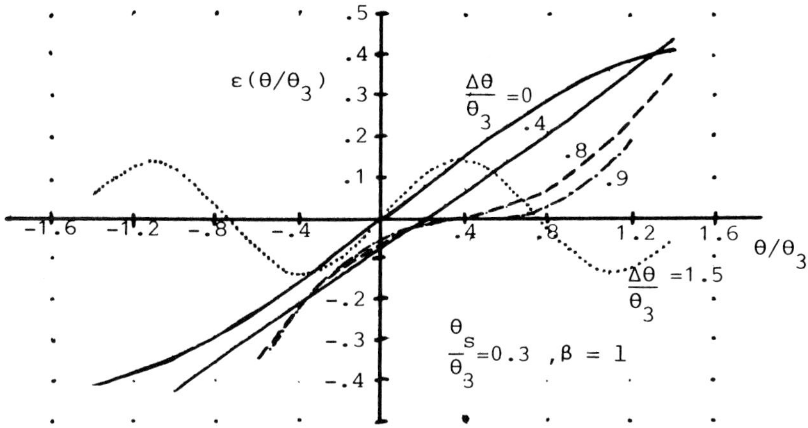

Fig. 9.3 Error voltage as a function of angle from target 1: Gaussian pattern.

sluggish or even unstable because of the low gain in the tracking loop. At slightly larger separations, the radar will select one or the other target. The angular resolution of the radar can then be defined [4, p. 191] as the angular separation $\Delta\theta_r$ such that

$$\frac{\partial \varepsilon_{az}}{\partial \theta}\bigg|_{\theta = \Delta\theta_r/2} = 0 \tag{9.16}$$

If the normalization of the error voltage by the AGC system is ignored, or equivalently if the sum- and difference-channel gains are assumed to be constant, the zeros of Eq. (9.6) still yield the stable track points. A plot of the voltage variation in the difference channel with angular error in tracking should be similar to Fig. 9.3 for small angles because of Eq. (9.10). Such a plot, given in Fig. 9.4, is identical to the plot presented by Shustov and Vakin [4, p. 190].

Tabulated in Table 9.1 are the magnitudes of the error, difference-channel, and sum-channel voltages that result when the track axis of a radar with a larger squint angle points away from a single target. These data show that a constant sum-channel output is a good assumption for large fractions of the radar beamwidth.

When the powers of the two sources are not equal, the radar will always track nearer the more powerful source [83, p. 9]. Shown in Fig. 9.5 is the angular displacement of the track axis from target 1 as a function of the power ratio of the two sources. The plot is generated by finding the zero of Eq. (9.6). The separation of the sources in angle is restricted to less than the resolving power of the radar. Without this restriction, one source or the other will be resolved and tracked at low power ratios; thus, a more accurate treatment of realistic radiation-pattern sidelobes must be performed to determine the direction of the track axis at high power ratios.

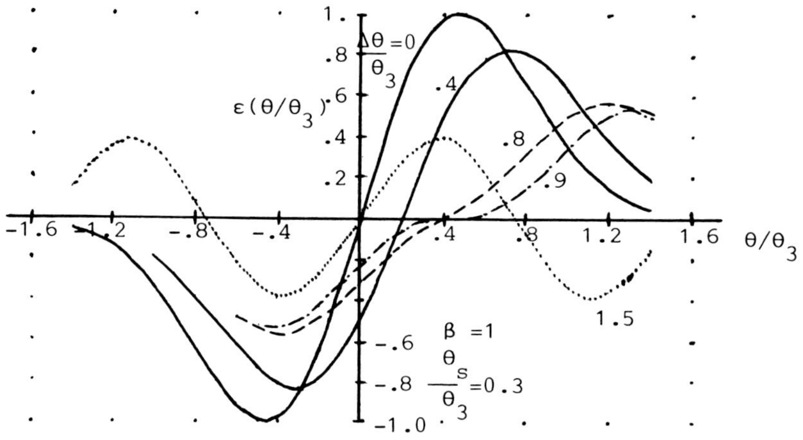

Fig. 9.4 Difference-channel voltage as a function of angle from target 1: no AGC, Gaussian pattern.

Table 9.1 Various monopulse voltages as a function of angle from a single target[a]

Angle offset θ/θ_3	Error voltage	Difference voltage	Sum voltage
0	0	0	8.000
0.1	0.069	0.546	7.932
0.2	0.135	1.04	7.720
0.3	0.197	1.451	7.375
0.4	0.252	1.734	6.881
0.5	0.300	1.875	6.250
0.6	0.341	1.875	5.505
0.7	0.374	1.753	4.682
0.8	0.402	1.540	3.832
0.9	0.424	1.275	3.008
1	0.441	0.996	2.258
1.1	0.455	0.735	1.617
1.1	0.465	0.513	1.103
1.3	0.474	0.339	0.716
1.4	0.480	0.212	0.441

[a]Normalized squint angle $\theta_s/\theta_3 = 0.5$.

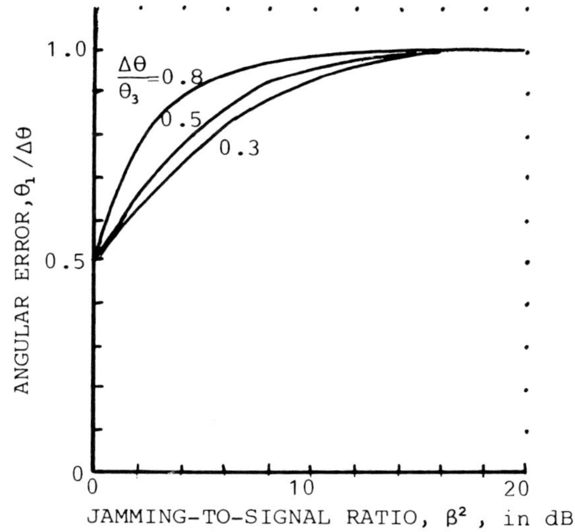

Fig. 9.5 Normalized tracking error from source 1 vs power ratio.

It is obvious that angular errors can make up a significant portion of the sum-pattern beamwidth. The next section applies this phenomenon in a countermeasure technique termed "terrain bounce."

Terrain Bounce

Chapter 3 briefly discussed terrain bounce, covering the problem geometry and relevant equations. This section presents the underlying assumptions related to that earlier development.

The geometry of the problem can be compared to the geometry in Fig. 9.6a showing a radiation source and two scatterers. One scatterer is the Earth. The radiation source simulates the radar and the other scatterer represents the target. The radiation field incident on the target can be found if the following are assumed: the Earth is flat and perfectly conducting and, after reflection from the Earth, the field scattered by the target has a negligible effect on the currents induced on the target by the original incident field. Under these assumptions, the image theory in electromagnetics [84, p. 103] allows replacement of the original problem in Fig. 9.6a by the equivalent problem in Fig. 9.6b for the purpose of computing the field incident on the scatterer.

After determining the incident field, one can find, in principle, the currents induced on the target and the resulting scattered field. That portion of the scattered field incident on the original source, as well as any field due to sources carried on the target, is found by constructing an image of both the induced currents on the target (scatterer) and its original sources on the Earth as shown in Fig. 9.7. Included in Fig. 9.7 is the original source and its

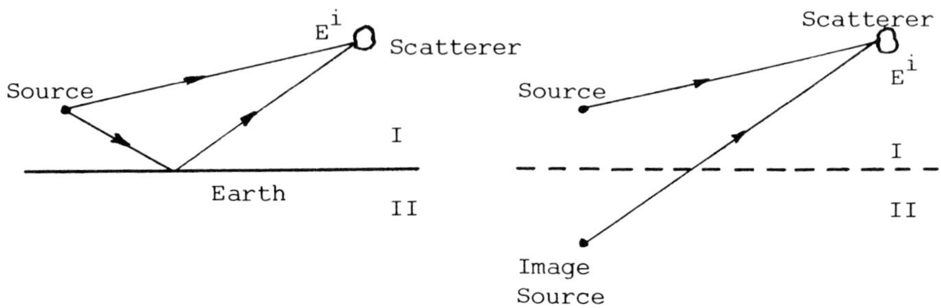

a) Original problem. b) Equivalent problem for E^i in region I.

Fig. 9.6 Equivalent scattering problems through image theory.

image. This model is valid only if the presence of the Earth does not significantly modify the currents on the radar or scatterer.

The surface of the Earth is not flat and the Earth is not a perfect conductor. A point source over a "rough" Earth has an extended image rather than a point image. The field incident at a point above the surface appears to come from an extended source or from an extended area of the surface of the Earth called the "glistening" area [71]. If image theory had not been used to replace the surface in Fig. 9.6a with the image source of Fig. 9.6b, the scattered field in Fig. 9.6a is also due to current sources extending over the entire surface of the Earth. These current sources obey a definite functional relationship over the surface. In modeling the rough Earth, one can break the surface into small areas on which the currents fluctuate in amplitude and phase around the value of the flat Earth. The total scattered field cannot be represented by the field from a point image. Usually, the total scattered field is representable by the sum of a coherent field generated by an image source

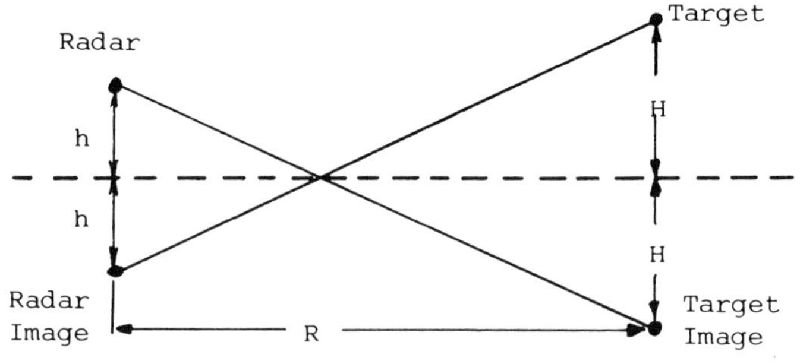

Fig. 9.7 Equivalent image problem for two sources over a flat Earth.

reduced in amplitude by ρ_s from the value of the flat Earth and a diffuse field created by sources within the glistening area. The ratio of the total diffuse field at a point above the surface to the field arriving directly from the original source defines a diffuse scattering coefficient ρ_d [71]. This diffuse field, however, is generated by an extended source.

The assumption here and in Chapter 3 is that the coherent scattered field is zero and that the diffuse scattered field is generated by a source at the image location. The true extended nature of the source of the diffuse field is ignored. The amplitude of the source of the diffuse field is reduced by $\rho = \rho_d$ from the original source. The phase of the source of the diffuse field varies rapidly in relation to the response time of the angle tracking loop of the radar. Strasser has achieved a more detailed development of this problem that accounts for the extended nature of the diffuse sources [83].

The geometry of the model of terrain bounce is shown in Fig. 9.8 where the gain of the radar antenna toward the target is G_{R1} and toward the ground is G_{R2}. The jammer is assumed to be a linear repeater and its radiated power depends on the effective area of the receive antenna of the jammer $\lambda^2 G_{JR}/4\pi$. The gain of the jammer system is G. The transmit antenna of the jammer is pointed at the ground, not directly at the radar, with a peak gain G_{J1}. Some value of gain G_{J2} must be assumed to be toward the radar.

Of the jammer power directed toward the ground, only a fraction ρ^2 is scattered toward the radar. The total intensity of diffuse power incident at the radar, including normal backscatter from the target, is

$$P_2 = \frac{P_R G_{R1}}{4\pi R^2}\left[\frac{\lambda^2}{4\pi} G_{JR} \frac{G G_{J1}}{4\pi R^2} + \frac{\sigma_T}{4\pi R^2}\right]\rho^2 \tag{9.17}$$

The power density incident on the radar from the direction of the target will include the power from the target of cross section σ_T added to the power radiated by the jammer toward the radar. This power density is

$$P_1 = \frac{P_R G_{R1}}{4\pi R^2}\left[\frac{\lambda^2}{4\pi} G_{JR} \frac{G G_{J2}}{4\pi R^2} + \frac{\sigma_T}{4\pi R^2}\right] \tag{9.18}$$

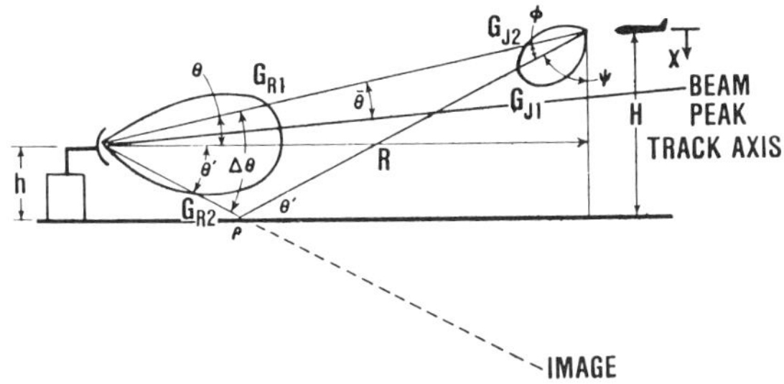

Fig. 9.8 Terrain bounce model.

Combining these results into the power centroid formula and again ignoring the target power scattered from the Earth results in a tracking point away from the real target of

$$\frac{\bar{\theta}}{\Delta\theta} = \frac{P_2}{P_1 + P_2} = \frac{\rho^2}{\dfrac{1}{(J/S)_o} + \dfrac{G_{J2}}{G_{J1}} + \rho^2} \tag{9.19}$$

where

$$\Delta\theta = \frac{2H}{R} \tag{9.20}$$

and, as defined in Chapter 3,

$$\left(\frac{J}{S}\right)_o = \frac{\lambda^2}{4\pi} \frac{G_{JR} G G_{J1}}{\sigma_T} \tag{9.21}$$

Alternatively, the power ratio β^2 of Eq. (9.11) can be defined as

$$\beta^2 = \frac{P_2}{P_1} = \rho^2 \frac{1 + \left(\dfrac{J}{S}\right)_o}{1 + \left(\dfrac{J}{S}\right)_o \dfrac{G_{J2}}{G_{J1}}} \tag{9.22}$$

and the error in tracking the target can be found in Fig. 9.5. A more convenient presentation of tracking error may be found in Fig. 9.9, with the abscissa directly expressed in $(J/S)_o$.

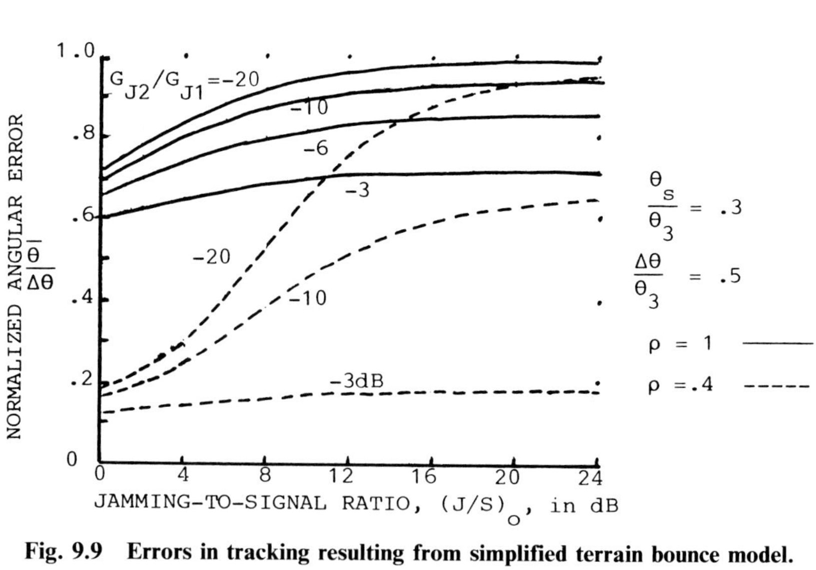

Fig. 9.9 Errors in tracking resulting from simplified terrain bounce model.

Two Coherent Sources

The monopulse radar will move its track axis until the voltage from the phase detector is zero and varies with angle in the proper sense. Thus, all angles at which Eq. (9.5) is zero are potential angles for location of the track axis and these zeros are now found for the case of coherent sources.

If one assumes that the two sources are located on one aircraft, it is desirable to calculate the error in the angle from the center of the two sources where, presumably, the aircraft fuselage is located. (This geometry is shown in Fig. 9.10.)

Using this new angular notation, the voltages from the two antennas become [see Eq. (9.2)]

$$V_1 = A_1 g(\theta_1 - \theta_s) + A_2 g(\theta_2 + \theta_s)e^{j\psi} \tag{9.23}$$

$$V_2 = A_1 g(\theta_1 + \theta_s) + A_2 g(\theta_2 - \theta_s)e^{j\psi} \tag{9.24}$$

Then, from Fig. 9.10,

$$V_1 = A_1 g\left(\frac{\Delta\theta}{2} - \theta - \theta_s\right) + A_2 g\left(\frac{\Delta\theta}{2} + \theta + \theta_s\right)e^{j\psi}$$

$$= A_1 g_1 + A_2 g_2 e^{j\psi} \tag{9.25}$$

$$V_2 = A_1 g\left(\frac{\Delta\theta}{2} - \theta + \theta_s\right) + A_2 g\left(\frac{\Delta\theta}{2} + \theta - \theta_s\right)e^{j\psi}$$

$$= A_1 g_3 + A_2 g_4 e^{j\psi} \tag{9.26}$$

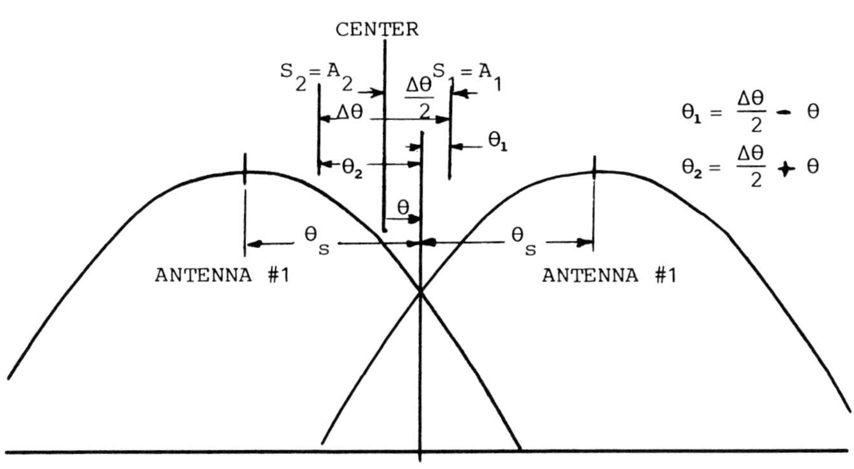

Fig. 9.10 Redefinition of angles for coherent problem.

If the source separation is small enough, a linear approximation to the gain variation of the antenna patterns should suffice. Using Eq. (9.13) results in

$$g_{\substack{1\\3}}(\theta) = g(\theta_s)\left[1 \pm k\left(\frac{\frac{\Delta\theta}{2} - \theta}{\theta_3}\right)\right] \tag{9.27}$$

$$g_{\substack{2\\4}}(\theta) = g(\theta_s)\left[1 \mp k\left(\frac{\frac{\Delta\theta}{2} + \theta}{\theta_3}\right)\right] \tag{9.28}$$

If again $\beta = A_2/A_1$, then Eq. (9.4) becomes

$$V_{\Sigma_o} = K_\Sigma 2A_1 g(\theta_s)(1 + \beta e^{j\psi}) \tag{9.29}$$

$$V_{\Delta_o} = K_\Delta 2A_1 g(\theta_s)\frac{k}{\theta_3}\left[\left(\frac{\Delta\theta}{2} - \theta\right) - \beta\left(\frac{\Delta\theta}{2} + \theta\right)e^{j\psi}\right] \tag{9.30}$$

and the zeros of Eq. (9.5) occur at a point on the side of the more powerful source at

$$\frac{\theta}{\Delta\theta} = \frac{1}{2}\frac{1 - \beta^2}{1 + 2\beta\cos\psi + \beta^2} \tag{9.31}$$

which is plotted in Fig. 3.45. The maximum error in angle, approximately $0.6\,\theta_3$, must be found by numerically solving for the zero of Eq. (9.5) [4, p. 207].

Alternative Development of Error Caused by Two Coherent Sources

Under the assumption that a tracking radar ideally aligns its track axis with the direction of the arrival of an approaching wavefront, the angular error induced by two coherent sources can again be developed. If the two sources are colinear as shown in Fig. 9.11, the radiation field at large distances, compared to the source separation L is

$$\bar{E}(\bar{r}) = \left(\frac{-j\omega\mu}{4\pi}\right)\left(\frac{e^{-jkr}}{r}\right)\int_{V'} [\bar{J} - (\bar{J}\cdot\hat{r})\hat{r}]e^{jk\bar{r}'\cdot\hat{r}}\,\mathrm{d}V' \tag{9.32}$$

For the two current elements shown, this far field becomes

$$\bar{E}(\bar{r}) = j\frac{\omega\mu}{2\pi r}I\Delta l\hat{\theta}\sin\theta e^{-jkr}(e^{j(\psi+\gamma)/2} + \beta e^{-j(\psi+\gamma)/2}) \tag{9.33}$$

where $\gamma = kL\sin\theta\cos\varphi$.

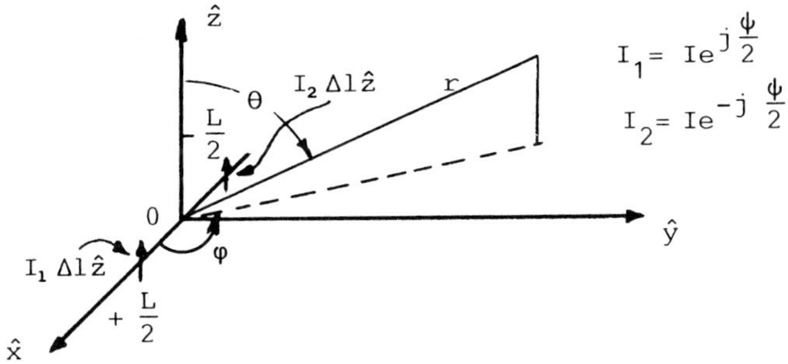

Fig. 9.11 Positions of coherent current sources.

Harrington [84, p. 85] shows that an electromagnetic wave can be expressed as

$$E(\bar{r}) = A(\bar{r})e^{j\Phi(\bar{r})} \tag{9.34}$$

where both A and Φ are real functions of position. The wavefront is defined as the surface over which $\Phi(\bar{r})$ is constant and the normal to the wavefront in the direction of propagation is given by $-\nabla\Phi(\bar{r})$. The total phase of Eq. (9.33) can be determined from Fig. 9.12 as

$$\Phi(\bar{r}) = -kr + \tan^{-1}\left[\frac{1-\beta}{1+\beta}\tan\left(\frac{\psi+\gamma}{2}\right)\right] = -kr + \xi \tag{9.35}$$

Under the original assumption, the direction in which the radar points is

$$\nabla\Phi(\bar{r}) = \hat{r}\left(\frac{\partial\Phi}{\partial r}\right) + \hat{\theta}\left(\frac{1}{r}\right)\left(\frac{\partial\Phi}{\partial\theta}\right) + \hat{\varphi}\left(\frac{1}{r\sin\theta}\right)\left(\frac{\partial\Phi}{\partial\varphi}\right) \tag{9.36}$$

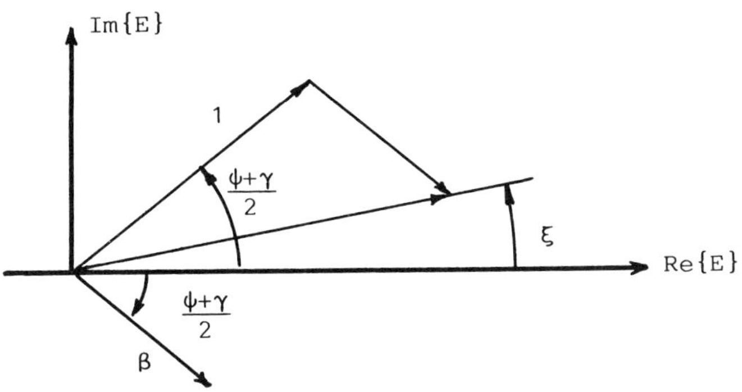

Fig. 9.12 Phase of far field from two current sources.

The second and third terms are easily evaluated by using

$$\frac{\partial\Phi}{\partial\theta} = \left(\frac{\partial\Phi}{\partial\gamma}\right)\left(\frac{\partial\gamma}{\partial\theta}\right) = \frac{\partial\xi}{\partial\gamma} kL\cos\varphi \tag{9.37}$$

and

$$\frac{\mathrm{d}\tan\xi}{\mathrm{d}\gamma} = \sec^2\xi \frac{\mathrm{d}\xi}{\mathrm{d}\gamma} \tag{9.38}$$

Then, as

$$\begin{aligned}\frac{\mathrm{d}\xi}{\mathrm{d}\gamma} &= \left(\frac{1}{\sec^2\xi}\right)\left(\frac{\mathrm{d}}{\mathrm{d}\gamma}\right)\left[\frac{1-\beta}{1+\beta}\tan\left(\frac{\psi+\gamma}{2}\right)\right] \\ &= \left(\frac{1}{2}\right)\left(\frac{1-\beta}{1+\beta}\right)\left(\frac{\sec^2[(\psi+\gamma)/2]}{\sec^2\xi}\right)\end{aligned} \tag{9.39}$$

and from Eq. (9.35)

$$\tan\xi = \frac{1-\beta}{1+\beta}\tan\frac{\psi+\gamma}{2} \tag{9.40}$$

it is found that

$$\frac{\partial\Phi}{\partial\theta} = \frac{kL}{2}\cos\theta\cos\varphi\left(\frac{1-\beta}{1+\beta}\right)\left[\frac{1+\tan^2\left(\frac{\psi+\gamma}{2}\right)}{1+\left(\frac{1-\beta}{1+\beta}\right)^2\tan^2\left(\frac{\psi+\gamma}{2}\right)}\right] \tag{9.41}$$

After a bit of algebra, Eq. (9.41) can be expressed as

$$\frac{\partial\Phi}{\partial\theta} = \frac{kL}{2}\cos\theta\cos\varphi\frac{1-\beta^2}{1+\beta^2+2\beta\cos(\psi+\gamma)} \tag{9.42}$$

and the radar points in the direction

$$\nabla\Phi(\bar{r}) = -k\hat{r} + \frac{1}{2r}\left(\cos\theta\cos\varphi\hat{\theta} - \frac{\sin\varphi}{\sin\theta}\hat{\varphi}\right)\frac{kL(1-\beta^2)}{1+\beta^2+2\beta\cos(\psi+\gamma)} \tag{9.43}$$

From Eq. (9.43), the magnitude of error in angle is contained in the components of $\nabla\Phi$ normal to $\hat{r}$. This error is, with $\gamma = kL\sin\theta\cos\varphi$,

$$\text{Pointing error} = \left(\frac{L}{2r}\right)\left[\frac{1-\beta^2}{1+\beta^2+2\beta\cos(\psi+\gamma)}\right]\left(\cos^2\theta\cos^2\varphi - \frac{\sin^2\varphi}{\sin^2\theta}\right)^{1/2} \tag{9.44}$$

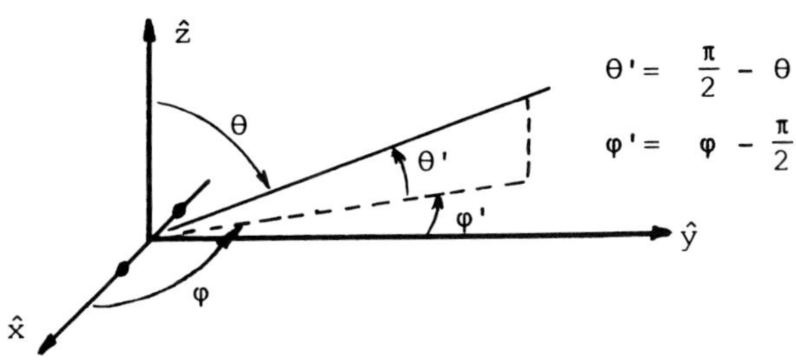

Fig. 9.13 Definition of angles from *y* axis.

If the two sources are out of phase, $\psi = \pi$. Then, in the x-y plane,

$$\text{Error} = \left(\frac{-L\sin\varphi}{2r}\right)\left[\frac{1-\beta^2}{1-2\beta\cos(kL\cos\varphi)+\beta^2}\right] \tag{9.45}$$

Equation (9.44) shows that the size of the angular tracking error is greatly affected by the polar angle θ, as well as the azimuth angle φ. If both angles are expressed with respect to the y axis, as shown in Fig. 9.13, the dominant error term becomes

$$\text{Error} = -\left(\frac{L}{2r}\right)\left[\frac{1-\beta^2}{1+\beta^2+2\beta\cos(\psi-kL\cos\theta'\sin\varphi')}\right] \tag{9.46}$$

Shustov and Vakin [4, p. 217] indicate that Eqs. (9.45) and (9.31) are valid for $\Delta\theta/\theta_3 < 0.1$ and $\beta < 0.9$ (or $\beta > 1.1$).

Combined Coherent and Incoherent Source

The impact of returns from the aircraft on coherent source jamming (cross-eye) or the specular ground return on incoherent jamming (terrain

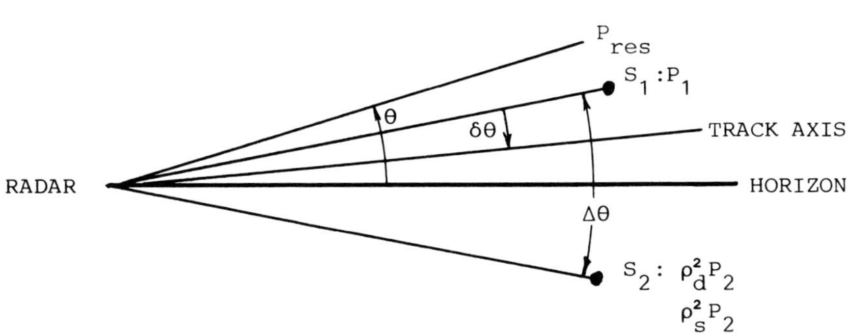

Fig. 9.14 Coherent source effect on terrain bounce.

bounce) can be determined by recognizing that the residue of the coherent sources is not coherent with the power centroid of the incoherent sources. Terrain bounce will be addressed here as the effect of target returns on cross-eye jamming was considered in Chapter 3.

In Fig. 9.14, source S_1 represents the power density of value P_1 radiated directly back to the radar. Source S_2, representing the power density reflected from the Earth, has two parts: an out-of-phase part of value $\rho_s^2 P_2$ and a part of a time-varying phase of power density $\rho_d^2 P_2$.

From Eqs. (9.17) and (9.18), ignoring any terrain scattered target return and assuming for a moment the total reflection from the Earth of all incident jamming signals, the two incident power densities are

$$P_1 = \frac{P_R G_{R1}}{4\pi R^2}\left(\frac{\lambda^2}{4\pi}\frac{G_{JR} G G_{J2}}{4\pi R^2} + \frac{\sigma_T}{4\pi R^2}\right) \tag{9.47}$$

$$P_2 = \frac{P_R G_{R1}}{4\pi R^2}\left(\frac{\lambda^2}{4\pi}\frac{G_{JR} G G_{J1}}{4\pi R^2}\right) \tag{9.48}$$

The error generated by the coherent sources is toward source S_1 at an angle from the horizon of

$$\theta = \frac{\Delta\theta}{2}\frac{1+\beta}{1-\beta} \tag{9.49}$$

where

$$\beta^2 = \rho_s^2 \frac{P_2}{P_1} = \frac{\rho_s^2}{\left(\dfrac{G_{J2}}{G_{J1}} + \dfrac{1}{(J/S)_o}\right)} \tag{9.50}$$

and the coherent source residue is

$$\begin{aligned} P_{res} &= P_1 + \rho_s^2 P_2 - 2\rho_s \sqrt{P_1 P_2} \\ &= P_1 + \rho_s^2 P_2 - 2\rho_s P_2 \left[\frac{G_{J2}}{G_{J1}} + \frac{1}{(J/S)_o}\right]^{1/2} \end{aligned} \tag{9.51}$$

The coherent residue and diffuse component are separated by $\theta + (\Delta\theta/2)$, resulting in an error from the target of

$$\frac{\delta\theta}{\Delta\theta} = -\frac{\beta}{1-\beta} + \frac{1}{1-\beta}\left[\frac{\rho_d^2}{\rho_d^2 + \rho_s^2 - 2\rho_s\left(\dfrac{G_{J_2}}{G_{J_1}} + \dfrac{1}{(J/S)_o}\right)^{1/2} + \left(\dfrac{G_{J_2}}{G_{J_1}} + \dfrac{1}{(J/S)_o}\right)}\right] \tag{9.52}$$

As $\rho_s \to 0$, $\beta \to 0$, Eq. (9.52) reduces to

$$\frac{\delta\theta}{\Delta\theta} = \frac{\rho_d^2}{\rho_d^2 + \dfrac{G_{J2}}{G_{J1}} + \dfrac{1}{(J/S)_o}} \tag{9.53}$$

which is identical (except for notation) to earlier results.

9.2 POLARIZATION JAMMING OF MONOPULSE

This section develops the initial response of a monopulse radar to a polarized incident signal containing random, time-varying components and describes incident signals that cause tracker instability. These results are useful in specifying the technical requirements or performance capabilities of a radar jammer.

Shustov and Vakin [4] do not consider the effects of incident signal polarization on radar tracking nor does a development in Leonov and Fomichev [26] address the detailed coupling of the azimuth and elevation channels. This section explores these issues and determines the necessary jammer characteristics for radar breaklock. It develops and simplifies the radar receiver error voltage resulting from receipt of a polarized signal through reasonable approximations of the monopulse antenna patterns near the boresight. The model of the radar tracker is simplified by decoupling the azimuth and elevation channels. Finally, it defines the properties of the incident signals necessary for the initiation of breaklock.

Monopulse Receiver Error Voltage

This section develops the error voltage of the monopulse receiver in sufficient generality to accommodate the effects of polarization. The steps include determining the antenna output signals, assuming a form of the monopulse receiver, and evaluating the receiver output. Since complex notation is used throughout, the actual time quantities are given by $\bar{\varepsilon}(\bar{r},t) = \mathrm{Re}\{\bar{E}(\bar{r})e^{j\omega_o t}\}$, where ω_o is the appropriate carrier.

The output voltage of an arbitrary antenna illuminated by an incident electromagnetic wave $\bar{E}_{in}$ is given by [34, p. 106] as a vector inner product

$$V_{oc} = k_1 \bar{E}_{in} \cdot \bar{E}_a \tag{9.54}$$

In Eq. (9.54), $\bar{E}_a$ is the field radiated by the antenna in the direction of the source of $\bar{E}_{in}$ when the antenna is driven. The quantity k_1 is a parameter accounting for the various constants in the formulation, such as range attenuation and actual power radiated.

Both sum $\bar{E}_\Sigma$ and difference $\bar{E}_\Delta$ antenna patterns exist in the monopulse radar, although only the sum pattern is radiated. Reference to Fig. 7.21 shows that $\bar{E}_a$ could be either $\bar{E}_\Sigma = \bar{E}_1 + \bar{E}_2$, the sum pattern, or $\bar{E}_\Delta = \bar{E}_1 - \bar{E}_2$, the difference pattern. Using Eq. (9.54) to express the input voltages to the IF amplifier with the constant k_1 suppressed,

$$V_{\Sigma i} = \bar{E}_{\Sigma} \cdot \bar{E}_{in} \tag{9.55}$$

$$V_{\Delta i} = \bar{E}_{\Delta} \cdot \bar{E}_{in} \tag{9.56}$$

and the inputs to the angle-error detector are

$$V_{\Sigma o} = K_{\Sigma} V_{\Sigma i} \tag{9.57}$$

$$V_{\Delta o} = K_{\Delta} V_{\Delta i} \tag{9.58}$$

where K is the IF channel gain. The output of the error detector is

$$\varepsilon_{az} = \tfrac{1}{2}\mathrm{Re}\{V_{\Sigma o} V_{\Delta o}^{*}\} \tag{9.59}$$

It is now necessary to make this error voltage independent of the target amplitude, a function accomplished by the automatic gain control (AGC) in the receiver. There are many approaches to AGC within the receiver. The one considered here is called instantaneous automatic gain control (IAGC) in which the output voltage is

$$V_{\Sigma o} = \frac{V_{\Sigma i} K_{\Sigma o}}{1 + \alpha k_2 |V_{\Sigma i}|} \tag{9.60}$$

$$V_{\Delta o} = \frac{V_{\Delta i} K_{\Delta o}}{1 + \alpha k_2 |V_{\Sigma i}|} \tag{9.61}$$

In this AGC model, the IF gain is inversely related to the magnitude of the sum input voltage through the term $\alpha k_2 |V_{\Sigma i}| \gg 1$ since this is the normal tracking condition. This condition is assumed even in the presence of jamming in full recognition of a small, but finite, probability that the jamming may almost cancel the target return and result in no sum-channel input voltage. The error voltage is then

$$\varepsilon_{az} = \frac{K_o^2}{2\alpha^2 k_2^2} \mathrm{Re}\left(\frac{V_{\Sigma i} V_{\Delta i}^{*}}{V_{\Sigma i} V_{\Sigma i}^{*}}\right) = K_1 \mathrm{Re}\left[\left(\frac{\bar{E}_{\Delta} \cdot \bar{E}_{in}}{\bar{E}_{\Sigma} \cdot \bar{E}_{in}}\right)^{*}\right] \tag{9.62}$$

in which it is assumed that $|K_{\Sigma o}| = |K_{\Delta o}| = K_o$ and that the voltage out of the difference channel is not so large as to become limited. Thus, the error voltage is stated in terms of radiation patterns.

Approximation to Monopulse Antenna Pattern near the Boresight

The general equation for receiver error voltage [Eq. (9.62)] can be simplified only if a description of the radar tracking error very close to the initial, correct tracking angle is desired. These limited objectives enable the use of a linear approximation to describe the variations in the antenna patterns with angle near the tracking axis (i.e., aperture normal). The discussion considers only a monopulse antenna with amplitude sum-difference processing and makes some assumptions concerning the cross-polarized portion of the radiation pattern.

An arbitrary radiation pattern $\bar{E}_a(\theta,\varphi)$ can be decomposed into a nominal component $\hat{e}_n$ and into a cross-polarized component $\hat{e}_p$ [85]

$$\bar{E}_a = (\bar{E}_a \cdot \hat{e}_n^*)\hat{e}_n + (\bar{E}_a \cdot \hat{e}_p^*)\hat{e}_p \tag{9.63}$$

where $\hat{e}_n \cdot \hat{e}_n^* = 1$, $\hat{e}_n \cdot \hat{e}_p^* = 0$. Herein, $\hat{e}_p^*$ is orthogonal to $\hat{e}_n$ but not necessarily to $\bar{E}_a$. If one assumes that $\hat{e}_n$ and $\hat{e}_p$ are constant over a restricted angular region (i.e., $\hat{e}_n = \hat{\theta}$, $\hat{e}_p = j\hat{\varphi}$), then a radiation pattern $\bar{E}_a$ can be expressed as

$$\bar{E}_a(\theta,\varphi) = |\bar{E}_a|[g_n(\theta,\varphi)\hat{e}_n + g_p(\theta,\varphi)\hat{e}_p] \tag{9.64}$$

where g_n is real and g_p is $g_p = |g_p|e^{j\psi}$ with ψ the assumed constant. The nominal and cross-polarized gain functions are linearized by expressing them as a Taylor series about an angle, θ_s,φ_s, as

$$g_n(\theta,\varphi) \approx g_n(\theta_s,\varphi_s) + \frac{\partial g_n}{\partial \theta}\bigg|_{\theta_s,\varphi_s} (\theta - \theta_s) + \frac{\partial g_n}{\partial \varphi}\bigg|_{\theta_s,\varphi_s} (\varphi - \varphi_s) \tag{9.65}$$

$$|g_p(\theta,\varphi)| \approx |g_p(\theta_s,\varphi_s)| + \frac{\partial |g_p|}{\partial \theta}\bigg|_{\theta_s,\varphi_s} (\theta - \theta_s) + \frac{\partial |g_p|}{\partial \varphi}\bigg|_{\theta_s,\varphi_s} (\varphi - \varphi_s) \tag{9.66}$$

Defining a set of error slope coefficients

$$k_{n\theta} = \frac{1}{g_n(\theta_s,\varphi_s)}\left(\frac{\partial g_n}{\partial \theta}\bigg|_{\theta_s,\varphi_s}\right) \tag{9.67}$$

$$k_{n\varphi} = \frac{1}{g_n(\theta_s,\varphi_s)}\left(\frac{\partial g_n}{\partial \varphi}\bigg|_{\theta_s,\varphi_s}\right) \tag{9.68}$$

$$k_{p\theta} = \frac{1}{|g_p(\theta_s,\varphi_s)|}\left(\frac{\partial |g_p|}{\partial \theta}\bigg|_{\theta_s,\varphi_s}\right) \tag{9.69}$$

$$k_{p\varphi} = \frac{1}{|g_p(\theta_s,\varphi_s)|}\left(\frac{\partial |g_p|}{\partial \varphi}\bigg|_{\theta_s,\varphi_s}\right) \tag{9.70}$$

allows Eq. (9.65) to be written as

$$g_n(\theta,\varphi) = g_n(\theta_s,\varphi_s)(1 + k_{n\theta}\Delta\theta + k_{n\varphi}\Delta\varphi) \tag{9.71}$$

$$|g_p(\theta,\varphi)| = |g_p(\theta_s,\varphi_s)|(1 + k_{p\theta}\Delta\theta + k_{p\varphi}\Delta\varphi) \tag{9.72}$$

Equations (9.71) and (9.72) describe the variation near an arbitrary point of the far-field pattern produced by a single antenna field. It is necessary to develop similar formulations for each of the four monopulse lobes and combine them to arrive at the overall variation in the sum or difference pattern. If the antenna is at the origin of spherical coordinates, the region into which power is radiated is as shown in Fig. 9.15. If feed 1 is displaced from the focal point in a manner that places the maximum gain of the associated nominal pattern in quadrant I, then Eq. (9.71) describes the gain variation of the nominal polarization about the tracking axis. Obviously, $k_{n\theta}$ and $k_{n\varphi}$ are negative. To simplify cataloging the gain variations for the other three symmetrical patterns, it is necessary to redefine k_{ni} and k_{pi} as positive; then Eq. (9.71) becomes

$$g_n(\theta,\varphi) = g_n(\theta_s,\varphi_s)(1 - k_{n\theta}\Delta\theta - k_{n\varphi}\Delta\varphi) \tag{9.73}$$

and the signs multiplying the error slope coefficients in the other patterns are as listed in Table 9.2 (see also Fig. 9.15).

In order to repeat the analysis for the cross-polarized component of the radiation pattern, this field component must either be known or assumed. The cross-polarized pattern of each feed, before it is offset, will be assumed to consist of four separate lobes, one in each quadrant. Further, the phase of each lobe will be assumed to be opposite of the phase of the adjacent lobes. Finally, the peak of the cross-polarized pattern will be assumed to lie in the quadrant opposite to the copolarized peak when the copolarized beam peak is squinted to form one of the four monopulse lobes. These assumptions are consistent with the literature concerning horn-fed paraboloids [86] and result

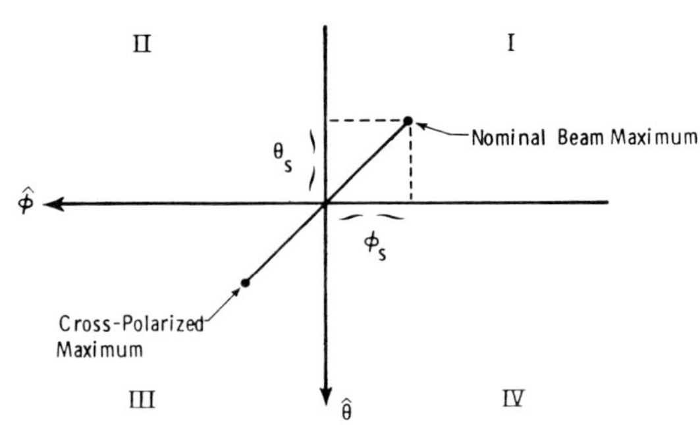

Fig. 9.15 Locations of beam maximum.

Table 9.2 Error slope coefficient multiplier

Antenna	Quadrant of copolarized peak	$k_{n\theta}$	$k_{n\varphi}$	$k_{p\theta}$	$k_{p\varphi}$
1	I	−1	−1	1	1
2	II	−1	1	1	−1
3	III	1	1	−1	−1
4	IV	1	−1	−1	1

in the location of the cross-polarized peak as shown in Fig. 9.15 and the sign multiplier of the error slope coefficient as shown in Table 9.2.

With these assumptions, the feed-horn arrangement shown in Fig. 9.16 (appropriate for a Cassagranian antenna), and $|g_p(\theta_s,\varphi_s)|e^{j\psi} = g_{p_s}$, the individual patterns are

$$\begin{aligned}
g_{n_1} &= g_{n_s}(1 - k_{n_\theta}\theta - k_{n_\varphi}\varphi) & g_{p_1} &= -g_{p_s}(1 + k_{p_\theta}\theta + k_{p_\varphi}\varphi) \\
g_{n_2} &= g_{n_s}(1 - k_{n_\theta}\theta + k_{n_\varphi}\varphi) & g_{p_2} &= g_{p_s}(1 + k_{p_\theta}\theta - k_{p_\varphi}\varphi) \\
g_{n_3} &= g_{n_s}(1 + k_{n_\theta}\theta + k_{n_\varphi}\varphi) & g_{p_3} &= -g_{p_s}(1 - k_{p_\theta}\theta - k_{p_\varphi}\varphi) \\
g_{n_4} &= g_{n_s}(1 + k_{n_\theta}\theta - k_{n_\varphi}\varphi) & g_{p_4} &= g_{p_s}(1 - k_{p_\theta}\theta + k_{p_\varphi}\varphi)
\end{aligned} \tag{9.74}$$

where angle is measured from the squint axis and all nominal (or cross-polarized) squint axis gains are assumed to be equal. Other assumptions in Eqs. (9.74) are that the copolarized and cross-polarized unit vectors for each pattern are identical at the squint angle. Finally, if $k_{n_\theta} = k_{n_\varphi} = k_n$ and $k_{p_\theta} = k_{p_\varphi} = k_p$, the resulting sum and difference patterns are

$$\bar{E}_\Sigma = \sum_{i=1}^{4} |\bar{E}_i|\{g_{n_i}\hat{e}_n + g_{p_i}\hat{e}_p\} = 4|\bar{E}_1|g_{n_s}\hat{e}_n \tag{9.75}$$

$$\bar{E}_{\Delta_{az}} = \bar{E}_2 + \bar{E}_3 - \bar{E}_1 - \bar{E}_4 \tag{9.76}$$

$$E_\Sigma = \begin{bmatrix} (E_{2+3} - E_{1+4})_n \\ (E_{2+3} - E_{1+4})_p \end{bmatrix} = |\bar{E}_1| \begin{bmatrix} 2g_{n_s}(1 + k_n\varphi) - 2g_{n_s}(1 - k_n\varphi) \\ 2g_{p_s}k_p\theta + 2g_{p_s}\theta k_p \end{bmatrix}$$

$$= 4|\bar{E}_1|\{g_{n_s}k_n\varphi\hat{e}_n + g_{p_s}k_p\theta\hat{e}_p\} \tag{9.77}$$

$$\bar{E}_{\Delta_{el}} = \bar{E}_3 + \bar{E}_4 - \bar{E}_1 - \bar{E}_2 = 4|\bar{E}_1|\{g_{n_s}k_n\theta\hat{e}_n + g_{p_s}k_p\varphi\hat{e}_p\} \tag{9.78}$$

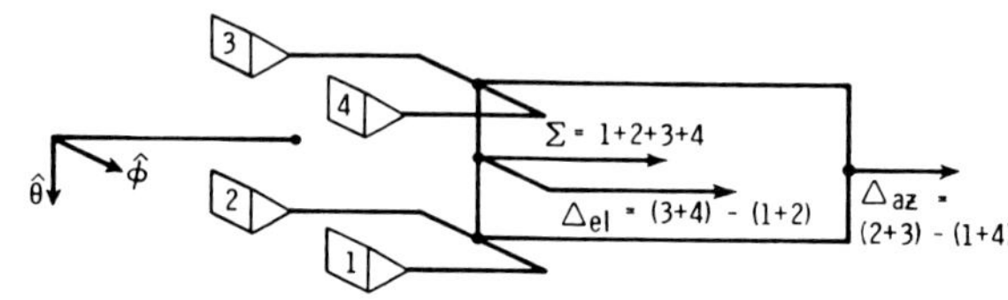

Fig. 9.16 Monopulse feed arrangement.

Example

It is assumed that the phase-sensing monopulse radar shown in Fig. 9.17 tracks in one plane only. Both antenna elements are short crossed dipoles, but each element has a different current density. The current densities are

$$\bar{J}_1 = V_1 G_a \Delta l \delta\left(\bar{r} - \frac{d}{2}\hat{x}\right)(\hat{x} + j\hat{z}) \tag{9.79}$$

$$\bar{J}_2 = V_2 \Delta l \delta\left(\bar{r} + \frac{d}{2}\hat{x}\right)(\hat{x} - j\hat{z}) \tag{9.80}$$

a) What is the radiated sum pattern, $\bar{E}_\Sigma(\theta,\varphi)$?

b) The nominal radiated field is $\bar{E}_\Sigma(\pi/2,\pi/2)$. What is $\hat{e}_n$?

c) What is ε_{az} if an incident field is expressed as $\bar{E}_{in} = E_i\hat{e}_n^*$? Find $\varepsilon_{az}(\varphi)$ at $\theta = \pi/2$.

d) What is $\varepsilon_{az}(\varphi)$ if the incident field is $\bar{E}_{in} = E_i\hat{e}_p^*$ at $\theta = \pi/2$ where $\hat{e}_n \cdot \hat{e}_p^* = 0$?

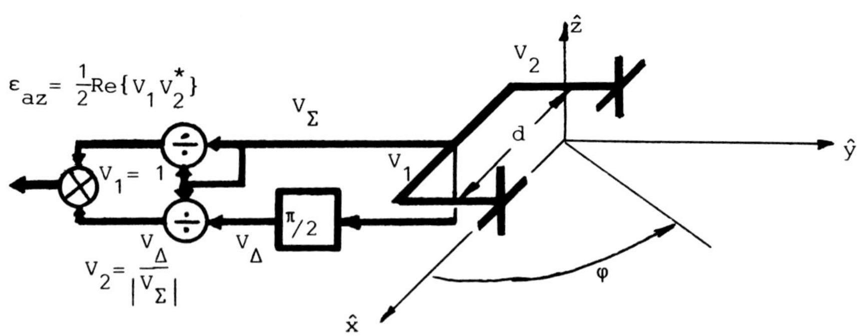

Fig. 9.17 Example of phase-sensing monopulse.

Solutions

a) The field of element 1 is

$$\bar{E}_1 = E_o \bar{h} e^{j\psi_o}$$

where

$$E_o = -j\frac{\omega\mu}{4\pi}\frac{V_\Sigma \Delta l G_a}{\sqrt{2}r} e^{-jkr}$$

$$\psi_o = e^{j(kd/2)\sin\theta\cos\varphi}$$

and $\bar{h}$ consists of the $\hat{\theta},\hat{\varphi}$ components of $\bar{J}_1$, or

$$\bar{h} = \hat{\theta}(\cos\theta\cos\varphi - j\sin\theta) - \hat{\varphi}\sin\varphi$$

Similarly, the field of element two is

$$\bar{E}_2 = E_o \bar{h}^* e^{-j\psi_o}$$

Thus, the sum pattern is

$$\bar{E}_\Sigma = \bar{E}_1 + \bar{E}_2 = 2E_o\{\hat{\theta}(\cos\theta\cos\varphi\cos\psi_o + \sin\theta\sin\psi_o) - \hat{\varphi}\sin\varphi\cos\psi_o\}$$

b) At $\theta = \varphi = \pi/2$, the nominal field is defined as $\bar{E}_\Sigma(\pi/2, \pi/2)$ or

$$\bar{E}_\Sigma\left(\frac{\pi}{2}, \frac{\pi}{2}\right) = -2E_o\hat{\varphi}$$

Therefore, $\hat{e}_n \approx \hat{\varphi}$.

c) If the incident field is $\bar{E}_{in} = E_i\hat{\varphi}$, then the sum output voltage is

$$V_\Sigma = -2E_o\sin\varphi\cos\left(\frac{kd}{2}\cos\varphi\right)$$

The difference antenna pattern is

$$\bar{E}_\Delta = \bar{E}_1 - \bar{E}_2 = 2E_o\{\hat{\theta}(j\sin\psi_o\cos\theta\cos\varphi - j\sin\theta\cos\psi_o) - j\hat{\varphi}\sin\varphi\sin\psi_o\}$$

yielding a difference channel voltage of

$$V_\Delta = -j2E_o\sin\varphi\sin\left(\frac{kd}{2}\cos\varphi\right)$$

The resultant error voltage out of the phase detector is

$$\varepsilon_{az}(\varphi) = \frac{1}{2}\tan\left(\frac{kd}{2}\cos\varphi\right)$$

d) If the incident field is orthogonal to $\bar{E}_\Sigma$ at $\theta = \varphi = \pi/2$, then $\bar{E}_{in} = E_i\hat{\theta}$, which is not orthogonal to $\bar{E}_\Sigma$ from other directions. The error voltage resulting from an error in azimuth angle in this case is found as follows:

$$V_\Sigma = 2E_o \sin\psi_o$$

$$V_\Delta = -j2E_o \cos\psi_o$$

$$\varepsilon_{az}(\varphi) = \frac{1}{2}\,\mathrm{Re}\left\{\frac{V_\Sigma V_\Delta^*}{|V_\Sigma|^2}\right\} = \frac{1}{2}\,\mathrm{ctn}\left(\frac{kd}{2}\cos\varphi\right)$$

Comment

The radar boresight is at $\varphi = 90$ deg; thus, the radar will track the source of the illumination in solution c. With $\varphi' = \varphi - (\pi/2)$, the error voltage in solution d becomes, for small φ',

$$\varepsilon_{az}(\varphi) = -\frac{1}{2}\,\mathrm{ctn}\left(\frac{kd}{2}\varphi'\right)$$

which is zero at an error of

$$\varphi' = \lambda/2d$$

It is now necessary to use the highly symmetric antenna patterns resulting from the many assumptions (note that the sum pattern is of constant polarization) as a part of a radar tracker and to continue with the original search for meaningful jammer requirements for breaklock.

Radar Tracker Model for Deterministic Inputs

This section first develops a linear radar tracker model consistent with the previous section. It then shows that a simpler model involving uncoupled axes is possible. Finally, it defines the properties of input signals that induce instability.

Let the incident wave be divided into a copolarized part $\bar{E}_n$ and a cross-polarized part $\bar{E}_p$, as

$$\bar{E}_{in} = \bar{E}_p + \bar{E}_n = E_p\hat{e}_p^* + E_n\hat{e}_n^* \tag{9.81}$$

Then, using Eqs. (9.62), (9.75), and (9.77) and suppressing the constant k_1, the complex detector voltage is

$$\varepsilon_{az} = \frac{\{g_{n_s}k_n\varphi\hat{e}_n + g_{p_s}k_p\theta\hat{e}_p\}\cdot\bar{E}_{in}}{g_{n_s}\hat{e}_n\cdot\bar{E}_{in}}$$

$$= \frac{g_{n_s} k_n \varphi E_n + g_{p_s} k_p \theta E_p}{g_{n_s} E_n}$$

$$= k_n \varphi + \left(\frac{g_{p_s}}{g_{n_s}} \right) k_p \left(\frac{E_p}{E_n} \right) \theta \tag{9.82}$$

Using Eq. (9.78), the error voltages resulting from angle errors in azimuth and elevation are (as k_n, k_p are real)

$$\varepsilon_{az} = k_n \varphi + k_p \mathrm{Re} \left\{ \frac{g_{p_s} E_p}{g_{n_s} E_n} \right\} \theta \tag{9.83}$$

$$\varepsilon_{el} = k_n \theta + k_p \ \mathrm{Re} \left\{ \frac{g_{p_s} E_p}{g_{n_s} E_n} \right\} \varphi \tag{9.84}$$

where we could define $|E_p / E_n| = \sqrt{J/S}$ when g_{p_s} is real.

The appropriate model of a radar tracker for Eqs. (9.83) and (9.84) is shown in Fig. 9.18 in which both the angular position of target $(\theta_{T_o}, \varphi_{T_o})$ and the radar boresight (θ_A, φ_A) are measured in relation to some arbitrary but fixed angle. The block labeled $h_1(t)$ represents the open-loop impulse response of the servo amplifier and motor driving the antenna.

The closed-loop servo response can be determined in a straightforward manner. The following substitutions will be used:

$$\bar{\varepsilon} = \begin{bmatrix} \varepsilon_{el} \\ \varepsilon_{az} \end{bmatrix}; \psi_T = \begin{bmatrix} \theta_{T_o} \\ \varphi_{T_o} \end{bmatrix}; \bar{\psi}_A = \begin{bmatrix} \theta_A \\ \varphi_A \end{bmatrix}; [K] = \begin{bmatrix} k_n & \gamma k_p \\ \gamma k_p & k_n \end{bmatrix} \tag{9.85}$$

$$[I] = \begin{bmatrix} 1 & 0 \\ 0 & 1 \end{bmatrix} \tag{9.86}$$

The error-voltage vector is

$$\bar{\varepsilon} = [K](\bar{\psi}_T - \bar{\psi}_A) = [K]\bar{\psi} \tag{9.87}$$

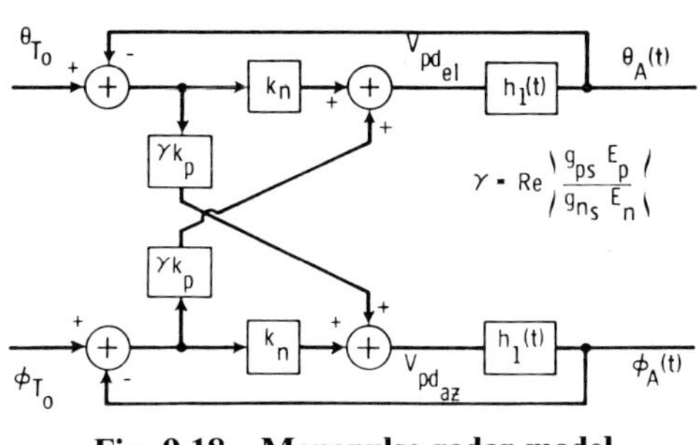

Fig. 9.18 Monopulse radar model.

and depends on the error in both axes. The eigenvectors and eigenvalues of the $[K]$ matrix can be used to define a new, uncoupled error voltage. These eigenvalues are

$$\lambda_{1,2} = k_n \pm \gamma k_p \tag{9.88}$$

and the eigenvectors are

$$[E] = [\bar{e}_1, \bar{e}_2] = \begin{bmatrix} 1 & 1 \\ 1 & -1 \end{bmatrix} \tag{9.89}$$

Then, defining new angular variables

$$\bar{\mu}_A = [E]^{-1}\bar{\psi}_A;\ \bar{\mu}_T = [E]^{-1}\bar{\psi}_T;\ \bar{\mu}_A = \begin{bmatrix} \mu_{A_1} \\ \mu_{A_2} \end{bmatrix} \tag{9.90}$$

Equation (9.87) becomes

$$[E]^{-1}\varepsilon = [E]^{-1}[K][E](\bar{\mu}_T - \bar{\mu}_A) = [\lambda](\bar{\mu}_T - \bar{\mu}_A) \tag{9.91}$$

as $[A][E] = [E][\lambda]$, where $[\lambda]$ is a diagonal matrix of the eigenvalues.

From Fig. 9.18,

$$\theta_A(t) = \int_0^t h_1(t-\tau)\varepsilon_{el}(\tau)\,\mathrm{d}\tau \tag{9.92a}$$

$$\varphi_A(t) = \int_0^t h_1(t-\tau)\varepsilon_{az}(\tau)\,\mathrm{d}\tau \tag{9.92b}$$

Adding and subtracting Eqs. (9.92a) and (9.92b) yields

$$\begin{aligned} \tfrac{1}{2}(\theta_A + \varphi_A) = \mu_{A_1} &= \tfrac{1}{2}\int_0^t h_1(t-\tau)[\varepsilon_{el}(\tau) + \varepsilon_{az}(\tau)]\,\mathrm{d}\tau \\ &= \int_0^t h_1(t-\tau)\lambda_1(\mu_{T_1} - \mu_{A_1})\,\mathrm{d}\tau \end{aligned} \tag{9.93}$$

$$\begin{aligned} \tfrac{1}{2}(\theta_A - \varphi_A) = \mu_{A_2} &= \tfrac{1}{2}\int_0^t h_1(t-\tau)[\varepsilon_{el}(\tau) - \varepsilon_{az}(\tau)]\,\mathrm{d}\tau \\ &= \int_0^t h_1(t-\tau)\lambda_2(\mu_{T_2} - \mu_{A_2})\,\mathrm{d}\tau \end{aligned} \tag{9.94}$$

or, for either newly defined axis,

$$\mu_{A_i} + \lambda_i \int_0^t h_1(t-\tau)\mu_{A_i}(\tau)\,\mathrm{d}\tau = \lambda_i \int_0^t h_1(t-\tau)\mu_{T_i}(\tau)\,\mathrm{d}\tau \qquad i = 1,2 \tag{9.95}$$

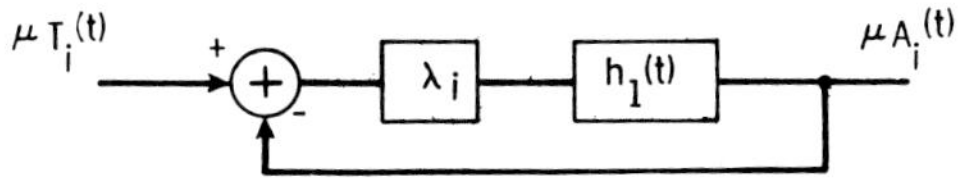

Fig. 9.19 Uncoupled monopulse model.

which is the same equation resulting from uncoupled channels but for the presence of the eigenvalues. The structure of the model for the decoupled axes system is shown in Fig. 9.19.

Equation (9.95) can easily be solved by Laplace transforms, yielding

$$\mu_{A_i}(s) = \frac{\lambda_i H_1(s)\mu_{T_i}(s)}{1 + \lambda_i H_1(s)} \qquad i = 1,2 \tag{9.96}$$

We are not interested in the unique solution corresponding to specific inputs $\mu_{T_i}(t)$, but we are concerned with the stability of the solution for an arbitrary initial offset μ_{T_o} and for a particular form of $H_1(s)$ corresponding to a typical radar servo [16, Chap. 9], such as

$$H_1(s) = \frac{K(s + \omega_2)}{s(s + \omega_1)} \tag{9.97}$$

With this servo response,

$$\mu_{A_i}(s) = \frac{\lambda_i K(s + \omega_2)\mu_{T_{o_i}}}{s^2 + (\omega_1 + \lambda_i K)s + \lambda_i K\omega_2} \tag{9.98}$$

and stability is governed by the roots of the denominator

$$s_{1,2} = -\frac{(\omega_1 + \lambda_i K) \pm \sqrt{\omega_1^2 + 2\lambda_i K(\omega_1 - 2\omega_2) + \lambda_i^2 K^2}}{2} \tag{9.99}$$

Instability results when either root has a positive real part, which is impossible if $\lambda_i > 0$. For $\lambda_i = 0$, one root is at the origin of the "s" plane. If $\lambda_i < 0$, one pole moves into the right half "s" plane and the servo is unstable.

Thus, instability requires

$$k_n \pm \gamma k_p = \lambda_i < 0$$

$$\left|\mathrm{Re}\left\{\frac{g_{p_s}E_p}{g_{n_s}E_n}\right\}\right| > \frac{k_n}{k_p} \tag{9.100}$$

If the nominal and cross-polarized patterns are in phase, Eq. (9.100) reduces to a somewhat expected result

$$\frac{J}{S} = \left|\mathrm{Re}\left\{\frac{E_p}{E_n}\right\}\right|^2 > \left(\frac{k_n g_{n_s}}{k_p g_{p_s}}\right)^2 = \left|\frac{\partial g_n/\partial\theta}{\partial|g_p|/\partial\theta}\right|_{\theta_s,\varphi_s} \tag{9.101}$$

One must recall that the ratio of co- to cross-polarization given in Eq. (9.100) is for the individual lobes of the pattern and *not* the total sum pattern.

Radar Tracker Model for Stochastic Inputs

The principal concern in this section is the determination of the relationship between the specifications and the performance of the jammer. The deterministic model in the last section permitted the solution of a problem similar to that considered in this section, but without the complication of stochastic signals. These more realistic signals must now be addressed, but the framework of the last section provides a logical path. Thus, we can derive a valid model for an incident, stochastic signal and provide the jammer properties for initiation of breaklock.

The incident wavefront is $\bar{E}_{in}$, consisting of a constant nominal part $\bar{E}_n$, a constant cross-polarized component $\bar{E}_p$, and a time-varying component, $\tilde{\bar{E}}_2$

$$\tilde{\bar{E}}_{in} = \bar{E}_n + \bar{E}_p + \tilde{\bar{E}}_2 \tag{9.102}$$

Then, as $\bar{E}_p \cdot \hat{e}_n^* = 0$, $\bar{E}_n \cdot \hat{e}_p^* = 0$, the signal $\tilde{\bar{E}}_s$ and jamming $\tilde{\bar{E}}_J$ inputs can be defined as

$$\tilde{\bar{E}}_s = (\tilde{\bar{E}}_{in} \cdot \hat{e}_n^*)\, \hat{e}_n^* = (E_n + \tilde{\bar{E}}_{2n})\hat{e}_n^* \tag{9.103}$$

$$\tilde{\bar{E}}_J = (\tilde{\bar{E}}_{in} \cdot \hat{e}_p^*)\, \hat{e}_p^* = (E_p + \tilde{\bar{E}}_{2p})\hat{e}_p^* \tag{9.104}$$

The complex error voltage of the radar receiver is still given by Eq. (9.62), but it is now a random process

$$\tilde{\varepsilon} \approx K_1 \left\{ \frac{\bar{E}_\Delta \cdot \tilde{\bar{E}}_{in}}{\bar{E}_\Sigma \cdot \tilde{\bar{E}}_{in}} \right\}^* \tag{9.105}$$

Errors introduced by the assumptions leading to Eq. (9.62), specifically $\alpha K_2 |V_{\Sigma i}| \gg 1$, are ignored.

Using approximations of the linear antenna pattern developed earlier in Eqs. (9.75–9.78) and Eq. (9.103) yields (suppressing K_1) the complex error voltage

$$\tilde{\varepsilon}_{az} = k_n \varphi + \frac{g_{p_s}}{g_{n_s}} \left(\frac{E_p + \tilde{E}_{2p}}{E_n + \tilde{E}_{2n}} \right) k_p \theta \tag{9.106}$$

If we define a new random process $\tilde{\gamma}(t)$,

$$\tilde{\gamma} = \mathrm{Re} \left\{ \frac{g_{p_s}}{g_{n_s}} \left(\frac{E_p + \tilde{E}_{2p}}{E_n + \tilde{E}_{2n}} \right) \right\} \tag{9.107}$$

then the real error voltages for the azimuth and elevation errors are

$$\tilde{\varepsilon}_{az} = k_n\varphi + \tilde{\gamma}k_p\theta \tag{9.108}$$

$$\tilde{\varepsilon}_{el} = \tilde{\gamma}k_p\varphi + k_n\theta \tag{9.109}$$

Figure 9.17 is still a valid model with the modification of the γk_p blocks to time-varying multipliers. The eigenvector decomposition is still the same, but the eigenvalues are time dependent.

$$\tilde{\lambda}_{1,2}(t) = k_n \pm \tilde{\gamma}(t)k_p \tag{9.110}$$

The decoupled model now appears in Fig. 9.20 and satisfies the integral equation,

$$\tilde{\mu}_{A_i}(t) + \int_0^t \tilde{\lambda}_i(\tau)h_1(t-\tau)\tilde{\mu}_{A_i}(\tau)\,d\tau = \int_0^t \tilde{\lambda}_i(\tau)h_1(t-\tau)\mu_{T_i}(\tau)\,d\tau \tag{9.111}$$

This equation will be solved only for $h_1(t)$, a perfect integrator. Then

$$\tilde{\mu}_{A_i}(t) + \int_0^t \tilde{\lambda}_i(\tau)\tilde{\mu}_{A_i}(\tau)\,d\tau = \int_0^t \tilde{\lambda}_i(\tau)\mu_{T_i}(\tau)\,d\tau \tag{9.112}$$

with the solution [87] for $\mu_{A_i}(o) = \mu_{A_{io}}$,

$$\tilde{\mu}_{A_i}(t) = \exp\left(-\int_0^t \tilde{\lambda}_i(\tau)\,d\tau\right)\left[\int_0^t \exp\left(\int_0^\tau \tilde{\lambda}_i(v)\,dv\right)\tilde{\lambda}_i(\tau)\mu_{T_i}(\tau)\,d\tau + \tilde{\mu}_{A_{io}}\right] \tag{9.113}$$

Since system stability without target maneuvers is the only concern, let $\mu_{T_i}(t) = 0$, yielding

$$\tilde{\mu}_{A_i}(t) = \tilde{\mu}_{A_{io}} \exp\left(-\int_0^t \tilde{\lambda}_i(\tau)\,d\tau\right) \tag{9.114}$$

In Eq. (9.114), the interpretation of $\tilde{\mu}_{A_i}(t)$ is the system response to one of an infinite variety of incident signals as represented by $\tilde{\lambda}_i(t)$. A stable system requires that $\lim_{t\to\infty} \tilde{\mu}_{A_i}(t)$ remain finite. Since there are many possible

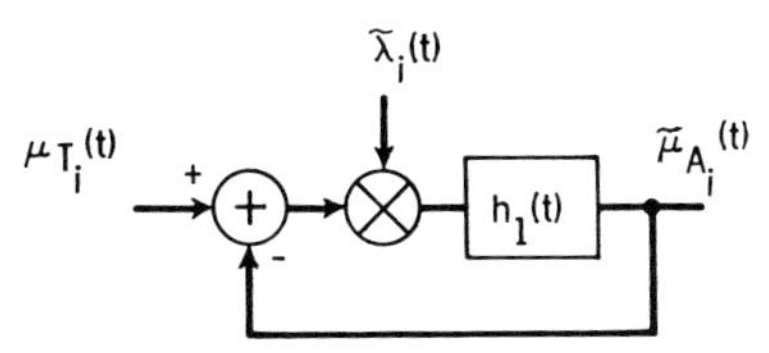

Fig. 9.20 Time-varying radar model.

responses, it is more meaningful to consider some average system response as the time increases. Specifically, we consider the mean square of the response and call a system unstable if

$$\lim_{t\to\infty} E\left\{\ell n\left|\left[\frac{\tilde{\mu}_{A_i}(t)}{\tilde{\mu}_{A_{io}}}\right]^2\right|\right\} = \lim_{t\to\infty} E\left\{-2\int_o^t \tilde{\lambda}_i(\tau)\, \mathrm{d}\tau\right\}$$

$$= \lim_{t\to\infty}\left[-2\int_o^t E\left\{\tilde{\lambda}_i(\tau)\right\}\mathrm{d}\tau\right]\to\infty \qquad (9.115)$$

Or, equivalently, a system is unstable if, assuming stationarity, one of the eigenvalues satisfies

$$E\{\tilde{\lambda}_i(t)\} < 0 \qquad (9.116)$$

Using Eq. (9.110) and as k_n,k_p are positive,

$$E\{\tilde{\lambda}\} = k_n - k_p E\{\tilde{\gamma}\} < 0 \qquad (9.117)$$

Using Eq. (9.107),

$$E\{\tilde{\gamma}\} = E\left\{\mathrm{Re}\left(\frac{g_{p_s}}{g_{n_s}}\frac{(E_p + \tilde{E}_{2p})}{(E_n + \tilde{E}_{2n})}\right)\right\} > \frac{k_n}{k_p} \qquad (9.118)$$

Some properties for the random components must be assumed. If the random components can be expressed as

$$E_{2n} = |E_{2n}|e^{j\xi n} \qquad (9.119)$$

$$E_{2p} = |E_{2p}|e^{j\xi p} \qquad (9.120)$$

where the probability density of the phase is

$$p(\xi_i) = 1/2\pi, \qquad \pi \le \xi \le \pi$$

$$= 0 \qquad \text{else} \qquad (9.121)$$

and the ξ_n and ξ_p are independent of each other and of the random amplitudes, then, it can be shown that Eq. (9.118) leads to

$$1 \ge \int_o^{E_n} p(|E_{2n}|)\,\mathrm{d}|E_{2n}| > \frac{k_n}{k_p}\frac{1}{\mathrm{Re}\left\{\dfrac{g_{ps}E_p}{g_{ns}E_n}\right\}} \qquad (9.122)$$

as the requirement for breaklock. For the special case of g_{p_s} real,

$$\int_o^{E_n} p(|E_{2n}|)d|E_{2n}| > \frac{k_n g_{n_s}}{k_p g_{p_s} \operatorname{Re}\{E_p/E_n\}} \tag{9.123}$$

or

$$\frac{k_p g_{p_s}}{k_n g_{n_s} \operatorname{Re}\{E_p/E_n\}} < 1 \tag{9.124}$$

as the left side of Eq. (9.123) is less than one. Again,

$$\frac{J}{S} \triangleq \left(\operatorname{Re}\left\{\frac{E_p}{E_n}\right\}\right)^2 > \left(\frac{k_n g_{n_s}}{k_p g_{p_s}}\right)^2 \tag{9.125}$$

as for the deterministic case. However, even if Eq. (9.125) is satisfied, the random portion of the incident wavefront, which is of the nominal polarization, must satisfy Eq. (9.123).

The ratio $\operatorname{Re}\{E_p/E_n\}$ is defined as the purity of the incident wavefront.

Summary and Conclusions on Polarization Jamming

This section has described an amplitude sum-difference monopulse with the desired polarization of the sum pattern defined by $\hat{e}_n$. It has assumed a nonzero cross polarization $\hat{e}_p$ in the equisignal direction for *each* of the four monopulse feeds. If the target is assumed constant and perfectly copolarized, $E_T = E_n$, while the jammer has a constant cross-polarized part, $E_J = E_p$, plus a fluctuating part E_2 breaklock requires a jamming-to-signal ratio of

$$\frac{J}{S} = \left(\operatorname{Re}\left\{\frac{E_J}{E_T}\right\}\right)^2 > \left(\frac{k_n g_{n_s}}{k_p g_{p_s}}\right)^2 \tag{9.126}$$

if it is also assumed that the copolarized and orthogonal components of the element patterns are in phase. (Note: $\hat{e}_n$ and $\hat{e}_p$ are complex vectors and are not necessarily in phase.) Additionally, the density of the copolarized random part of the jammer emission must be such that

$$\int_o^{E_T} p(|E_{2n}|)\, \mathrm{d}|E_{2n}| > \frac{k_n g_{n_s}}{k_p g_{p_s}} \frac{1}{\operatorname{Re}|\{E_J/E_T\}|} \tag{9.127}$$

This latter requirement states that the random copolarized jammer output, $\tilde{E}_{2n}$ must be very small relative to the signal E_T if the J/S requirement given in Eq. (9.126) is met with only a small excess (i.e., the right side of Eq. (9.127) is then nearly unity as can be seen from observation of Eq. (9.126).

As a specific example of the meaning of Eq. (9.127), we assume that the jammer's random amplitude component has a Rayleigh distribution. Then, with

$$\tilde{z} = |\tilde{E}_{2n}| \tag{9.128}$$

$$2\sigma^2 = E\{|\tilde{E}_{2n}|^2\}, \qquad \mathrm{Re}\left|\left\{\frac{E_J}{E_T}\right\}\right| = \sqrt{\frac{J}{S}} \tag{9.129}$$

$$p(|E_{2n}|) = p(z) = \frac{z}{\sigma^2} e^{-z^2/2\sigma^2} \tag{9.130}$$

it follows that

$$\int_o^{E_T} p(z)\, \mathrm{d}z = 1 - e^{-E_T{}^2/2\sigma^2} > \frac{k_n g_{n_s}}{(k_p g_{p_s})\sqrt{J/S}} \tag{9.131}$$

Equation (9.131) can be manipulated to define a normalized second-moment for the jammer of

$$\frac{E\{|\tilde{E}_{2n}|^2\}}{E_T^2} < \frac{-1}{\ell n \left\{1 - \dfrac{k_n g_{n_s}}{k_p g_{p_s}} \dfrac{1}{\sqrt{J/S}}\right\}} \tag{9.132}$$

which holds only when g_{p_s} is in phase with g_{n_s}. Equation (9.132) is plotted in Fig. 9.21 under the condition of $k_n = k_p$. In Fig. 9.21, the abscissa value is the amount by which the constant cross-polarized part of the jammer output, relative to the constant copolarized target, exceeds the squint-angle purity of each of the radar beams. The ordinate could also be labeled the ratio of the nominal part of the random jammer power to the target power. (The total random jammer power must include the cross-polarized random part.)

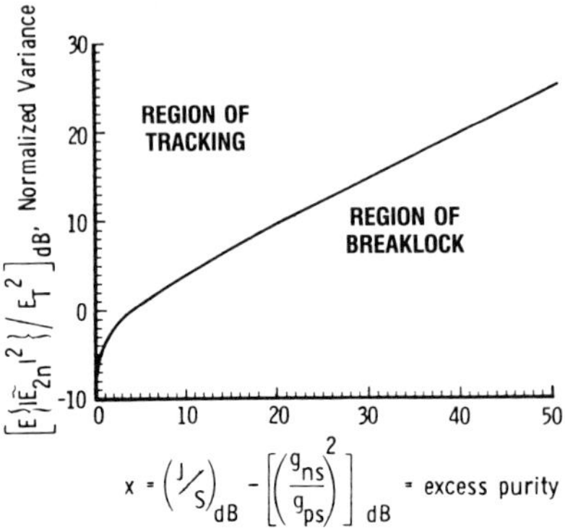

Fig. 9.21 Requirements on jammer second moment for breaklock.

From Fig. 9.21, one can conclude that, if the incident wavefront ratio of copolarization to cross-polarization (purity) is equal to the purity of one of the radar beams, then there must be no random component. If the incident purity exceeds the purity of the radar by 8 dB, then the jammer can have a Rayleigh fluctuating part with a root-mean-square value equal to the target return. (This assumes equal nominal and crossed random parts.) Since a perfect jammer cannot be built, the purity of the incident signal must obviously exceed the purity of any one radar beam.

Alternatively, if the constant nominal return is considered as a constant jammer signal E_{J_n} of incorrect polarization, then Fig. 9.21 still holds provided that this signal is large enough to activate the automatic gain control. The jammer requirement is then

$$\left(\mathrm{Re}\left\{\frac{g_{p_s}E_{J_p}}{g_{n_s}E_{J_n}}\right\}\right)^2 > \left(\frac{k_n}{k_p}\right)^2 \tag{9.133}$$

subject to the nominal part of the random component, which is either signal or jamming, satisfying

$$\int_o^{E_{J_n}} p(|E_n|)\,\mathrm{d}E_n > \frac{1}{\mathrm{Re}\left\{\dfrac{g_{p_s}E_{J_p}}{g_{n_s}E_{J_n}}\right\}} \tag{9.134}$$

One should note that, although the results above are presented in terms of the radiation pattern of a single element [(Eq. (9.126)], the required ratio is a measurable quantity. From Eqs. (9.75–9.78), nominal and orthogonal polarizations are defined at the peak of the sum pattern, located most accurately from the null direction. Upon rotating the monopulse antenna through a known angle in both azimuth and elevation from the illuminator, and then switching the polarization of the illumination between orthogonals, the required ratio $g_{ns}k_n/g_{ps}k_p$ can be found.

10 ACTIVE ECM VS AUTOMATIC GAIN CONTROL

Previous chapters stressed the importance of the proper functioning of the AGC mechanism in a radar. When the AGC system is operative, the angle servo of the radar is driven by an error voltage related through the constant K_{D_0} to the angular tracking error and the gain of the angle servo is at its designed (presumed optimum) value. The signals analyzed in this chapter reduce the gain of the angle-servo loop through their influence on the radar AGC voltage. Small reductions in gain result in sluggish response; larger reductions may result in oscillations or instability.

10.1 EFFECT OF IMPULSIVE SIGNALS ON THE VOLTAGE OUT OF IF AMPLIFIERS

The equivalent circuit of an IF amplifier, with gain controlled by an AGC loop, is shown in Fig. 10.1. This equivalent circuit is identical to the circuit presented in Fig. 7.9, with the exception of the AGC delay voltage V_d. The purpose of the delay voltage is to prevent any reduction in amplifier gain at low levels of the received signals. This analysis is attributed to Bakut et al. [88, pp. 95–96, 126–128].

We make the following assumptions regarding the operation of the circuit in Fig. 10.1. The gain $K(V_r)$ of the amplifier responds instantly to changes in the control voltage V_r. Because of the averaging of the filter $h(t)$ in the feedback circuit, the intermediate frequency of the input and output of the IF amplifier has no impact on the analysis of the variations in IF gain. The input and output voltages can thus be regarded as the envelope of the associated IF frequency signals. Finally, if the output becomes less than the delay voltage, the AGC filter input becomes open circuited.

With a sufficiently large increase in input voltage, the output voltage of the amplifier will become limited at some value V_L, as shown in Fig. 10.2. (In Fig. 10.2 the bipolar IF frequency of the signal is retained for visual clarity.) Such limiting usually results from a physical limitation within the IF amplifier. The dotted lines represent the variation of the IF output voltage in the absence of limiting. The output of the IF amplifier contains no encoded information about the error in tracking (assuming a scanning radar) while it is limited or clipped. The duration of the limited condition, measured from the application of the input pulse, is T_1.

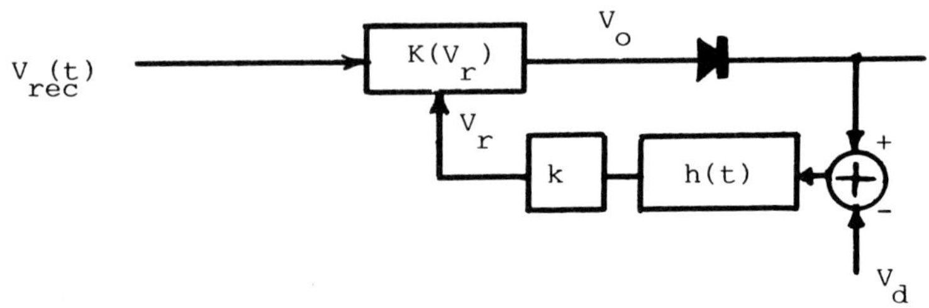

Fig. 10.1 Equivalent circuit of a gain controlled amplifier.

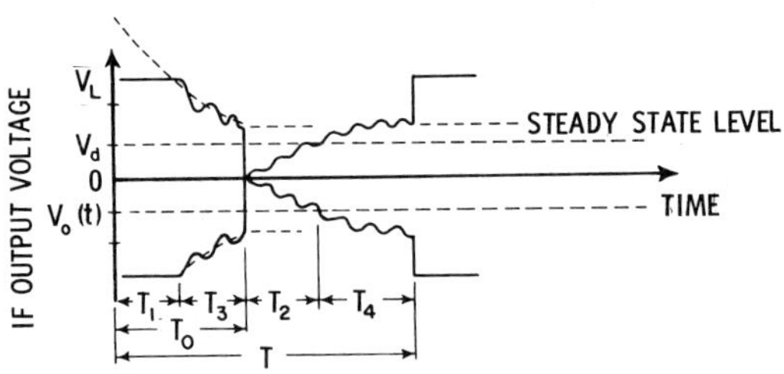

Fig. 10.2 Amplifier output voltage variations caused by impulsive input voltage.

Upon removal of the large input signal, the existing low value of the gain results in the suppression of any useful target signal. Gradually, the AGC voltage will decrease, the IF gain will increase, and the amplitude of the output voltage will reach a desired voltage, here set equal to the delay voltage. The time required to achieve this desired output is T_2 and we assume that the IF output is only partially useful over the period from elimination of the jammer input until T_2. (Bakut et al. [88] assumes that the error derived from the signal existing over the interval T_2 has no use.)

Under the influence of periodic jamming pulses of period T, the radar receiver produces a degraded error voltage for a time $T_1 + T_2$. These two times will be derived for a particular variation in IF amplifier gain with control voltage.

10.2 LINEAR IF GAIN VARIATION WITH CONTROL VOLTAGE

If a linear variation in IF gain with control voltage is assumed, the IF gain varies with control voltage as described in Chapter 7, Fig. 7.10. Specifically,

$$K(t) = K_o - \alpha V_r(t) \tag{10.1}$$

$$K \geq 0 \tag{10.2}$$

If a constant input voltage V_1 is applied, the steady-state output voltage in the absence of a delay voltage [Eq. (7.7)] is

$$V_o(\infty) = KV_1 = \frac{K_o V_1}{1 + \alpha k V_1} \tag{10.3}$$

Because of the incorporation of a delay voltage, in the steady state, $V_r = k(V_o(\infty) - V_d)$ and the output voltage becomes

$$V_o(\infty) = V_1 \frac{K_o + \alpha k V_d}{1 + \alpha k V_1} = V_3 \tag{10.4}$$

We assume that the AGC filter has the impulse response

$$h(t) = \frac{1}{T_f} e^{-t/T_f} \tag{10.5}$$

If the input V_1 has been applied long enough for the steady-state condition in Eq. (10.4) to hold and a new input V_2 is applied at $t = 0$ causing the output of the IF amplifier to become limited, the input voltage to the AGC filter varies as shown in Fig. 10.3 for $t < T_1$. The regulating voltage varies as

$$V_r(t) = k(V_3 - V_d) + k(V_L - V_3)(1 - e^{-t/T_f}), \qquad 0 \leq t \leq T_1 \tag{10.6}$$

By definition, at $t = T_1$, the output voltage of the IF amplifier is V_L or

$$V_L = K[V_r(T_1)]V_2 \tag{10.7}$$

Inserting Eq. (10.6) into Eq. (10.7) and solving to T_1 yields

$$T_1 = -T_f \ell n \left[\frac{V_L - V_d}{V_L - V_3} - \frac{K_o - (V_L/V_2)}{\alpha k(V_L - V_3)} \right] \tag{10.8}$$

If the large input continues beyond T_1, the amplifier output will decrease to some constant value. If the interference now disappears, the input of the IF amplifier will decrease to V_1, if the target is still present, and the output

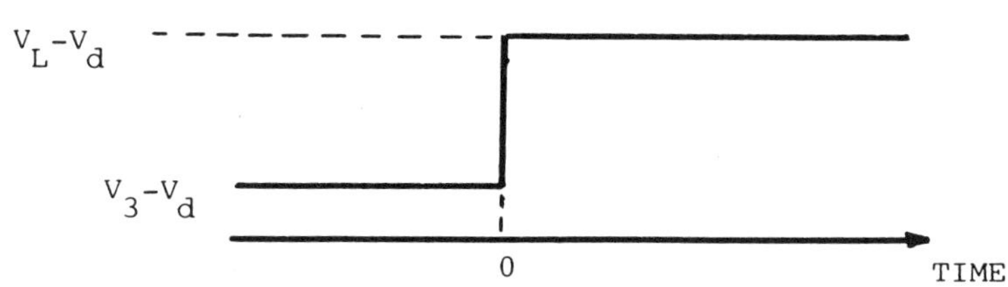

Fig. 10.3 AGC filter input while output of amplifier is limited.

of the IF amplifier will drop below V_d, open-circuiting the input to the AGC filter. The AGC voltage will decrease from the initially large value at a rate controlled by the time constant of the filter T_f.

At the time of the change in the input voltage (set time of change to zero), the AGC voltage is a solution of

$$V_2 \frac{K_o + \alpha k V_d}{1 + \alpha k V_2} = [K_o - \alpha V_r(0)] V_2 \tag{10.9}$$

or

$$V_r(0) = \frac{k(K_o V_2 - V_d)}{1 + \alpha k V_2} \tag{10.10}$$

if the larger input signal V_2 is of sufficient duration for the output of the IF amplifier to stabilize.

The output of the IF amplifier will achieve a level of V_d at a time T_2, measured from the time of removal of input V_2. The time T_2 is a solution of

$$V_o(T_2) = V_d = V_1[K_o - \alpha V_r(0) e^{-T_2/T_f}] \tag{10.11}$$

Using Eq. (10.9), the time T_2 is

$$T_2 = T_f \ell n \left[\left(\frac{\alpha k V_2}{1 + \alpha k V_2} \right) \frac{K_o - (V_d/V_2)}{K_o - (V_d/V_1)} \right] \tag{10.12}$$

Shown in Fig. 10.4 are the impulsive input signal, AGC voltage, IF output voltage, and error voltage available to the angle servo. Because of the servo bandwidth, only the average error voltage is effective in driving the angle servo. Since the error voltage is not continuous, there results a reduction in the coefficient K_D (see Fig. 7.3) relating physical angular tracking error θ_T to error voltage ε_{az},

$$\varepsilon_{az} = K_{D_o} \left(1 - \frac{T_3 + T_4 + x T_2}{T} \right) \theta_T = K_D \theta_T \tag{10.13}$$

The total jammer period T is still unknown because the time interval required for the IF output voltage to stabilize under large or small input signal conditions is unknown. The following is an analysis of the gain-controlled amplifier in Fig. 10.1 to determine these time intervals, subject to the gain variation of Eq. (10.1) and the AGC filter response of Eq. (10.5).

After the output voltage of the IF amplifier drops below the limiting voltage, it satisfies the following equation:

$$V_o(t) + \alpha k V_2 V_o(t) * h(t) = V_2[K_o - \alpha k V_r(0) e^{-t/T_f} + \alpha k V_d (1 - e^{-t/T_f})] \tag{10.14}$$

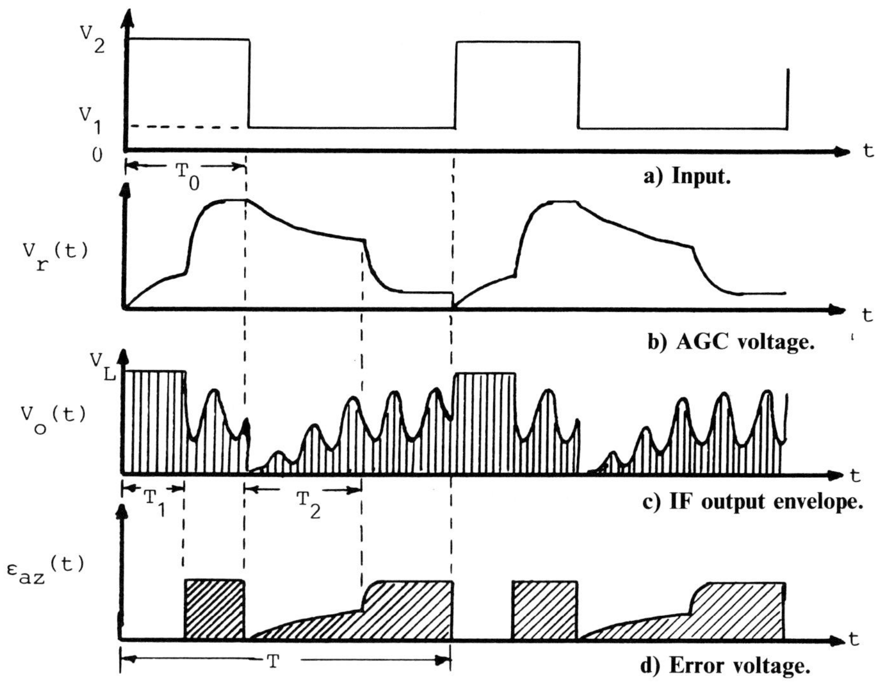

Fig. 10.4 Effects of impulsive jamming on radar systems employing slow AGC.

In Eq. (10.14), time begins at the point limiting ends; thus,

$$V_o(0) = V_L \tag{10.15}$$

$$V_r(0) = \frac{1}{\alpha}\left(K_o - \frac{V_L}{V_2}\right) \tag{10.16}$$

The output voltage varies as

$$V_o(t) = \frac{V_2(K_o + \alpha k V_d)}{1 + \alpha k V_2}\,[1 - e^{-(1+\alpha k V_2)(t/T_f)}] + V_L e^{-(1+\alpha k V_2)(t/T_f)} \tag{10.17}$$

If $\alpha k V_2 \gg 1$, the output voltage very rapidly approaches the desired level near $(K_o/\alpha k) + V_d$, as also claimed by Bakut [88, p. 126], and a proper tracking error voltage is available for the remainder of the jammer on-time T_o. Upon termination of the jammer signal, the IF gain is

$$K = \frac{K_o + \alpha k V_d}{1 + \alpha k V_2} \tag{10.18}$$

and the output voltage has the initial amplitude,

$$V_o = KV_1 = \frac{K_o + \alpha k V_d}{1 + \alpha k V_2} V_1 \approx \left(\frac{K_o}{\alpha k} + V_d\right)\frac{V_1}{V_2} \tag{10.19}$$

Certainly, if $V_1 \ll V_2$, the resulting error voltage is initially unusable. At what point, however, does the output voltage reach a usable level? Previously, the output was claimed usable after it reached the delay voltage V_d. The output must be allowed to rise or the receiver will not saturate on the next jamming pulse. Once the IF output reaches V_d, the output will vary as

$$V_o(t) = \frac{V_1(K_o + \alpha k V_d)}{1 + \alpha k V_1}[1 - e^{-(1 + \alpha k V_1)(t/T_f)}] + V_d e^{-(1 + \alpha k V_1)(t/T_f)} \tag{10.20}$$

and will again quickly reach the desired value if $\alpha k V_1 \gg 1$, as suggested by a strong target. Because of this rapid variation in output voltage once the AGC loop is closed, one is justified in ignoring both transition periods.

Equating the on-time of the jammer to slightly more than the time the receiver output is limited and the off-time to the time required for the IF output to reach the delay voltage immediately appears optimum, for there is no constant error voltage. However, closer examination discloses that the usefulness of the error voltage during the off-time of the jammer must be considered. Once the jammer turns off, the IF output increases as

$$V_o(t) = V_1[K_o - \alpha V_r(0)e^{-t/T_f}] \qquad 0 < t < T_2 \tag{10.21}$$

which has an average value due to jamming of

$$E\{V_o\}_j = \frac{1}{T_2}\int_0^{T_2} V_o(t)\,dt \tag{10.22}$$

Normalizing this average value to the value that would exist for an IF input voltage always at V_1 (no pulsing jammer) results in an effective reduction in the error signal over the interval T_2 of

$$\frac{E\{V_o\}_j}{E\{V_o\}_s} = X = \frac{1 + \alpha k V_1}{K_o + \alpha k V_d}\left[K_o + \frac{\dfrac{K_o + \alpha k V_d}{1 + \alpha k V_2} - \dfrac{V_d}{V_1}}{\ell n\left[\left(\dfrac{\alpha k V_2}{1 + \alpha k V_2}\right)\left(\dfrac{K_o - (V_d/V_2)}{K_o - (V_d/V_1)}\right)\right]}\right] \tag{10.23}$$

The reduction in the gain of the servo loop can now be computed as

$$\frac{K_D}{K_{D_o}} = \frac{XT_2 + T_3 + T_4}{T_1 + T_2 + T_3 + T_4} \tag{10.24}$$

Table 10.1 Effect of impulsive jamming on receiver gain when signal level is 11 mV

Step	Jammer input, V	Receiver output, V	K_D/K_{D_o}	X	Transition times, s			
					T_1	T_3	T_2	T_4
2	0.01	1.0989	0.640075	0.515959	1.00004E-04	0.002997	0.0943107	0.02970
3	0.0316228	1.09965	0.598326	0.482572	8.67777E-04	9.48384E-04	0.101877	0.02970
4	0.1	1.09989	0.585168	0.472025	1.11064E-03	2.9997E-04	0.10426	0.02970
5	0.316228	1.09997	0.581005	0.468683	1.18742E-03	9.48653E-05	0.105013	0.02970
6	1	1.09999	0.579693	0.467631	1.21172E-03	2.99997E-05	0.10525	0.02970
7	3.16228	1.1	0.579284	0.467306	1.21938E-03	9.48681E-06	0.105326	0.02970
8	10	1.1	0.57915	0.467198	1.22187E-03	0.000003	0.10535	0.02970
9	31.6228	1.1	0.579108	0.467164	1.22261E-03	9.48683E-07	0.105357	0.02970
10	100	1.1	0.579088	0.467145	0.0012229	0.0000003	0.105359	0.02970

Table 10.2 Effect of impulsive jamming on receiver gain when signal level is 10 mV

Step	Jammer input, V	Receiver output, V	K_D/K_{D_o}	X	Transition times, s			
					T_1	T_3	T_2	T_4
2	0.1	1.09989	0.638285	0.505828	1.11144E-05	2.9997E-04	8.94986E-03	0.00299
3	0.316228	1.09997	0.594666	0.470884	8.78806E-05	9.48653E-05	9.70237E-03	0.00299
4	1	1.09999	0.581079	0.460072	1.12249E-04	2.99997E-05	9.94026E-03	0.00299
5	3.16228	1.1	0.576616	0.456428	1.19961E-04	9.48681E-06	0.0100155	0.00299
6	10	1.1	0.576478	0.456944	1.22321E-04	0.000003	0.0100394	0.00299
7	31.6228	1.1	0.575184	0.455467	1.23112E-04	9.48683E-07	0.0100468	0.00299
8	100	1.1	0.574889	0.455151	1.23354E-04	0.0000003	0.0100492	0.00299
9	316.228	1.1	0.575352	0.455783	1.23475E-04	9.48683E-08	0.01005	0.00299
10	1000	1.1	0.575126	0.455492	1.23475E-04	3E-08	0.0100502	0.00299

The previous formulations for the various time constants and the reduction in gain have been applied for a receiver with the following parameters: $K_o = 10^4$, $T_f = 1$ s, $\alpha k = 10^5/\text{V}$, $V_d = 1$ V, and $V_L = 10$ V.

Table 10.1 shows the resulting transition times T_i as well as the overall gain reduction K_D/K_{D_o}. The reduction in gain appears to asymptotically approach a value of 0.58, essentially achieving it when the jammer-to-signal power ratio (J/S) is 60 dB. At this J/S level, the jammer duty cycle can be defined as

$$\text{Duty cycle} = \frac{T_1 + T_3}{T} = 0.009 \tag{10.25}$$

Thus, the average J/S ratio is only 39.5 dB. From Eq. (10.20), the transition time T_4 should decrease if the signal level increases or if the jammer provides a "floor." Table 10.2 presents the results of a tenfold increase in the signal. All transition times but T_3 have decreased by a factor of 10, but the overall reduction in gain is still close to 0.58.

10.3 STABILITY OF ANGLE-SERVO LOOPS AT LOWERED GAIN

Barton [16, pp. 296–297] describes two typical angle servos. One servo is the improved type I, including the high-frequency phase-lag term discussed in Chapter 7. The servo open-loop transfer function is

$$G(s) = \frac{K\omega_3}{\omega_2} \frac{(s + \omega_2)}{s(s + \omega_1)(s + \omega_3)} \tag{10.26}$$

yielding a phase shift of

$$< G(j\omega) = -\frac{\pi}{2} - \tan^{-1}\left(\frac{\omega}{\omega_1}\right) + \tan^{-1}\left(\frac{\omega}{\omega_2}\right) - \tan^{-1}\left(\frac{\omega}{\omega_3}\right) \tag{10.27}$$

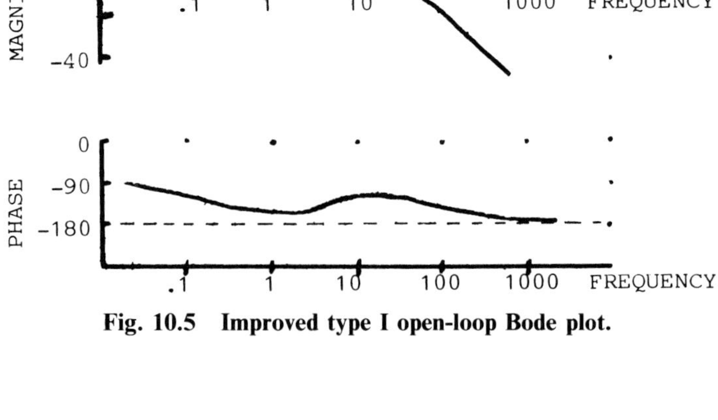

Fig. 10.5 Improved type I open-loop Bode plot.

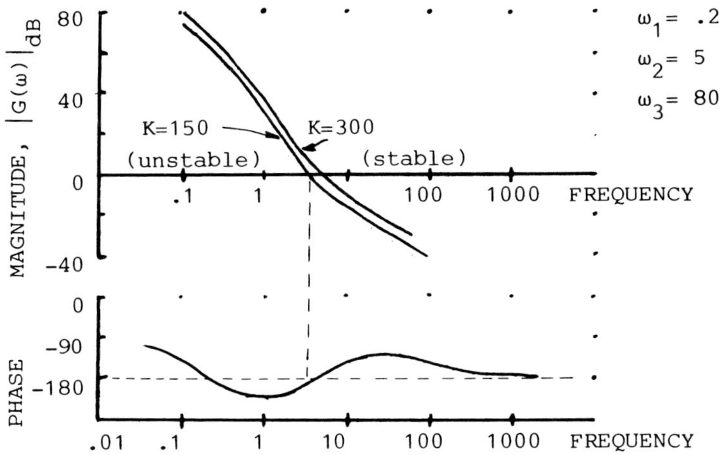

Fig. 10.6 First-order servo with double integration.

Since $\omega_3 > \omega_2 > \omega_1$, the total phase shift is less than 180 deg and the closed loop is always stable. The Bode plot for this servo is shown in Fig. 10.5; reductions of one-half in gain will clearly not result in instability.

The second servo described by Barton has the open-loop transfer function

$$G(s) = \frac{K\omega_1\omega_2}{\omega_2^2} \frac{(s+\omega_2)^2}{s(s+\omega_1)^2(s+\omega_3)} \tag{10.28}$$

If the coefficients of this transfer function are changed from the coefficients suggested by Barton, the result is the Bode plot shown in Fig. 10.6. This Bode plot is similar to the plot shown in Shustov and Vakin [4, p.176] where it is called a "leak-shaped" amplitude-phase characteristic. A closed-loop servo employing this second type of transfer function with the coefficients selected here will become unstable if the loop gain is reduced by one-half.

10.4 SENSITIVITY OF JAMMING PARAMETERS

The effectiveness of the jamming technique just described depends to a great extent on the accuracy of the assumed receiver model. Two variations are considered: variation of jammer parameters with a fixed receiver and relocation of the position within the receiver of the delay voltage.

If the peak J/S ratio is fixed, the transition times in Table 10.2 are also fixed. If the jammer on time T_o is longer than $T_1 + T_3$ as shown in Table 10.2, then a good error voltage is derived for the time $T_o - T_1$. Similarly, if the off time is extended, a useful error voltage exists for a time $T - T_o - T_2 + XT_2$.

The effective reduction in gain is then

$$\frac{K_D}{K_{D_o}} = \frac{T - T_1 - (1 - X)T_2}{T} \tag{10.29}$$

The jammer period given by Table 10.2 is optimum. As the jammer period increases, the reduction in gain will decrease (i.e., $K_D/K_{D_o} \rightarrow 1$). As the jammer period decreases from the optimum value, the level of reduction in receiver gain will also decrease as stated by Shustov and Vakin. This reduced effectiveness results from failure of the IF gain to decrease to a stable value before the next jammer pulse. The output voltage of the receiver is not clipped as heavily and the time T_1 is reduced. From Table 10.2 for a 40 dB J/S ratio, the jammer period is 0.0131 s. The variation in loop gain for an increasing jammer period is shown in Fig. 10.7.

A variation of AGC mechanization places the delay voltage after the low-pass filter as shown in Fig. 10.8. This system responds differently to impulsive jamming because the filter input is never open-circuited. An analysis would show the time duration of limiting of the output T_1 to be identical to the time duration of the previous mechanization, or Eq. (10.8). The time T_2, however, is no longer governed by the discharge time of an undriven

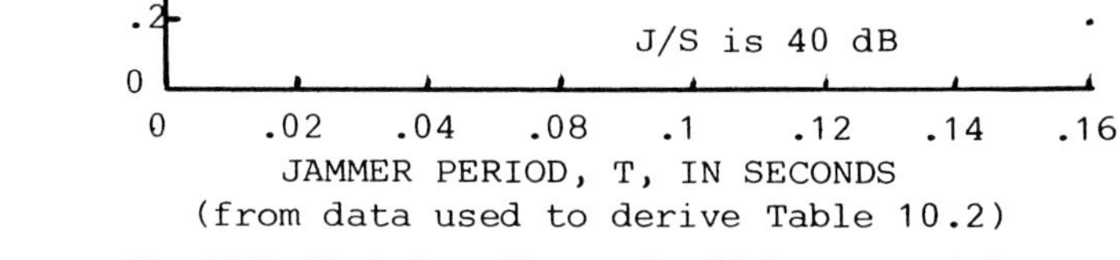

Fig. 10.7 Variation of loop gain with jammer period.

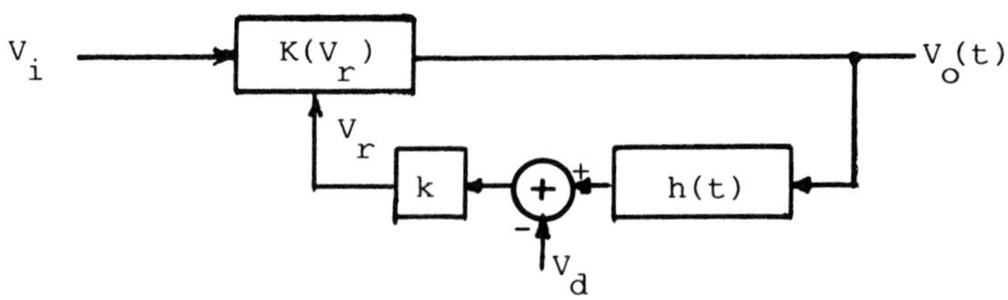

Fig. 10.8 Gain controlled amplifier with delay voltage after filter.

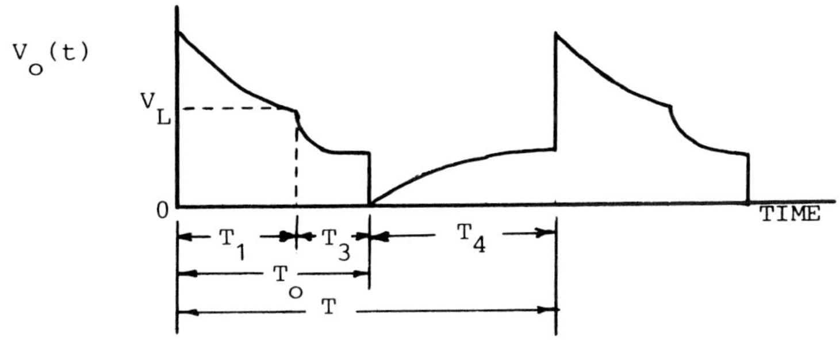

Fig. 10.9 Definition of transition times for modified AGC system.

AGC filter. The regulating voltage now varies as

$$V_r(t) = \frac{k(K_o V_1 - V_d)}{1 + \alpha k V_1}[1 - e^{-(1+\alpha k V_1)(t/T_f)}] + V_r(o)e^{-(1+\alpha k V_1)(t/T_f)} \quad (10.30)$$

and the time to achieve an IF output voltage level of V_d is now

$$T_2 = \frac{T_f}{1 + \alpha k V_1} \ell n \left\{ \alpha k V_1 \frac{K_o + \alpha k V_d}{1 + \alpha k V_2} \frac{V_2 - V_1}{K_o V_1 - V_d} \right\} \quad (10.31)$$

For this second AGC system, the time to achieve an output voltage level of V_d is not important. The control voltage varies continuously according to Eq. (10.30) and the total time required to achieve a stable output can be approximated at three times the time constant in Eq. (10.30), which is identical to the T_4 defined previously.

The time constant T_3 defined earlier will also remain the same. Thus, Tables 10.1 or 10.2 can be used to determine T_1, T_4, and T_3 as defined in Fig. 10.9. The maximum reduction in gain at a 30 dB J/S ratio, from the values of Table 10.1, is then

$$\frac{K_D}{K_{D_o}} = \frac{T_3 + 0.67T_4}{T_1 + T_3 + T_4} = 0.64 \quad (10.32)$$

However, if the optimum duty cycle for the previous AGC system were used against this AGC system, the reduction in the gain of the radar servo-loop is only

$$\frac{K_D}{K_{D_o}} = 0.92 \quad (10.33)$$

This small reduction in gain results from the rapid adjustment of the AGC during the relatively long off-time of the jammer. Clearly, this form of the ECM technique is sensitive to the parameters of the radar because a small reduction in gain does not result in instability of the radar.

APPENDIX A
GLOSSARY OF TERMS

AAA	antiaircraft artillery
AFIT	Air Force Institute of Technology
AGC	automatic gain control
AM	amplitude modulation
BWO	backward-wave oscillator
C^3CM	command, control, and communications countermeasures
CFA	crossed-field amplifier
CFAR	constant false alarm rate
CO	copolarized maximum
CONSCAN	conical scan
COSRO	conical scan on receive only
CW	continuous wave
DC	direct current
DINA	direct noise amplification
EC	electronic combat
ECCM	electronic counter-countermeasures
ECM	electronic countermeasures
ELINT	electronic intelligence
EM	electromagnetic
ENR	excess noise ratio
ERP	effective radiated power
ESM	electronic warfare support measure
EW	early warning; electronic warfare
EWO	electronic warfare officer
FEBA	forward edge of the battle area
FM/SN	frequency modulation by sine wave plus noise
FTC	fast-time constant
GCI	ground-controlled intercept
HF	height finding
HOJ	home-on-jam
Hz	cycle/second (kHz = 1000 Hz, MHz = 10^6 Hz, GHz = 10^9 Hz
IAGC	instantaneous automatic gain control
IF	intermediate frequency
IFM	instantaneous frequency measurement
IR	infrared
ITU	International Telecommunications Union

LORO	lobe on receive only
ML	maximum likelihood
MOPA	master oscillator-power amplifier
MTI	moving target indication
MWR	multisensor warning receiver
NF	noise figure
PD	pulse Doppler
PPI	plan position indicator
PPM	periodic permanent magnet
PRF	pulse-repetition frequency
PRI	pulse-repetition interval
RAM	radar-absorbing material
RCS	radar cross section
RF	radio frequency
RGPO	range gate pull off
rms	root mean square
rpm	revolutions per minute
RWR	radar warning receiver
SAM	surface-to-air missile
S&H	sample-and-hold
SCV	subclutter visibility
SEAD	suppression of enemy air defense
SLB	sidelobe blanking
SLC	sidelobe canceller or cancellation
SOJ	standoff jammer
STC	sensitivity time control
SWC	scan with compensation
TWS	track while scan
TWT	traveling wave tube
VCO	voltage-controlled oscillator
VGWO	velocity gate walk-off
VSWR	voltage standing wave ratio

A	physical area of antenna
A_e	effective area of antenna
B_R	receiver bandwidth
f_d	Doppler frequency
f_{PRF}	pulse-repetition frequency
f_R	frequency of oscillation, cycles/s
f_s	scan frequency
G_R	radar antenna gain
J/S	jamming-to-signal ratio
L_a	horizontal aperture dimension
L_e	vertical aperture dimension
N_{INT}	receiver noise power watts

N_o	ideal receiver noise power density, W/Hz, assumed to be − 144 dBW/MHz
P_J	jammer power
P_R	radar transmit power
P_r	power received by radar
P_{r_J}	jammer power received
P_{r_T}	target power received
R	range
ΔR	range resolution
R_J	range to jammer
R_T	range to target
$(S/C)_p$	signal-to-clutter ratio per pulse
$(S/N)_{\min}$	minimum signal-to-noise ratio for proper system performance
$(S/N)_p$	S/N per pulse at IF
V	3×10^8 m/s, speed of light in vacuum
η	aperture efficiency
η^o	volume clutter reflectivity, m^2/m^3
θ_{az}	azimuth 3 dB beamwidth
θ_{el}	elevation 3 dB beamwidth
θ_s	squint angle from boresight
θ_T	target angle from boresight
θ_3	3 dB beamwidth
λ	wavelength
σ_T	target average cross section
σ^o	area clutter reflectivity, m^2/m^2
τ	pulse width
ω	frequency, rad/s

APPENDIX B
MATHEMATICAL BASICS

Presented here is a brief review of the mathematical techniques used in the text. Topics include exponentials, logarithms, trigonometry, decibels, and some geometric formulas.

Exponentials

In the formulation $A^2 = A \times A$, the factor 2 is an exponential: it is the power to which A is raised. Rules for use of exponentials are

$$A^a A^b = A^{a+b}$$

$$(A^a)^b = A^{ab}$$

Thus,

$$A^{1/3} \times A^{1/3} \times A^{1/3} = A \quad \text{or} \quad A^{1/3} \text{ is the cube root of } A$$

Logarithms

If the letter A above is equal to 10, we can write

$$10^a = X \text{ and } 10^b = Y$$

Then, from the rules for using exponents,

$$XY = 10^a 10^b = 10^{a+b} = Z$$

The factor a is called the logarithm of X to the base 10,

$$a = \log_{10} X$$

From the above, the logarithm of the product of two numbers is equal to the sum of the logarithms of the individual numbers.

The base A need not be 10. If the base is taken as e (approximately defined as the number 2.71828), then the exponents are called natural logarithms, as

$$e^c = Q \qquad \ell n_e Q = c$$

Certain useful rules are

$$\log \frac{X}{Y} = \log X - \log Y$$

$$\log \frac{1}{Y} = \log 1 - \log Y = -\log Y$$

Decibels

A number is expressed in decibel form by taking the logarithm of the number to base 10 and multiplying the result by 10, as

$$A_{dB} = 10 \times \log_{10} A$$

The utility of this operation can be seen by noting that

$$(A \times B)_{dB} = 10 \log(AB) = 10 \log A + 10 \log B = A_{dB} + B_{dB}$$

or an expression involving products becomes one involving sums. Since log (0) is undefined, this definition is restricted to numbers greater than zero and, by convention, the quantities converted are power or power ratios. Some examples are given in Table B1.

Table B1 Example of conversions of power

Power, W	Power in dB relative to 1 W, dBW	Power in dB relative to 1 mW, dBm
1	0	30
2	3	33
10	10	40
100	20	50
1000	30	60

Power ratio A	dB
1	0
10	10
100	20

Trigonometry

Formulas used are listed below and are related to Fig. B1. Both radian and degree units are used where 1 rad = 57.3 deg. Definitions are

$$\sin\theta = Y_o/r \qquad \cos\theta = X_o/r \qquad \tan\theta = Y_o/X_o$$

Important relationships are

$$\sin 90^\circ = \sin (\pi/2) = 1$$

$$\cos 0^\circ = \cos 0 = 1$$

$$\sin 30^\circ = \sin (\pi/6) = \tfrac{1}{2}$$

$$\sin 45^\circ = \sin (\pi/4) = 1/\sqrt{2}$$

$$\sin 60^\circ = \sin (\pi/3) = \sqrt{3}/2$$

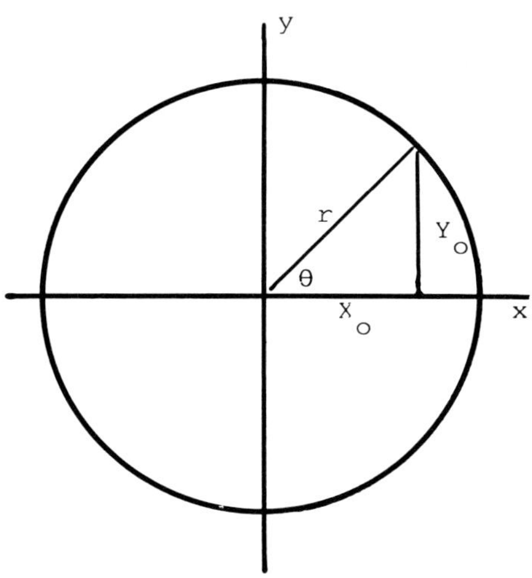

Fig. B1 Trigonometry variables

Useful approximations for $\theta \ll 1$ in radians is

$$\sin\theta \approx \theta$$

$$\cos\theta \approx 1 - \theta^2/2$$

The double-angle formulas are

$$\sin(\theta \pm \varphi) = \sin\theta\cos\varphi \pm \cos\theta\sin\varphi$$

$$\cos(\theta \pm \varphi) = \cos\theta\cos\varphi \mp \sin\theta\sin\varphi$$

Using the double-angle formulas, we can show that

$$\cos(\theta - \varphi) + \cos(\theta + \varphi) = 2\cos\theta\cos\varphi$$

$$\cos(\theta - \varphi) - \cos(\theta + \varphi) = 2\sin\theta\sin\varphi$$

Geometric Formulas

Area of a circle, radius r, is πr^2
Area of a sphere, radius r, is $4\pi r^2$
Volume of a sphere, radius r, is $(4/3)\pi r^3$

APPENDIX C
ANSWERS TO PROBLEMS

Chapter 2

2.2a	37.5 m
2.2b	40.5 n.mi.
2.2c	3.3 cm
2.3a	52 dBW
2.3b	0.03 W/m^2
2.3c	150 MW
2.4a	180 n.mi.
2.4b	−132 dBW
2.5a	6667
2.5b	13.7 dB
2.6	4.4 deg beamwidth/31.4 dB antenna gain
2.7	2.7 MW
2.8	Phase sum-difference
2.9	61.6 dB
2.10a	8.24 km
2.10b	110 m
2.10c	38.2 dB

Chapter 3

3.1	Active, passive
3.2	CONSCAN and CONSCAN with SWC
3.3	60 dB
3.4	12.3 dB
3.5a	56 n.mi., 7.5 min
3.5b	4600 ft
3.5c	19.7 n.mi.
3.5d	19.7 n.mi.
3.6	19.5 n.mi.
3.7	6.7 n.mi.
3.8	J/S = 18.75 dB at 1 n.mi.
3.9	J/S = 35.8 dB

3.10

θ, deg	R, n.mi.	θ, deg	R, n.mi.
0	1.15	60	3.98
15	1.06	70	8.86
30	1.03	90	∞
45	1.15		

3.11a $R = \dfrac{L}{2\tan(0.45\theta_3)}$

3.11b Miss distance $= \dfrac{L}{2} - \dfrac{a_n L^2}{8V^2\tan^2(0.45\theta_3)}$

$a_n = 200 \text{ m/s}^2,\ V = 2000 \text{ m/s},\ \theta_3 = 10 \text{ deg}$

3.11c $L = 248$ m, miss distance $= 62$ m

Chapter 4

4.1 No
4.2 Yes
4.3 True
4.4 SLB
4.5 True
4.6a No
4.6b Yes
4.7 All to some extent

Chapter 6

6.6 $V_r = -\dfrac{j\omega\mu E_{in} I \Delta l}{2\pi} \cos[k d \sin\theta \cos\varphi]\cos\theta\cos\varphi$

REFERENCES

[1]U.S. Department of the Air Force, *Electronic Warfare Principles*, Air Force Pamphlet 51-3, Government Printing Office, Washington, DC, Sept. 1, 1978.

[2]Boyd, J. A., Harris, D. B., King, D. D., and Welch, H. W., Jr., *Electronic Countermeasures*, University of Michigan, Ann Arbor, 1961, pp. 14-17 (AD 357877). (Also available from Peninsula Publishing, Los Altos, CA 94022.)

[3]Smith, H. F., "EW History," *Electronic Warfare Magazine*, Vol. 1, No. 4, 1969, p. 18.

[4]Shustov, L. N. and Vakin, S. A., *Principles of Jamming and Electronic Reconnaissance*, Vol. 1, Soviet Radio, Moscow, 1968, p. xii (AD 692642, 692643).

[5]*Encyclopedia Americana*, Vol. 24, American Corp., New York, 1979, p. 45.

[6]Locke, A. S., *Guidance*, D. Van Nostrand Co., Princeton, NJ, 1955, p. 39.

[7]Porter, J. W. (interview), "Porter Scores EW Cost Effectiveness," *Electronic Warfare Magazine*, Vol. 5, No. 1, Feb. 1973, p. 11.

[8]Alberts, D. J., "A Call from the Wilderness," *Air University Review*, Air University, Maxwell AFB, AL, 1976, p. 35.

[9]"Yom Kippur Fighting Underscores EW Importance," *Electronic Warfare*, Vol. 6, No. 1, Jan./Feb. 1974, p. 24.

[10]U. S. Department of the Air Force, *Basic Aerospace Doctrine of the United States Air Force*, Air Force Manual 1-1, Government Printing Office, Washington, DC, pp. 3–6.

[11]Hall, W. M. and Ward, H. R., "Signal-to-Noise Loss in Moving Target Integrators," *IEEE Proceedings*, Vol. 56, Feb. 1968, pp. 233–234.

[12]Eustace, H. F. and Schoniger, K. R., *The International Countermeasures Handbook*, EW Communications, Palo Alto, CA, 1975, p. 124.

[13]Wolff, E. A., *Antenna Analysis*, Wiley, New York, 1967.

[14]Eustace, H. F. (ed.), *The International Countermeasures Handbook*, EW Communications, Palo Alto, CA, 1976.

[15]Skolnik, M. (ed.), *Radar Handbook*, McGraw-Hill, New York, 1970.

[16]Barton, D., *Radar Systems Analysis*, Artech House, Dedham, MA, 1976, Chap. 1.

[17]Blake, L. V., *Radar Range Performance Analysis*, Lexington Books, 1980.

[18]Skolnik, M., *Introduction to Radar Systems*, McGraw-Hill, New York, 1980, p. 139.

[19]Nathanson, F. E., *Radar Design Principles*, McGraw-Hill, New York 1969.

[20]Schleher, D. C., "Radar Detection in Weibull Clutter," *IEEE Transactions on Aerospace and Electronic Systems*, Vol. AES-12, Nov. 1976, pp. 736-743.

[21]DeVore, C., "AEGIS: Shield of the Fleet," *IEEE-AES Newsletter*, Vol. 13, No. 12, Dec. 1978.

[22]Faraone, J. N., "Automatic Range Tracking Accuracy," Illinois Institute of Technology, Chicago, Tech. Note 75-RS-101, Nov. 1974.

[23]Faraone, J. N., "Angle Tracking Accuracy of a Conical Scan Radar," Illinois Institute of Technology, Chicago, Tech. Note 75-RS-103, Oct. 1974.

[24]Barton, D. and Ward, H., *Handbook of Radar Measurements*, Prentice Hall, Englewood Cliffs, NJ, 1969.

[25]Rhodes, D. R., *Introduction to Monopulse*, McGraw-Hill, New York, 1959.

[26]Leonov, A. I. and Fomichev, K. I., *Monopulse Radar*, Soviet Radio, Moscow, 1962 (AD 742696).

[27]Walsh, T., "Military Radar Systems: History, Current Position, and Future Forecast," *Microwave Journal*, Vol. 21, Nov. 1978, p. 87.

[28]Van Brunt, L. B., *Applied ECM*, Vol. 1, EW Engineering, Inc., 1978, p. xiv.

[29]Turner, F. M., "Noise Quality Optimizes Jammer Performance," *Electronic Warfare*, Vol. 9, No. 6, Nov. 1977, p. 117 (reprinted in Johnston, S. L., *Radar Electronic Counter-Countermeasures*, Artech House, Dedham, MA, 1980).

[30]Schlesinger, R. J., *Principles of Electronic Warfare*, Prentice-Hall, Englewood Cliffs, NJ, 1961, p. 125.

[31]"Electronic Countermeasures Calculator," General Electric Co., Utica, N.Y., Rept. GEN-188A, 1959.

[32]Turner, F. M., "Deflection-versus Intensity-Modulated CRT's as Anti-Jam Radar Display," *Electronic Warfare*, Vol. 3, No. 1, Winter 1971, p. 6.

[33]Golden, A., "Amplitude Modulated Repeater Jammer J/S Requirements—A Derivation," *25th Annual Joint EW Conference Proceedings*, Naval Postgraduate School, Monterey, CA, May 1980.

[34]Collin, R. F. and Zucker, F. J., *Antenna Theory, Part I*, McGraw-Hill, New York, 1969, p. 135.

[35]Thurbon, M. T., "The Nature of Electronic Warfare," *Journal of Electronic Defense*, Vol. 3, No. 3, May/June 1980, p. 31.

[36]Hoffman, J. F., "Radar ESM Systems and Technology," *Introduction to Electronic Warfare*, U.S. Air Force Institute of Technology, Wright-Patterson AFB, OH, 1982.

[37]Davis, R., "Anatomy of a Jammer," *Microwave System News*, Vol. 5, No. 3, July 1977.

[38]Fusfield, R., "High Power Airborne Jammers," *Electronic Progress*, Vol. XVII, No. 2, Fall 1975, p. 7.

[39]Jones, E. M. T., "Paraboloid Reflector and Hyperboloid Lens Antennas," *IRE Transactions on Antennas and Propagation*, Vol. AP-2, No. 4, July 1954, p. 119.

[40]Balanis, C. A., *Antenna Theory—Analysis and Design*, Harper & Row, New York, 1982.

[41]Cikalo, J. and Greenbaum, M., "Radar/ECM Computer Modeling," *Radar Electronic Counter-Countermeasures*, Artech House, Dedham, MA, 1980, pp. 513-519.

[42]Crispin, J. W. and Maffett, A. L., "Radar Cross-Section Estimation for Simple Shapes," *IEEE Proceedings*, Vol. 53, Aug. 1965, pp. 833-847.

[43]Pelton, E. L., "A Streamlined Metallic Radome," *IEEE Transactions on Antennas and Propagation*, Vol. 62, Nov. 1974, p. 799.

[44]Mahaffey, M., "Electrical Fundamentals of Countermeasure Chaff," *The International Countermeasures Handbook*, edited by H. F. Eustace, EW Communications, Palo Alto, CA, 1976, p. 512.

[45]King, R. W. P., *The Theory of Linear Antennas*, Harvard University Press, Cambridge, MA 1956, p. 508.

[46]Coleman, E. R., "Chaff Meteorology," *The International Countermeasures Handbook*, edited by H. F. Eustace, EW Communications, Palo Alto, CA, 1976, p. 518.

[47]Silverman, G., "Chaff Dispersion and Polarization," *Journal of Electronic Defense*, Vol. 1, No. 3, Nov./Dec. 1978, p. 39.

[48]Hoisington, D. B., *Electronic Warfare*, Naval Post Graduate School, Monterey, CA, April 1980, p. 3L-6.

[49]Maksimov, M. V. et al., *Radar Anti-Jamming Techniques*, Artech House, Dedham, MA, 1979.

[50]"How Chaff Protects Aircraft," *Microwave Systems News*, Vol. 6, No. 5, Oct./Nov. 1976, p. 69.

[51]Bang, S. B., *The Effectiveness of Self-Protective Chaff Against Radar MTI, Resolution and Control Servos*, Master's Thesis, School of Engineering, U.S. Air Force Institute of Technology, Wright-Patterson AFB, OH, Dec. 1975, p. 11.

[52]Ruck, G. T., Barrick D. E. et al., *Radar Cross Section Handbook*, Plenum Press, New York, 1970.

[53]Bissegger, C. A., "VCO Subsystems: What to Test, How to Test It," *Microwaves*, Vol. 16, May 1977, p. 60.

[54]*VCO Definition Workshop*, U.S. Air Force Avionics Laboratory, Wright-Patterson AFB, OH, Tech. Rept. AFAL-TR-75-187.

[55]Muskus, C. Z., "Solid State Noise Generators for Radar Performance Monitoring," *Microwave Journal*, Vol. 22, Oct. 1979, p. 69.

[56]*Coaxial and Waveguide Catalog*, Hewlett Packard Co., 1976.

[57]Advertisement by Microwave Semiconductor, Inc., Somerset, NJ, *Microwave Journal*, Vol. 23, April 1980, p. 30.

[58]*Klystron Twystron TWT BWO Component and Amplifier Catalog*, Varian Co., Palo Alto, CA, Brochure 3725 10M, Aug. 1977.

[59]Hughes TWT and TWTA Handbook, Hughes Aircraft Co., Torrence, CA.

[60]Mendel, J. T., "Helix and Coupled-Cavity Traveling-Wave Tubes," *IEEE Proceedings*, Vol. 61, March 1973, p. 280.

[61]Skowron, J. F., "The Continuous-Cathode (Emitting-Sole) Crossed-Field Amplifier," *IEEE Proceedings*, Vol. 61, March 1973, p. 330.

[62]*The Gridded Crossed-Field Amplifier*, Warnecke Electron Tubes, Inc., Des Plaines, IL.

[63]Fink, D. G. (ed.), *Electronics Engineer's Handbook*, McGraw-Hill, New York, 1975, pp. 9–40.

[64]Yaw, D., "Electronic Warfare Antenna Systems—Past and Present," *Microwave Journal*, Vol. 24, Sept. 1981, p. 22.

[65]Jasik, H. (ed.), *Antenna Engineering Handbook*, McGraw-Hill, New York, 1961, p. 10-7.

[66]Archer, D. H. and Black, A. A., "Higher ERP with Lens Fed Multibeam Array," *Journal of Electronic Defense*, Vol. 5, March 1982, p. 51.

[67]Simpson, M., "High ERP Phased Array ECM Systems," *Journal of Electronic Defense*, Vol. 5, March 1982, p. 41.

[68]Johnston, S. L., *Radar Electronic Counter-Countermeasures*, Artech House, Dedham, MA, 1980, p. 11.

[69]Burrows, C. R. and Atwood, S. S., *Radio Wave Propagation, Consolidated Summary Technical Report of the Committee on Propapation NDRC*, Academic Press, Inc., New York, 1949, p. 219.

[70]Rubin, W. L. and Kamen S. K., "SCAMP—A Single-Channel Monopulse Radar Signal Processing Technique," *IEEE Transactions on Military Electronics*, Vol. MIL-6, No. 2, April 1962, pp. 146–152.

[71]Barton, D. K., "Low-Angle Radar Tracking," *IEEE Proceedings*, Vol. 62, June 1974, p. 687.

[72]Letter to the Editor, *Electronic Warfare Magazine*, Vol. 6, No. 2, March/April 1974, p. 79.

[73]Editorial, *Electronic Warfare Magazine*, Vol. 5, No. 6, Nov./Dec. 1973, p. 96.

[74]Papoulis, A., *The Fourier Integral and Its Applications*, McGraw-Hill, New York, 1962, p. 44.

[75]Ridenous, L. N. (ed.), *Theory of Servomechanisms*, Radiation Laboratory Series No. 25, McGraw-Hill, New York, 1947, p. 25.

[76]Shinners, S. M., *Control System Synthesis*, Wiley, New York, 1964, Chap. 4.

[77]Truxal, J. G., *Control System Synthesis*, McGraw-Hill, New York, 1955, pp. 80–84.

[78]Hartman, R., "Multisensor Warning Receiver Tackles Growing Threat," *Defense Electronics*, May 1979, p. 79.

[79]Hodgman, C. D. (ed.), *C. R. C. Standard Mathematical Tables*, 12th ed., Chemical Rubber Co., Cleveland, OH, 1963.

[80]Skolnik, M., *Introduction to Radar Systems*, McGraw-Hill, New York, 1962, p. 171.

[81]Abramowitz, M. and Stegun, I. A., *Handbook of Mathematical Functions*, Dover Publications, New York, 1965, Eq. 8.6.35.

[82]Whalen, A. D., *Detection of Signals in Noise*, Academic Press, New York, 1971.

[83]Strasser, N. C., "Investigation of Terrain Bounce Electronic Countermeasure," Master's Thesis, School of Engineering, U.S. Air Force Institute of Technology, Wright-Patterson AFB, OH, Doc. AFIT/GE/EE/80D-41, 1980.

[84]Harrington, R. F., *Time-Harmonic Electromagnetic Fields*, McGraw-Hill, New York, 1961.

[85]Ludwig, A. C., "The Definition of Cross-Polarization," *IEEE Transactions on Antennas and Propagation*, Vol. AP-21, Jan. 1973, pp. 116–119.

[86]Sandler, S. S., "Paraboloidal Reflector Patterns for Off-Axis Feed," *IRE Transactions on Antennas and Propagation*, Vol. AP-8, July 1960, p. 368.

[87]Murray, D. A., *Introductory Course in Differential Equations*, Longmans, Green, London, 1961.

[88]Bakut, P. A. et al., *Questions of the Statistical Theory of Radar* (Vol. I), Soviet Radio, Moscow, 1964 (AD 645 775).

[89]Hoffman, C. B. and Baron, A. R., "Wideband ESM Receiving Systems," *Microwave Journal*, Vol. 23, Sept. 1980, pp. 25–34 and Vol. 24, Feb. 1981, pp. 57–61.

[90]D'Azzo, J. J. and Houpis, C. C., *Principles of Electrical Engineering*, McGraw-Hill, New York.

[91]Jones, R. V., *The Wizard War*, Coward, McCann & Geoghegan, Inc., New York, 1978.

[92]Price, A., *Instruments of Darkness*, William Kimber, London, 1967, MacDonald and Jones, London, 1977.

[93]"Performing Electronic Countermeasures in the United States and Canada," U.S. Air Force Regulation 55-44 (C1), July 3, 1980.

[94]Sakamoto, H. and Pebbles, P. Z., Jr.,"Conopulse Radar," *IEEE Transactions on Aerospace and Electronic Systems*, Vol. AE S-14, Jan. 1978, pp. 199–208.

[95]Johnston, S. L., "Tracking Radar Electronic Counter-Countermeasures Against Inverse Gain Jammers," *Radar* 82, IEEE Conf. Pub. 216, Oct. 1982, pp. 444–447.

[96]Steer, D. J. and Abram, T. M., "Airborne Microwave ECM," *Proceedings of the Military Microwaves Conference*, London, Oct. 1978 (reprinted in Johnston, S. L., *Radar Electronic Counter-Countermeasures*, Artech House, Dedham, MA, 1980).

[97]Davenport, W. D. and Root, W. L., *Random Signals and Noise*, McGraw-Hill, New York, 1958.

[98]Gardner, F. M., *Phaselock Techniques*, Wiley, New York, 1979.

[99]Garnell, P. and East, D. J., *Guided Weapon Control Systems*, Pergamon Press, Elmsford, NY, 1977.

[100]Felsenthal, H. D., Jr., "Radar System," U.S. Patent 3,947,847, March 30, 1976.

[101]Pifer, P. M., "Automatic Tracking System Utilizing Coded Scan Rate," U.S. Patent 3,419,867, Dec. 31, 1968.

[102]Clayton, L., Jr., "Automatic Tracking System Utilizing Coded Scan Sequence," U.S. Patent 3,419,868, Dec. 31, 1968.

[103]Jaffe, R. and Rechtin, E., "Design and Performance of Phase-Lock Loops Capable of Near-Optimum Performance Over a Wide Range of Input Signal and Noise Levels," *IRE Trans.*, IT-1, pp. 66–76, March 1955.

[104]Transco Products, Inc. Advertisement, *Microwaves*, Vol. 15, No. 8, August 1976, p. 13.

Index